Pitman Research Notes in Mathematics Series

Submission of proposals for consideration
Suggestions for publication, in the form of outlines and representative samples, are invited by the Editorial Board for assessment. Intending authors should approach one of the main editors or another member of the Editorial Board, citing the relevant AMS subject classifications. Alternatively, outlines may be sent directly to the publisher's offices. Refereeing is by members of the board and other mathematical authorities in the topic concerned, throughout the world.

Preparation of accepted manuscripts
On acceptance of a proposal, the publisher will supply full instructions for the preparation of manuscripts in a form suitable for direct photo-lithographic reproduction. Specially printed grid sheets are provided and a contribution is offered by the publisher towards the cost of typing. Word processor output, subject to the publisher's approval, is also acceptable.

Illustrations should be prepared by the authors, ready for direct reproduction without further improvement. The use of hand-drawn symbols should be avoided wherever possible, in order to maintain maximum clarity of the text.

The publisher will be pleased to give any guidance necessary during the preparation of a typescript, and will be happy to answer any queries.

Important note
In order to avoid later retyping, intending authors are strongly urged not to begin final preparation of a typescript before receiving the publisher's guidelines and special paper. In this way it is hoped to preserve the uniform appearance of the series.

Longman Scientific & Technical
Longman House
Burnt Mill
Harlow, Essex, UK
(tel (0279) 26721)

Longman Scientific & Technical
Churchill Livingstone Inc.
1560 Broadway
New York, NY 10036, USA
(tel (212) 819-5453)

Titles in this series

1 Improperly posed boundary value problems
A Carasso and A P Stone
2 Lie algebras generated by finite dimensional ideals
I N Stewart
3 Bifurcation problems in nonlinear elasticity
R W Dickey
4 Partial differential equations in the complex domain
D L Colton
5 Quasilinear hyperbolic systems and waves
A Jeffrey
6 Solution of boundary value problems by the method of integral operators
D L Colton
7 Taylor expansions and catastrophes
T Poston and I N Stewart
8 Function theoretic methods in differential equations
R P Gilbert and R J Weinacht
9 Differential topology with a view to applications
D R J Chillingworth
10 Characteristic classes of foliations
H V Pittie
11 Stochastic integration and generalized martingales
A U Kussmaul
12 Zeta-functions: An introduction to algebraic geometry
A D Thomas
13 Explicit *a priori* inequalities with applications to boundary value problems
V G Sigillito
14 Nonlinear diffusion
W E Fitzgibbon III and H F Walker
15 Unsolved problems concerning lattice points
J Hammer
16 Edge-colourings of graphs
S Fiorini and R J Wilson
17 Nonlinear analysis and mechanics: Heriot-Watt Symposium Volume I
R J Knops
18 Actions of fine abelian groups
C Kosniowski
19 Closed graph theorems and webbed spaces
M De Wilde
20 Singular perturbation techniques applied to integro-differential equations
H Grabmüller
21 Retarded functional differential equations: A global point of view
S E A Mohammed
22 Multiparameter spectral theory in Hilbert space
B D Sleeman
24 Mathematical modelling techniques
R Aris
25 Singular points of smooth mappings
C G Gibson
26 Nonlinear evolution equations solvable by the spectral transform
F Calogero
27 Nonlinear analysis and mechanics: Heriot-Watt Symposium Volume II
R J Knops
28 Constructive functional analysis
D S Bridges
29 Elongational flows: Aspects of the behaviour of model elasticoviscous fluids
C J S Petrie
30 Nonlinear analysis and mechanics: Heriot-Watt Symposium Volume III
R J Knops
31 Fractional calculus and integral transforms of generalized functions
A C McBride
32 Complex manifold techniques in theoretical physics
D E Lerner and P D Sommers
33 Hilbert's third problem: scissors congruence
C-H Sah
34 Graph theory and combinatorics
R J Wilson
35 The Tricomi equation with applications to the theory of plane transonic flow
A R Manwell
36 Abstract differential equations
S D Zaidman
37 Advances in twistor theory
L P Hughston and R S Ward
38 Operator theory and functional analysis
I Erdelyi
39 Nonlinear analysis and mechanics: Heriot-Watt Symposium Volume IV
R J Knops
40 Singular systems of differential equations
S L Campbell
41 N-dimensional crystallography
R L E Schwarzenberger
42 Nonlinear partial differential equations in physical problems
D Graffi
43 Shifts and periodicity for right invertible operators
D Przeworska-Rolewicz
44 Rings with chain conditions
A W Chatters and C R Hajarnavis
45 Moduli, deformations and classifications of compact complex manifolds
D Sundararaman
46 Nonlinear problems of analysis in geometry and mechanics
M Atteia, D Bancel and I Gumowski
47 Algorithmic methods in optimal control
W A Gruver and E Sachs
48 Abstract Cauchy problems and functional differential equations
F Kappel and W Schappacher
49 Sequence spaces
W H Ruckle
50 Recent contributions to nonlinear partial differential equations
H Berestycki and H Brezis
51 Subnormal operators
J B Conway

52 Wave propagation in viscoelastic media
F Mainardi
53 Nonlinear partial differential equations and their applications: Collège de France Seminar. Volume I
H Brezis and J L Lions
54 Geometry of Coxeter groups
H Hiller
55 Cusps of Gauss mappings
T Banchoff, T Gaffney and C McCrory
56 An approach to algebraic K-theory
A J Berrick
57 Convex analysis and optimization
J-P Aubin and R B Vintner
58 Convex analysis with applications in the differentiation of convex functions
J R Giles
59 Weak and variational methods for moving boundary problems
C M Elliott and J R Ockendon
60 Nonlinear partial differential equations and their applications: Collège de France Seminar. Volume II
H Brezis and J L Lions
61 Singular systems of differential equations II
S L Campbell
62 Rates of convergence in the central limit theorem
Peter Hall
63 Solution of differential equations by means of one-parameter groups
J M Hill
64 Hankel operators on Hilbert space
S C Power
65 Schrödinger-type operators with continuous spectra
M S P Eastham and H Kalf
66 Recent applications of generalized inverses
S L Campbell
67 Riesz and Fredholm theory in Banach algebra
B A Barnes, G J Murphy, M R F Smyth and T T West
68 Evolution equations and their applications
F Kappel and W Schappacher
69 Generalized solutions of Hamilton-Jacobi equations
P L Lions
70 Nonlinear partial differential equations and their applications: Collège de France Seminar. Volume III
H Brezis and J L Lions
71 Spectral theory and wave operators for the Schrödinger equation
A M Berthier
72 Approximation of Hilbert space operators I
D A Herrero
73 Vector valued Nevanlinna Theory
H J W Ziegler
74 Instability, nonexistence and weighted energy methods in fluid dynamics and related theories
B Straughan
75 Local bifurcation and symmetry
A Vanderbauwhede
76 Clifford analysis
F Brackx, R Delanghe and F Sommen
77 Nonlinear equivalence, reduction of PDEs to ODEs and fast convergent numerical methods
E E Rosinger
78 Free boundary problems, theory and applications. Volume I
A Fasano and M Primicerio
79 Free boundary problems, theory and applications. Volume II
A Fasano and M Primicerio
80 Symplectic geometry
A Crumeyrolle and J Grifone
81 An algorithmic analysis of a communication model with retransmission of flawed messages
D M Lucantoni
82 Geometric games and their applications
W H Ruckle
83 Additive groups of rings
S Feigelstock
84 Nonlinear partial differential equations and their applications: Collège de France Seminar. Volume IV
H Brezis and J L Lions
85 Multiplicative functionals on topological algebras
T Husain
86 Hamilton-Jacobi equations in Hilbert spaces
V Barbu and G Da Prato
87 Harmonic maps with symmetry, harmonic morphisms and deformations of metrics
P Baird
88 Similarity solutions of nonlinear partial differential equations
L Dresner
89 Contributions to nonlinear partial differential equations
C Bardos, A Damlamian, J I Díaz and J Hernández
90 Banach and Hilbert spaces of vector-valued functions
J Burbea and P Masani
91 Control and observation of neutral systems
D Salamon
92 Banach bundles, Banach modules and automorphisms of C*-algebras
M J Dupré and R M Gillette
93 Nonlinear partial differential equations and their applications: Collège de France Seminar. Volume V
H Brezis and J L Lions
94 Computer algebra in applied mathematics: an introduction to MACSYMA
R H Rand
95 Advances in nonlinear waves. Volume I
L Debnath
96 FC-groups
M J Tomkinson
97 Topics in relaxation and ellipsoidal methods
M Akgül
98 Analogue of the group algebra for topological semigroups
H Dzinotyiweyi
99 Stochastic functional differential equations
S E A Mohammed

100 Optimal control of variational inequalities
V Barbu
101 Partial differential equations and dynamical systems
W E Fitzgibbon III
102 Approximation of Hilbert space operators. Volume II
C Apostol, L A Fialkow, D A Herrero and D Voiculescu
103 Nondiscrete induction and iterative processes
V Ptak and F-A Potra
104 Analytic functions – growth aspects
O P Juneja and G P Kapoor
105 Theory of Tikhonov regularization for Fredholm equations of the first kind
C W Groetsch
106 Nonlinear partial differential equations and free boundaries. Volume I
J I Díaz
107 Tight and taut immersions of manifolds
T E Cecil and P J Ryan
108 A layering method for viscous, incompressible L_p flows occupying R^n
A Douglis and E B Fabes
109 Nonlinear partial differential equations and their applications: Collège de France Seminar. Volume VI
H Brezis and J L Lions
110 Finite generalized quadrangles
S E Payne and J A Thas
111 Advances in nonlinear waves. Volume II
L Debnath
112 Topics in several complex variables
E Ramírez de Arellano and D Sundararaman
113 Differential equations, flow invariance and applications
N H Pavel
114 Geometrical combinatorics
F C Holroyd and R J Wilson
115 Generators of strongly continuous semigroups
J A van Casteren
116 Growth of algebras and Gelfand–Kirillov dimension
G R Krause and T H Lenagan
117 Theory of bases and cones
P K Kamthan and M Gupta
118 Linear groups and permutations
A R Camina and E A Whelan
119 General Wiener–Hopf factorization methods
F-O Speck
120 Free boundary problems: applications and theory, Volume III
A Bossavit, A Damlamian and M Fremond
121 Free boundary problems: applications and theory, Volume IV
A Bossavit, A Damlamian and M Fremond
122 Nonlinear partial differential equations and their applications: Collège de France Seminar. Volume VII
H Brezis and J L Lions
123 Geometric methods in operator algebras
H Araki and E G Effros
124 Infinite dimensional analysis–stochastic processes
S Albeverio
125 Ennio de Giorgi Colloquium
P Krée
126 Almost-periodic functions in abstract spaces
S Zaidman
127 Nonlinear variational problems
A Marino, L Modica, S Spagnolo and M Degiovanni
128 Second-order systems of partial differential equations in the plane
L K Hua, W Lin and C-Q Wu
129 Asymptotics of high-order ordinary differential equations
R B Paris and A D Wood
130 Stochastic differential equations
R Wu
131 Differential geometry
L A Cordero
132 Nonlinear differential equations
J K Hale and P Martinez-Amores
133 Approximation theory and applications
S P Singh
134 Near-rings and their links with groups
J D P Meldrum
135 Estimating eigenvalues with *a posteriori/a priori* inequalities
J R Kuttler and V G Sigillito
136 Regular semigroups as extensions
F J Pastijn and M Petrich
137 Representations of rank one Lie groups
D H Collingwood
138 Fractional calculus
G F Roach and A C McBride
139 Hamilton's principle in continuum mechanics
A Bedford
140 Numerical analysis
D F Griffiths and G A Watson
141 Semigroups, theory and applications. Volume I
H Brezis, M G Crandall and F Kappel
142 Distribution theorems of L-functions
D Joyner
143 Recent developments in structured continua
D De Kee and P Kaloni
144 Functional analysis and two-point differential operators
J Locker
145 Numerical methods for partial differential equations
S I Hariharan and T H Moulden
146 Completely bounded maps and dilations
V I Paulsen
147 Harmonic analysis on the Heisenberg nilpotent Lie group
W Schempp
148 Contributions to modern calculus of variations
L Cesari
149 Nonlinear parabolic equations: qualitative properties of solutions
L Boccardo and A Tesei
150 From local times to global geometry, control and physics
K D Elworthy

151 A stochastic maximum principle for optimal control of diffusions
U G Haussmann
152 Semigroups, theory and applications. Volume II
H Brezis, M G Crandall and F Kappel
153 A general theory of integration in function spaces
P Muldowney
154 Oakland Conference on partial differential equations and applied mathematics
L R Bragg and J W Dettman
155 Contributions to nonlinear partial differential equations. Volume II
J I Díaz and P L Lions
156 Semigroups of linear operators: an introduction
A C McBride
157 Ordinary and partial differential equations
B D Sleeman and R J Jarvis
158 Hyperbolic equations
F Colombini and M K V Murthy
159 Linear topologies on a ring: an overview
J S Golan
160 Dynamical systems and bifurcation theory
M I Camacho, M J Pacifico and F Takens
161 Branched coverings and algebraic functions
M Namba
162 Perturbation bounds for matrix eigenvalues
R Bhatia
163 Defect minimization in operator equations: theory and applications
R Reemtsen
164 Multidimensional Brownian excursions and potential theory
K Burdzy
165 Viscosity solutions and optimal control
R J Elliott
166 Nonlinear partial differential equations and their applications. Collège de France Seminar. Volume VIII
H Brezis and J L Lions
167 Theory and applications of inverse problems
H Haario
168 Energy stability and convection
G P Galdi and B Straughan
169 Additive groups of rings. Volume II
S Feigelstock
170 Numerical analysis 1987
D F Griffiths and G A Watson
171 Surveys of some recent results in operator theory. Volume I
J B Conway and B B Morrel
172 Amenable Banach algebras
J-P Pier
173 Pseudo-orbits of contact forms
A Bahri
174 Poisson algebras and Poisson manifolds
K H Bhaskara and K Viswanath

Pseudo-orbits of contact forms

A Bahri

École Polytechnique, Palaiseau

Pseudo-orbits of contact forms

Longman Scientific & Technical

Copublished in the United States with
John Wiley & Sons, Inc., New York

Longman Scientific & Technical
Longman Group UK Limited
Longman House, Burnt Mill, Harlow
Essex CM20 2JE, England
and Associated Companies throughout the world.

Copublished in the United States with
John Wiley & Sons, Inc., 605 Third Avenue, New York, NY 10158

First published 1988

ISSN 0269-3674

British Library Cataloguing in Publication Data
Bahri, A.
Pseudo-orbits of contact forms.—
(Pitman research notes in mathematics
series; ISSN 0269–3674; 173).
1. System analysis
I. Title
003 QA402
ISBN 0-582-01991-5

Library of Congress Cataloging-in-Publication Data
Bahri, A.
Pseudo-orbits of contact forms.
(Pitman research notes in mathematics series,
ISSN 0269-3674 ; 173)
Bibliography: p.
1. Variational inequalities (Mathematics)
2. Critical point theory (Mathematical analysis)
3. Compact spaces. I. Title. II. Series.
QA316.B243 1988 515′.64 87-33912
ISBN 0-470-21065-6 (USA only)

Printed and bound in Great Britain by
Biddles Ltd, Guildford and King's Lynn

Contents

Preface and introduction

In this book, we present a study of a variational problem related to contact form geometry.

The work was initiated with the study of the Weinstein conjecture about contact vector fields on compact orientable manifolds. This conjecture states that any contact vector field ξ on a $(2n+1)$-dimensional orientable manifold M has a periodic orbit, provided $H^1(M;\mathbb{Z}) = 0$. It was formulated by A. Weinstein [O1] after P. H. Rabinowitz's [O5] results about periodic orbits for Hamiltonian vector fields on star-shaped surfaces in $\mathbb{R}^{2n}$. In the $\mathbb{R}^{2n}$ and cotangent framework, it is now solved under suitable hypotheses (see C. Viterbo [O6]; H. Hofer, C. Viterbo [O7]). The full conjecture is still open. Motivated by this conjecture, we examine here a variational problem on a submanifold of the loop-space of M. The features of this variational problem are interesting to us from many points of view: it is a "non compact" problem, in the sense which is given to this expression today, i.e. it does not satisfy the Palais-Smale condition. Starting from here, we have been led to study this failure of the Palais-Smale condition and to try to understand its incidence on the variational problem itself: we analyse in this book the "critical points at infinity" of this variational problem, that is the (pseudo)-gradient lines which do not end up at a critical point (a periodic orbit of ξ). This has a corresponding aspect in contact form geometry: we are led to introduce a notion of conjugate points adapted to this special geometry; with this notion, we are able to describe geometric curves, together with a parametrization, which are the "critical points at infinity" of this variational problem.

In fact, through this failure of Palais-Smale, a full new variational problem arises "at infinity", involving essentially the geometric features of the problem.

Lastly, this study has also a corresponding aspect in dynamical systems: the dynamics of the contact form along a vector field v of its kernel is essential to understanding these critical points at infinity. It is expressed in a pendulum equation, which is described in great detail below.

From the analysis point of view these critical points at infinity arise through a singular-perturbation equation. This singular-perturbation equation describes the way the tangent vector to curves approximating these critical points at infinity behave. It is quite surprising, while there is no control at all on the length of such curves, to be able to derive convergence theorems for these curves in graph. The geometric nature of the problem somewhat restrains the oscillations allowed by the singular-perturbation equation, or makes them compensate, leading to a complete control on these curves.

Lastly, one can attribute to these curves a Morse index allowing an extension of Morse Theory. In order to make the presentation of this book clearer, we have included in the introduction a summary, following immediately this quick presentation. This summary is based on [O8].

Acknowledgements

Special thanks are due to Professor H. Brezis: he gave me very warm encouragement throughout this work and thought it worth publishing. Without his help, I do not think that I would have been able to write this work or to publish it.

Thanks are also due to Professor F. Browder and Professor L. Nirenberg. Both have allowed me to work quietly on this project, without the pressure of fast publication.

I wish also to thank all my collaborators and friends: Professor Michel Demazure of the Ecole Polytechnique; Alain Chenciner; Daniel Bennequin; H. Berestycki; J. M. Coron; P. L. Lions.

Finally, many thanks are due to Mrs Marie-Jo Lécuyer for her original typing of such a difficult manuscript.

1. THE DATAS - THE FUNCTIONALS

Given (M, α), ξ is the Reeb vector field of α, i.e.

$$\alpha(\xi) \equiv 1; \quad d\alpha(\xi, \cdot) \equiv 0 . \tag{I1.1}$$

Let $H^1(S^1;M)$ be the space of H^1-loops on M. On $H^1(S^1;M)$, there are some "natural" functionals whose critical points are orbits of ξ; namely:

$$\int_0^1 \alpha_x(\dot{x})\,dt; \quad \int_0^1 \alpha_x(\dot{x})^2 dt . \tag{I1.2}$$

The gradient of these functionals along a variation z in $T_x H^1(S^1;M)$ is:

$$\int_0^1 d\alpha_x(\dot{x}, z)\,dt; \quad 2\int_0^1 \alpha_x(\dot{x})\left(\frac{d}{dt}\alpha_x(z) - d\alpha_x(\dot{x}, z)\,dt\right. \tag{I1.3}$$

and we have the following Proposition.

Proposition O.1: Critical points of $\int_0^1 \alpha_x(\dot{x})^2 dt$ of strictly positive energy are periodic orbits to ξ. Critical points of $\int_0^1 \alpha_x(\dot{x})\,dt$ of non-zero energy are, after reparametrization, periodic orbits to ξ.

Such a proposition leaves some hope for studying the related variational problems on $H^1(S^1;M)$ and trying to find an existence mechanism from Morse theory.

Nevertheless, the variational problems are very much ill-posed and, in some senses, are useful in showing that a functional is not enough to derive the existence of critical points. Indeed, both functionals we are considering have very bad features, which may be summed up as follows:

1. The gradients are not Fredholm.
2. The second variation at a critical point has an infinite Morse index.
3. The level sets have the same homotopy type.
4. The Palais-Smale condition is not satisfied.

In order to see 1., one writes for instance the gradients in local coordinates. 2. also is almost immediate. 4. will be discussed in the sequel, at least for the restriction of these variational problems to a submanifold of the loop space. The arguments are more complicated to demonstrate 3. Nevertheless, there is a result by S. Smale [O2], which shows 3. partly:

Theorem (Smale [O2]): Let α be a contact form on M. Let $\mathcal{L}_\alpha = \{x \in H^1(S^1;M) \text{ s.t. } \alpha_x(\dot{x}) \equiv 0\}$. The injection of $\mathcal{L}_\alpha$ in $H^1(S^1;M)$ is then a weak homotopy equivalence

Denoting now:

$$J(x) = \int_0^1 \alpha_x(\dot{x})^2 dt, \tag{I1.4}$$

$$J^a = \{x \in H^1(S^1;M) \text{ s.t. } J(x) \le a\}. \tag{I1.5}$$

Smale's theorem asserts that J^∞ and J^0 are homotopy equivalent. Although this does not prove 3., it is very reasonable after having studied the problem, to conjecture that $J^a = J^b$ for any a and b. Such a result, which indeed holds, has nevertheless not very much interest, but in giving another bad feature of these functionals.

With 1., 2., 3. and 4., no variational theory is in fact possible and some thought shows that either of these forbids the use of variational arguments:

Indeed, if the gradients (or pseudo-gradients) are not Fredholm (at least in some weak sense), no Morse lemma is provided; and therefore one does not know any more whether or not a given critical point induces a difference of topology in the level sets.

If the Morse index of a critical point is infinite, and if the gradient is Fredholm, it is known that this critical point does not induce a difference of topology at least if the space of variations (here $H^1(S^1;m)$) is modelled on a Hilbert or a reflexive Banach space. As the gradient is not Fredholm here, we cannot derive such a result.

The homotopy type of the level sets is another key-feature. Usually, in order to find critical points, one tries to find variations in the homotopy type of these level sets. Here, there are none; and if there were some, we would not be able, for instance because of 1., or 2., or 4., to derive any existence result.

Lastly, the Palais-Smale condition is also a key tool as can be seen on very simple functions from $\mathbb{R}$ to $\mathbb{R}$, for instance.

Therefore, Proposition 1 is useless.

The remaining possibilities are either to find other functionals in order to study the problem, other methods (see, for instance, Paul Rabinowitz's approach when M is star-shaped in $\mathbb{R}^{2n}$, or more recent developments by C. Viterbo); or to restrict the variations, i.e. to study the variational problems on suitable submanifolds of the loop space[(★)].

Following Smale's theorem, there are "natural" submanifolds of the loop space where it is tempting to consider the variational problem. Namely, considering:

v a vector field in $\ker \alpha$; which we assume to be non-singular. (I1.6)

We introduce:

$$\beta = d\alpha(v, \cdot), \tag{I1.7}$$

$$\mathcal{L}_\beta = \{x \in H^1(S^1; M) \text{ s.t. } \beta_x(\dot{x}) \equiv 0\}, \tag{I1.8}$$

$$C_\beta = \{x \in H^1(S^1; M) \text{ s.t. } \beta_x(\dot{x}) \equiv 0; \; \alpha_x(\dot{x}) = c^t > 0\}. \tag{I1.9}$$

From now on, we denote:

$J(x) = \int_0^1 \alpha_x(\dot{x})^2 dt$, if the space of variations is $\mathcal{L}_\beta$, (I1.10)

$J(x) = \alpha_x(\dot{x}) = \int_0^1 \alpha_x(\dot{x})\, dt$, if the space of variations is C_β. (I1.11)

There are arguments in favor of such a choice of restricted variations, which are summed up in the following proposition:

(★) The variational framework presented here is a joint work of D. Bennequin and the author.

Proposition O.2:

1) Generically on v, the singularities of $\mathcal{L}_\beta/M$ lie on finitely many geometric curves. If β is a contact form, $\mathcal{L}_\beta/M$ is a submanifold of the loop space.
2) C_β is, generically on v, a submanifold of the loop space.
3) The critical points of J on $\mathcal{L}_\beta/M$, of finite Morse index, are periodic orbits of ξ. The same holds true on C_β.

Proposition O.2 leaves some hope that we will be able to find those periodic orbits by restricting the variational problem to $\mathcal{L}_\beta$ or C_β. In fact, when β is chosen to be a contact form (which imposes restrictions on v), the topology of $\mathcal{L}_\beta$ is known, through Smale's theorem, to be that of the loop space on M. To relate this to a more classical framework, we simply say here that if M is a convex hypersurface in $\mathbb{R}^{2n}$, then such a choice of v is possible.

We will rediscuss points 1 to 4 for these new variational problems later on. For the moment, we start a description of the dynamics of α along v.

2. THE PENDULUM EQUATION

From now on, we assume, for sake of simplicity, that β is itself a contact form. By transversality of β and α, we then have:

$$(\beta \wedge d\beta)(\alpha \wedge d\alpha) > 0. \tag{I2.1}$$

We normalize v so that:

$$\beta \wedge d\beta = \alpha \wedge d\alpha. \tag{I2.2}$$

Let w be the Reeb vector field of β. We set:

$$\alpha(w) = \bar{\mu}; \quad d\alpha([\xi,[\xi,v]],[\xi,v]) = \tau \tag{I2.3}$$

and we have:

$$\left.\begin{aligned}\bar{\mu} &= d\alpha(v,[v,[\xi,v]])\\ d\bar{\mu}(\xi) &= \bar{\mu}_{\xi} = d\alpha([\xi,v],[v,[\xi,v]])\end{aligned}\right\} \qquad (I2.4)$$

where $d\bar{\mu}(\xi)$ is the value of the differential of $\bar{\mu}$ on ξ.

There is a geometric property of the dynamics of α along v which we single out now:

Let x_0 be a point of M and let φ_s be the one-parameter group generated by v. We choose at x_0 two vectors in $T_{x_0}(M)$, denoted $e_1(0)$ and $e_2(0)$ such that

$$\alpha \wedge d\alpha(v, e_1(0), e_2(0)) < 0. \qquad (I2.5)$$

In particular, $(v, e_1(0), e_2(0))$ is a basis of $T_{x_0}(M)$. Let:

$$e_1(s) = D\varphi_s(e_1(0)); \quad e_2(s) = D\varphi_s(e_2(0)). \qquad (I2.6)$$

The plane $P_s = \{z \in T_{x_s}(M) \text{ s.t. } \alpha_{x_s}(z) = 0\}$, where $x_s = \varphi_s(x_0)$, intersects transversally the plane $\text{span}(e_1(s), e_2(s))$ along a direction:

$$u(s) = \alpha_{x_s}(e_1(s))\,e_2(s) - \alpha_{x_s}(e_2(s))\,e_1(s). \qquad (I2.7)$$

We then have the following:

<u>Proposition O.3</u>: $u(s)$ rotates monotonically (strictly) from $e_1(s)$ to $e_2(s)$ when s increases.

Proposition O.3 expresses exactly the fact that α is a contact form. Due to this rotation property, we introduce the two following definitions:

<u>Definition O.1</u>: $x_s = \varphi_s(x_0)$ is a <u>coincidence point</u> of x_0 if $u(s)$ has completed, in the $(e_1(s), e_2(s))$ frame, k rotations, $k \in \mathbb{Z}$, between time 0 and s. x_s is an <u>oriented coincidence point</u> if $k \equiv 0 \ (2)$.

Therefore, if x_s is a coincidence point of x_0,

$$(\varphi_s^{\star}\alpha)_{x_0} = \lambda(s, x_0)\,\alpha_{x_0}; \qquad \lambda \neq 0. \qquad (I2.8)$$

$\lambda(s, x_0)$ is positive if x_s is an oriented coincidence point of x_0.

Definition O.2: Let x_s be a coincidence point of x_0. x_s is said to be conjugate to x_0 if $\lambda(s, x_0) = 1$.

As can be checked, these notions of coincidence and conjugate points are intrinsic, i.e. they depend only on α and v.

Starting from here, we have a two-fold discussion. On the one hand, there is a discussion on the existence of coincidence and conjugate points which we will complete later on. On the other hand, there are the analytical aspects of these definitions. To summarize these analytical aspects, we first point out the following second order differential equation:

Denoting L_v the Lie-derivative of forms along v, we have three one-forms $\alpha, L_v\alpha, L_v^2\alpha$ which vanish on v. They must therefore be dependent and we have the differential equation:

$$a(x)(L_v^2\alpha)_x + b(x)(L_v\alpha)_x + c(x)\alpha_x \equiv 0. \qquad (D0)$$

Recalling that:

$$\left.\begin{aligned} &\beta\wedge d\beta(\xi, v, w) = d\beta(\xi, v) = (L_v^2\alpha)(\xi) = \alpha\wedge d\alpha(\xi, v, w)\\ &= d\alpha(v, w) = \beta(w) = 1,\\ &d\alpha(v, \xi) \equiv 0. \end{aligned}\right\} \qquad (I2.9)$$

We derive:

$$c(x) \equiv -a(x). \qquad (I2.10)$$

Recalling that:

$$\left.\begin{aligned} &\alpha(w) = \bar{\mu},\\ &(L_v\alpha)(w) = \beta(w) = 1,\\ &(L_v^2\alpha)(w) = d\beta(v, w) = 0. \end{aligned}\right\} \qquad (I2.11)$$

We have:

$$b(x) + c(x)\bar{\mu} \equiv 0. \tag{I2.12}$$

Therefore, (D0) yields:

$$L_v^2 \alpha + \alpha - \bar{\mu} L_v \alpha \equiv 0. \tag{D1}$$

This is the first pendulum equation that we encounter. In case $M = S^3$ and $\alpha = \alpha_0$ is the standard contact form on S^3 with v defining a Hopf fibration of S^3 over S^2, $\bar{\mu} \equiv 0$ and we have the exact pendulum equation.

Next, we may write (D1) as a transport equation of forms along v, in a selected basis. Namely, introducing:

$$\gamma = d\beta(v, \cdot). \tag{I2.13}$$

We may write along a v-orbit x_s the transport equation of forms in the frame $(\alpha_{x_0}, -\gamma_{x_0}, \beta_{x_0})$. Let

$$\left.\begin{array}{ll} u^\star(0) = \begin{bmatrix} A^\star(0) \\ B^\star(0) \\ C^\star(0) \end{bmatrix} & \text{be the coordinates of a one-form at } x_0 \\ u^\star(s) = \begin{bmatrix} A^\star(s) \\ B^\star(s) \\ C^\star(s) \end{bmatrix} & \text{be the coordinates of } \phi^\star_{-s} u^\star(0) \text{ at } x_0. \end{array}\right\} \tag{I2.14}$$

We then have:

<u>Proposition O. 4:</u>

$$\left.\begin{array}{l} \dot{u}^\star(s) = -{}^t\Gamma(x_s)\, u^\star(s) \\ u^\star(0) = \begin{bmatrix} A^\star(0) \\ B^\star(0) \\ C^\star(0) \end{bmatrix} \end{array}\right\}$$

where

$$\Gamma(x) = \begin{bmatrix} 0 & 0 & -1 \\ 0 & 0 & \bar{\mu}_\xi \\ 1 & 0 & \bar{\mu} \end{bmatrix}$$

or else

$$\left.\begin{aligned} \dot{A}^\star(s) &= C^\star(s) + \bar{\mu}_\xi B^\star(s) \\ \dot{C}^\star(s) &= -\dot{A}^\star(s) - \bar{\mu} C^\star(s) \\ \dot{B}^\star(s) &= 0. \end{aligned}\right\}$$

In particular, if $B^\star(0) = 0$, i.e. if the initial data belong to $\mathrm{span}\{\alpha_{x_0}, \beta_{x_0}\}$, then $B^\star(s) = 0$ and the differential equation reduces to:

$$\left.\begin{aligned} \dot{A}^\star(s) &= C^\star(s) \\ \dot{C}^\star(s) &= -A^\star(s) - \bar{\mu}(x_s)\, C^\star(s) \end{aligned}\right\} \tag{D2}$$

once again the pendulum equation. Coincidence and conjugate points to x_0 are then described as follows:

Proposition O.5: Considering the initial data

$$\begin{bmatrix} 1 \\ 0 \\ 0 \end{bmatrix},$$

x_s is a coincidence point of x_0 if

$$u^\star(s) = \begin{bmatrix} A^\star(s) \\ 0 \\ 0 \end{bmatrix};$$

x_s is conjugate to x_0 if

$$u^\star(s) = \begin{bmatrix} 1 \\ 0 \\ 0 \end{bmatrix}.$$

Therefore, conjugate points are related to periodic solutions of (D2).

Through these analytical descriptions of coincidence and conjugate points, we will encounter them in the previously defined functional framework, while studying the Palais-Smale condition.

We discuss now the existence of coincidence and conjugate points: There are two opposite situations which are relevant to this discussion, which we summarize in two opposite hypotheses:

Hypothesis A1: α turns well along v, i.e. every x_0 on M has a coincidence point distinct from itself. In this case, we say that (α, v) is elliptic.

Hypothesis A2: v belongs to a codimension one foliation transverse to α.

It is proved later that a large class of contact forms (presumably all of them) admit a transverse foliation $\tilde{\gamma}$. If v belongs then to $\ker \tilde{\gamma} \cap \ker \alpha$, no point x_0 of M can possibly admit another coincidence point distinct from itself. (A complete rotation forces $\tilde{\gamma}$ and α to coincide somewhere in between.)

In this situation, the variational problem, as well as the topology of C_β, have drastically different features from the case in hypothesis A1. This will be described in a later publication.

When (A1) holds, there are, of course, infinitely many coincidence points to each given point x_0 of M and the question which is left open is the existence of conjugate points. In this framework, we introduce:

Definition O.3: Under (A1), let $\gamma^i : M \to \mathbb{R}$ be the function mapping $x_0 \in M$ to the ith-time $s = \gamma^i(x_0)$ such that x_s is a coincidence point of $x_0 (i \in \mathbb{Z})$. Let $f^i : M \to M$ be the diffeomorphism of M mapping x_0 on $x_{\gamma^i(x_0)}$. Let $\mu_i(x_0)$ be the coefficient of $(\varphi^\star_{\gamma^i(x_0)} \alpha)_{x_0}$ on α_{x_0}, i.e.:

$$(\varphi^\star_{\gamma^i(x_0)} \alpha)_{x_0} = \mu_i(x_0)\, \alpha_{x_0}.$$

Conjugate points are defined by the equation:

$$\mu_i(x_0) = 1. \tag{I2.15}$$

Therefore, points admitting a conjugate point distinct from themselves lie generically on unions of hypersurfaces $\Sigma_i = \{x \in M \text{ s.t. } \mu_i(x) = 1\}$. An interesting situation is provided here by $(S^3, \alpha = \lambda \alpha_0)$ where $\lambda \in C^\infty(S^3, \mathbb{R}^{+\star})$; then all points admitting a conjugate point distinct from themselves lie on

$\Sigma = \{x \in S^3$ s.t. $\lambda(x) = 1\}$. If x belongs to Σ, the conjugate points of x are x and $-x$. It is also interesting to notice that the existence of conjugate points is related to the behavior of the function:

$$(\varphi_s^\star \alpha)_{x_0} \wedge \varphi_s^\star(d\alpha)_{x_0} / (\alpha \wedge d\alpha)_{x_0} \tag{I2.16}$$

in particular when s goes to $\pm\infty$. Therefore, the asymptotic behavior or v, the fact that it expands or contracts volume is relevant to the conjugate points.

Before concluding this section, we introduce:

$$\Gamma = \{x_0 \in M \text{ s.t. } \lambda(s, x_0) \geq 1 \text{ for any } s \text{ such that } x_s \text{ is a coincidence point of } x_0\}, \tag{I2.17}$$

and in the framework of (A1), the following hypotheses about v and the dynamics of α along v. We choose a distance on M, denoted d and a related norm for differentials, denoted $\| \ \|$:

(H2) v has a periodic orbit;

(H3) for a vector field v_1, non-singular and collinear to v, we have: $\exists\, k_1 > 0$ such that $\|D\varphi_s^1\| \leq k\ \forall s \in \mathbb{R}$, where φ_s^1 is the one-parameter group of v_1;

(H4) $\exists\, k_2$ and $k_3 > 0$ such that $\forall\, i \in \mathbb{Z}$, we have:

$$\exists\, k_2 d(x,y) \leq d(f^i(x), f^i(y)) \leq k_3 d(x,y)\ \forall x, y \in M;$$

(H5) $\exists\, k_4 > 0$ such that $|\mu_i(x) - \mu_i(y)| \leq k_4 d(x,y)\ \forall x, y \in M$;

(H6) $\exists\, \rho > 0$ such that $\forall x \in M$, the set $C_\rho(x) = \{f^i(x) / |\mu_i(x) - 1| < \rho;\ i \in \mathbb{Z}\}$ is finite.

(H2)-(H6) are heavy hypotheses, under which clean statements about critical points at infinity can be made. Nevertheless, these hypotheses are necessary only at the end of the analysis of the phenomenon; our feeling is that they can be considerably weakened.

3. TOPOLOGICAL ASPECTS OF THE VARIATIONAL PROBLEMS

3.1 The topology of $\mathcal{L}_\beta$ and C_β

We limit our discussion to the case when (A1) holds. We have already pointed out Smale's theorem, which states that, if β is a contact form, the injection of $\mathcal{L}_\beta = \{x \in H^1(S^1;M)$ s.t. $\beta_x(\dot{x}) \equiv 0\}$ into $H^1(S^1;M)$ is a weak homotopy equivalence. Smale proves this theorem by showing that the projection:

$$\begin{aligned} &\mathcal{L}_\beta \to M \\ &x \to x(0) \end{aligned} \tag{I3.1}$$

is a Serre-fibration.

For this, he needs the hypothesis that β is a contact form and he gives an example where this is no longer true, when β is no longer a contact form. There is a slight improvement of Smale's theorem that we have completed. Introducing:

$$F = \{x \in M \mid (\beta \wedge d\beta)_x = 0\} \tag{I3.2}$$

we have:

Proposition O.6: Assume F is a codimension one submanifold of M and assume that β is transverse to F at any point x of F. $\mathcal{L}_\beta \hookrightarrow H^1(S^1;M)$ is then a weak homotopy equivalence.

Corollary: Generically on β, $\pi_0(\mathcal{L}_\beta) = \pi_1(M)$.

The proof of Proposition O.6 follows the method of Stephen Smale; nevertheless, as β is no more a contact form, we need some other reduction of β in the neighbourhood of a point of F. The Corollary follows immediately Proposition O.6, through a transversality argument.

Remark: When β is not transverse to F, the trace of β on F has singularities. Near the elliptic ones, β has a nice reduction (see D. Bennequin [O3]) which was shown by D. Bennequin, in a celebrated theorem, to provide exotic contact structures on $\mathbb{R}^3$.

In fact, the same paper by S. Smale gives some insight on the topology of C_β under (A1). Namely, S. Smale proves in [O.2] the following theorem, later developed in many directions:

<u>Theorem</u> (S. Smale [O2]): Let N be a compact surface without boundary. Let $Imm(S^1;N)$ be the space of immersions of S^1 in N and $\Omega(ST^\star N)$ be the space of continuous closed curves in the cotangent S^1-bundle over N. Then the inclusion $Imm(S^1, N) \hookrightarrow \Omega(ST^\star N)$ is a weak homotopy equivalence.

This theorem is related to the topology of C_β as follows.

Consider α_0 the standard contact form on $ST^\star N$ and let α be a constant form obtained by multiplying α_0 by a C^∞ positive function λ. Let v be the vector field defining the natural fibration of $ST^\star N$ over N.

It is not difficult to see that, in this context, the form α_0, hence the form α, makes one (and only one) complete revolution along the fiber (in the sense of rotation introduced in Proposition O. 3; see Definitions O. 1, O. 2 and O. 3).

Suppose now $\beta = d\alpha(v,\cdot)$ is also a contact form, which means, from the Hamiltonian point of view, the convexity of the kinetic energy in the cotangent variable. As $\beta(v)$ is zero, β is also tangent to the fibers. Being transverse to α, β also makes one (and only one) complete revolution along the fiber.

Let x be an immersion of S^1 in N; $x(t_0)$ be a fixed point on the curve and $\dot{x}(t_0)$ be its tangent vector.

Let $p : ST^\star N \to N$ be the natural projection.

As β turns once along the fiber, ξ, which belongs to the kernel of β, also turns once along this fiber. This implies the existence and the unicity of a point $(x(t_0), \theta)$, projecting on $x(t_0)$, in $ST^\star N$ such that:

$$dp_{(x(t_0),\theta)}(\xi_{(x(t_0),\theta)}) = \mu\dot{x}(t_0); \quad \mu > 0, \tag{I3. 3}$$

where $dp_{(x(t_0),\theta)}$ is the differential of p at $(x(t_0), \theta)$ and $\xi_{(x(t_0),\theta)}$ is the value of ξ at $(x(t_0), \theta)$.

When t_0 runs in S^1, this device allows us to lift the immersion x in a closed curve $y(\cdot) = (x(\cdot), \theta(\cdot))$ belonging to $\Omega(ST^{\star}N)$. This curve y is defined by the following equation:

$$\left.\begin{aligned} & p(y(t)) = x(t) \\ & dp(\dot{y}(t)) = \dot{x}(t)\ ;\ dp_{y(t)}(\xi_{y(t)}) = \mu(t)\,\dot{x}(t)\ ;\ \mu > 0. \end{aligned}\right\} \tag{I3.4}$$

Equation (3.4) yields that $\dot{y}(t) = \frac{1}{\mu(t)} \xi_{y(t)} + \lambda(t)\, v_{y(t)}$. Hence:

$$\dot{y}(t) = a_1 \xi_{y(t)} + b_1 v;\quad a_1 > 0. \tag{I3.5}$$

As a_1 is continuous and positive, one can reparameterize the curve $y(\cdot)$; hence obtaining $z(\cdot)$ such that:

$$\dot{z}(t) = a \xi_{z(t)} + bv_{z(t)};\quad a = \text{Cst.} > 0. \tag{I3.6}$$

We defined in this way a map h:

$$\mathrm{Imm}(S^1, N) \xrightarrow{h} C_\beta \tag{I3.7}$$

Conversely, if y is a given curve in C_β, $p(y)$ belongs to $\mathrm{Imm}(S^1, N)$. Indeed:

$$\dot{y} = a\xi + bv;\quad a = \text{positive constant.} \tag{I3.8}$$

Hence

$$dp(\dot{y}) = \dot{\overline{p(y)}} = a\,dp(\xi). \tag{I3.9}$$

As $dp(\xi)$ is not zero, $p(y)$ belongs to $\mathrm{Imm}(S^1, N)$. We thus have another map:

$$\hat{h} : C_\beta \to \mathrm{Imm}(S^1, N). \tag{I3.10}$$

One can check that $\hat{h} \circ h(x)$ is a reparameterization of $x(\cdot)$ while $h \circ \hat{h}(y)$ is a reparameterization of y.

This, together with some approximation argument appropriate for dealing with the H^1-topology, proves that C_β has the same topology as $\mathrm{Imm}(S^1, N)$; hence as $\Omega(ST^{\star}N)$ by Smale's theorem.

The same argument works if M is a finite covering of $ST^\star N$; $Imm(S^1, N)$ and $\Omega(ST^\star N)$ are in this context to be replaced by $Imm_M(S^1, N)$ and $\Omega_M(ST^\star N)$ where these indexed sets denote the results of curves which lift to M as closed curves. S. Smale's theorem suggests the following result, which may be derived intuitively through complicated variational arguments:

Conjecture: Assume β is a contact form. Under (A1), the injection of C_β in $H^1(S^1;M)$ is a weak homotopy equivalence.

3.2 The change of topology induced by J

We first introduce the following definitions; d is a distance on M.

Definition O.4: Let v_1 be a nonsingular vector field collinear to v ($v_1 = \lambda v$; $\lambda > 0$) and let φ^1_s be the one-parameter group generated by v_1. v is said to be ξ-conservative if there exist v_1 and a constant K such that

$$\|D\varphi^1_s(\xi)\| \leq K \quad \forall s \in \mathbb{R}. \tag{I3.11}$$

Here $\| \ \|$ stands for any norm of differentiable maps from M to M.

Definition O.5: v is said to be α-nonresonant if there exists a non-singular one-differential form θ, transverse to α and tangent to v, such that:

> there exists $\varepsilon_4 > 0$ such that for any x in M and s_1 in $\mathbb{R}$ satisfying $d(\varphi_{s_1}(x), x) < \varepsilon_4$, $(\varphi^\star_{s_1}\alpha)_x$ is transverse to θ_x. (I3.12)

Definition O.6: C_β is said to be branched on v if there exists $C > 0$ such that for any $\varepsilon > 0$, there exists x in C_β, with $\dot{x} = a\xi + bv$, $0 < a < \varepsilon$ and $\left|\int_0^1 b(t)\,dt\right| \geq C$.

Here are some comments about these definitions:

First, if v is ξ-conservative, there is no hyperbolic invariant set to v. Indeed, on such a set, Equation (I3.11) cannot hold.

Hence, v might be called "elliptic" in this situation.

Second, we consider a point x in M which is recurrent to v. Let f_x be the local Poincaré map on a section transverse to v at x.

As α rotates monotonically along v from x to $f_x(x)$, we can define:

$$\tilde{\varphi}(x) = \text{total rotation of } \alpha \text{ from } x \text{ to } f_x(x) \qquad (I3.13)$$

and

$$n_x = \left[\frac{\tilde{\varphi}(x)}{\pi} \right]. \qquad (I3.14)$$

If v is α-nonresonant, n_x is constant on the connected components of the v-recurrent set.

Third, C_β is branched on v if there exists curves of C_β close to recurrent orbits of v. We then have:

Lemma O.1: If v has a periodic orbit and if (A_1) holds, C_β is branched on v.

Finally, we notice that in case v is the fiber vector field of an S^1-fibration over a surface N, then v is ξ-conservative, α-nonresonant and C_β is branched on v. With these definitions, we have the following Proposition:

Proposition O.7: Assume (A_1) holds and v is ξ-conservative, α-nonresonant and C_β is branched on v. Then, for $\varepsilon > 0$ small enough, $\pi_0(J^\varepsilon)$ is infinite, where

$$J^\varepsilon = \{x \in C_\beta \text{ s.t. } J(x) \leq \varepsilon\}.$$

It is difficult to sum up the results of this section.

The conjecture we formulated may be proven through very complicated variational problems (one studies on $\mathcal{L}_\beta$ the functional $\left(\int_0^1 \alpha_x(\dot{x})^2 dt\right)^{\frac{1}{2}} - \int_0^1 \alpha_x(\dot{x})\, dt)$). The methods are nevertheless too complicated and another approach would be welcome, in order to build a clean proof.

Nevertheless, Smale's theorem shows that the conjecture holds for $(ST^\star N, \lambda\alpha_0)$, $\lambda \in C^\infty(ST^\star N, \mathbb{R}^{+\star})$ and also for all finite coverings of $ST^\star N$, (M, α)

such that $q_\star(\ker \alpha) = \ker \alpha_0$, where q is the projection $M \to ST^\star N$. The simplest example of such coverings is, of course $(S^3, \lambda \alpha_0)$, α_0 being the standard contact form on S^3. In this case, with v defining a Hopf fibration of S^3 over S^2, $\pi_0(C_\beta) = 0$; while $\pi_0(J^\varepsilon)$ is infinite.

This difference of topology is drastic. Some thought shows that it cannot only be due to the periodic orbits of ξ.

Indeed, we may assume, by transversality arguments on λ, that given $a_0 > 0$, there are only finitely many of these orbits of length less than a_0 and that they all are nondegenerate.

On the other hand, the Morse index of J on C_β at these critical points increases to $+\infty$ with the length of these periodic orbits. Although the gradient is not Fredholm, a perturbation of J near these critical points, whose gradient is Fredholm, induces then a finite difference of topology in the indices 0 and 1; therefore a contradiction. This shows that the only possibility is for the Palais-Smale condition to fail and that it is precisely this failure that will induce such a heavy difference of topology. Starting from here, we see that we are directly led to the idea of critical points at infinity, which have to explain such a drastic difference of topology. Therefore, these critical points at infinity cannot be mere sequences.

4. A COMPACTIFYING DEFORMATION LEMMA

We point out here that the compactifying deformation lemma we are presenting provides us with a control on <u>each flow-line</u> of the vector fields we will introduce. These vector fields, which depend on a parameter $\varepsilon > 0$, and which decrease the functional J, <u>do not</u> satisfy the Palais-Smale condition <u>on sequences,</u> unless we specify that these sequences belong to the same flow-line.

As we will see later on in this introduction, there are intuitive reasons which imply that J does not satisfy the Palais-Smale condition or else any weaker compacity criterion. This is already hinted at by the study of the topology of C_β and the differences of topology induced by J, which are far too drastic to be due only to the ξ periodic orbits.

For the time being, we can get a first understanding of the problem: along (pseudo)-gradient lines, one knows that $J(x)$ remains bounded; which provides a bound on the ξ-component of $\dot{x} = a\xi + bv$. (As x belong to C_β, $\dot{x}$ splits on (ξ, v) with a constant component on ξ.)

But we have no control on the v-component and it might happen that this v-component becomes infinite on a gradient line.

On the other hand, if $\partial J(x)$ goes to zero, we get: ($\partial J(x)$ is the gradient of J)

$$\partial J(x) \cdot z = -\int_0^1 b\eta \, dt \to 0 \quad (z = \lambda\xi + \mu v + \eta w). \tag{I4.1}$$

Hence

$$\left| \int_0^1 b\eta \, dt \right| \leq \varepsilon \, |z|_{H^1}. \tag{I4.2}$$

The H^1-norm of z contains an L^2-norm of $\dot{\eta}$ and other terms also. Hence (I4.2) will give a very weak information on b and anyway, no bound whatsoever (there are b-terms in $|z|_{H^1}$); hence the failure of the Palais-Smale condition.

In this situation, we construct a special deformation lemma, which will compactify the situation, by introduction of a certain "viscosity term" which allows us to control the b-component. The deformation depends on a certain parameter $\varepsilon > 0$ and is induced by a vector field Z_ε such that the scalar product $\partial J(x) \cdot Z_\varepsilon(x)$ is positive. However, Z_ε can be zero while $\partial J(x)$ is not zero; which will bring the differential equation giving rise to critical points at infinity. Let

$$I(x) = \int_0^1 b^2 dt \quad (\dot{x} = a\xi + bv). \tag{I4.3}$$

I is a C^2 functional on C_β on which we wish to have control along deformation lines of the level sets of J.

For this purpose, we introduce the following vector field on C_β:

$$Z = |\partial J| \, \partial I + |\partial I| \, \partial J, \tag{I4.4}$$

∂J and ∂I are the gradients of I and J and $|\partial I|$, $|\partial J|$, are their norms which respect to the H^1-metric of C_β (inherited for the H^1-metric on $H^1(S^1, M)$). We will note (,) the H^1 scalar product on the tangent space to C_β. The idea underlying the deformation lemma is to use Z selectively: where $\partial J(x)$ is small (in fact, some other quantity dominating $|\partial J(x)|$), we use $-Z$ to deform. Otherwise, we use $-\partial J(x)$.

Let $\varepsilon > 0$ be given. We consider:

$$\left.\begin{array}{l} \omega_\varepsilon : \bar{\mathbb{R}} \to [0,1] \\ \omega_\varepsilon(x) = 1 \text{ if } x \geq \varepsilon \\ \omega_\varepsilon(x) = 0 \text{ if } x \leq \varepsilon/2 \\ \omega_\varepsilon \in C^\infty, \end{array}\right\} \tag{I4.5}$$

$$\left.\begin{array}{l} l_\varepsilon : C_\beta \to [0,1];\ l_\varepsilon \in C^\infty \\ l_\varepsilon = 0 \text{ on an } \varepsilon/2\text{-neighborhood of the} \\ \qquad \text{critical points of } J \\ l_\varepsilon = 1 \text{ outside an } \varepsilon\text{-neighborhood of these points.} \end{array}\right\} \tag{I4.6}$$

$$\left.\begin{array}{l} \varphi : C_\beta / \{\text{critical points of } J\} \to \bar{\mathbb{R}} \\ \varphi(x) = \dfrac{|\partial J|(x)}{|\partial I|(x)} I(x) \text{ if } |\partial I|(x) \text{ is nonzero} \\ \varphi(x) = +\infty \text{ is } |\partial I|(x) = 0. \end{array}\right\} \tag{I4.7}$$

We introduce then the following vector field on C_β:

$$Z_\varepsilon(x) = l_\varepsilon(x)\,(\omega_\varepsilon(\varphi(x))\,\partial J(x) + Z(x)). \tag{I4.8}$$

This formula of Z_ε is <u>a priori</u> defined only on $C_\beta/\{\text{critical points of } J\}$ as φ is only defined on this set. But it clearly extends to all of C_β by setting $Z_\varepsilon = 0$ on a critical point of J. In fact, Z_ε is zero on an $\varepsilon/2$-neighborhood of these critical points, by definition of l_ε.

The vector field Z_ε will provide us with the compactifying deformation lemma that we state now:

We consider the differential equation on C_β

$$\left.\begin{array}{l}\frac{\partial x}{\partial s} = -Z_{\varepsilon}(x) \\ x(0) \ \text{given.}\end{array}\right\} \qquad (I4.9)$$

<u>Compactifying Deformation Lemma:</u>

1. On an integral curve of (I4.8) $x(s)$, $J(x(s))$ decreases and $J(x(s))$ remains bounded by a constant depending of ε, $J(x(0))$ and $I(x(0))$, for all time $s \geq 0$ such that $J(x(s))$ is positive.

2. Let $0 < a_0 < a_1$. Suppose $J^{a_0} = \{x \in C_\beta | J(x) \leq a_0\}$ is not retracted by deformation of $J^{a_1} = \{x \in C_\beta | J(x) \leq a_1\}$.

Then there exists $\varepsilon_0 > 0$ such that for any $0 < \varepsilon < \varepsilon_0$, there is a point x_ε in C_β with:

$$a_0 \leq J(x_\varepsilon) \leq a_1 \text{ and either } \partial J(x_\varepsilon) = 0 \text{ or} \qquad (I4.10)$$

$$Z(x_\varepsilon) = 0; \ \varphi(x_\varepsilon) \leq \varepsilon.$$

5. ANALYTICAL ASPECTS OF THE CRITICAL POINTS AT INFINITY

5.1 The equation of critical points at infinity

In this section, we study those sequences, given by the compactifying lemma, which satisfy:

$$Z(x_\varepsilon) = 0; \ \varphi(x_\varepsilon) \leq \varepsilon; \ a_0 \leq J(x_\varepsilon) \leq a_1. \qquad (I5.1)$$

For the moment, we are interested in making explicit the equation satisfied by these sequences. For this purpose, we introduce:

$$\dot{x}_\varepsilon = a\xi + bv; \quad \omega = \frac{|\partial I|(x_\varepsilon)}{|\partial J|(x_\varepsilon)}. \qquad (I5.2)$$

We will drop for sake of simplicity the subscripts ε in the variables we will use. a is a constant. b is an $L^2(S^1)$-function. We will assume:

$$\int_0^1 b^2 dt \to +\infty \quad \text{when } \varepsilon \to 0. \qquad (I5.3)$$

This is indeed the interesting case, when there is no compactness.

We then have the following proposition which gives the equation satisfied by b.

Proposition O. 8: Under (I5. 3), (I5. 1) is equivalent to:

$$\left.\begin{aligned} &\ddot{b} + b\left(-\frac{\omega a}{2} + \frac{b^2}{2} - \int_0^1 \frac{b^2}{2}\right) + a^2 b\tau - ab^2\bar{\mu}_\xi + b\dot{b}\bar{\mu} = 0 \\ &b(0) = b(1)\,; \quad \dot{b}(0) = \dot{b}(1) \\ &\int_0^1 \frac{b^2}{\omega} \to 0. \end{aligned}\right\} \qquad (\text{I5. 4})$$

5.2 Geometric interpretation of the equation of critical points at infinity

We write down the equation satisfied by the critical points at infinity in a matricial form. Let

$$\left.\begin{aligned} A^\star_1 &= 1 - \frac{b^2}{2\omega a} + \int_0^1 \frac{b^2 dt}{2\omega a}, \\ B^\star_1 &= -\frac{b}{\omega}, \\ C^\star_1 &= -\frac{\dot{b}}{\omega a}. \end{aligned}\right\} \qquad (\text{I5. 5})$$

Then we have:

$$\left.\begin{aligned} \dot{A}^\star_1 &= -\frac{b\dot{b}}{\omega a} = -bC^\star_1, \\ \dot{B}^\star_1 &= -\frac{\dot{b}}{\omega}, \\ \dot{C}^\star_1 &= -\frac{\ddot{b}}{\omega a} = bA^\star_1 - \frac{ab\tau}{\omega} - b\bar{\mu}_\xi B^\star_1 - b\bar{\mu}C^\star_1 \end{aligned}\right\} \qquad (\text{I5. 6})$$

or else

$$\dot{Z}^\star_1 = \begin{bmatrix} \dot{A}^\star_1 \\ \dot{B}^\star_1 \\ \dot{C}^\star_1 \end{bmatrix} = b \begin{bmatrix} 0 & 0 & -1 \\ 0 & 0 & 0 \\ 1 & -\bar{\mu}_\xi & -\bar{\mu} \end{bmatrix} \begin{bmatrix} A^\star_1 \\ B^\star_1 \\ C^\star_1 \end{bmatrix} + \begin{bmatrix} 0 \\ -\frac{\dot{b}}{\omega} \\ -\frac{ab\tau}{\omega} \end{bmatrix} \qquad (\text{I5. 7})$$

$$= b \begin{bmatrix} 0 & 0 & -1 \\ 0 & 0 & 0 \\ 1 & -\bar{\mu}_\xi & -\bar{\mu} \end{bmatrix} Z^\star_1 + \begin{bmatrix} 0 \\ -\frac{\dot{b}}{\omega} \\ \frac{ab\tau}{\omega} \end{bmatrix}.$$

The matrix

$$\begin{bmatrix} 0 & 0 & -1 \\ 0 & 0 & 0 \\ 1 & -\bar{\mu}_{\xi} & -\bar{\mu} \end{bmatrix}$$

is ${}^t\Gamma$ where Γ is defined in Proposition O.4. Hence (I5.7) is also:

$$\dot{Z}_1^\star = -b\,{}^t\Gamma Z_1^\star + \begin{bmatrix} 0 \\ -\frac{\dot{b}}{\omega} \\ -\frac{ab\tau}{\omega} \end{bmatrix}. \tag{I5.8}$$

Equation (I5.8), and Equation (D2) which gives the dynamics of α along v, have to be thought of together.

Indeed these equations are very close:

In Equation (D2), $\dot{Z}^\star$ is the derivative of $Z^\star(s)$ with respect to s or to v, as s represents the time on the v-orbit.

In Equation (I5.8) $\dot{Z}_1^\star$ is the derivate of $Z_1^\star$ with respect to time t along x(t) or with respect to $\dot{x} = a\xi + bv$.

Let us rewrite (I5.8) with respect to:

$$\frac{\partial}{\partial s} = \frac{1}{b}\frac{\partial}{\partial t} = \frac{1}{b}\xi + v. \tag{I5.9}$$

We then find:

$$\frac{\partial Z_1^\star}{\partial s} = -{}^t\Gamma Z_1^\star + \begin{bmatrix} 0 \\ -\frac{\dot{b}}{\omega b} \\ -\frac{a\tau}{\omega} \end{bmatrix}. \tag{I5.10}$$

Now, when b is very large, a/b is small; hence $\partial/\partial s$ looks like v. As ω goes to infinity, $a\tau/\omega$ goes to zero.

Hence (I5.10) is very close to the equation governing the dynamics of α along v, provided b is very large and $\dot{b}/\omega b$ is small.

Under these conditions, which will amount later on to the fact that b is large, (I5.10) acquires a geometric significance: it is very close to the transport equations of forms along v.

This is a key point.

6. THE CONVERGENCE THEOREM; THE GEOMETRICAL CURVES

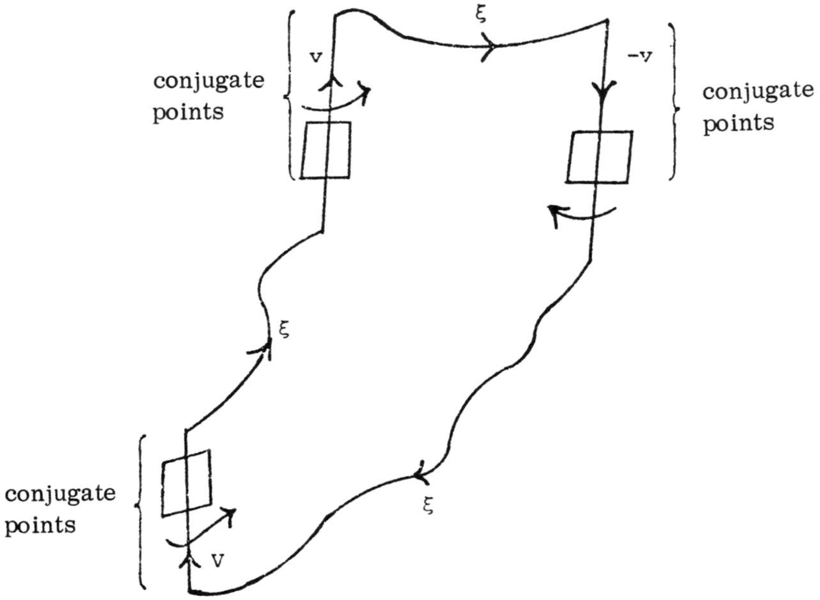

(I6.1)

(6.1) is a geometrical description of the critical points at infinity. To understand this description, we set in (I5.4), $\omega = 1/\varepsilon$. x is a curve in C_β with:

$$\dot{x} = a\xi + bv; \text{ a being a constant; } b \in L^2(S^1, \mathbb{R}) \tag{I6.2}$$

b and a satisfying (I5.4).

To understand qualitatively the phenomenon, we analyse the convergence process.

Due to (I5.4), in particular to the fact that $\int_0^1 b^2 dt/\omega \to 0$, we are able to distinguish on the curves $x(\omega)$ or x_ε, two types of pieces:

1. The pieces rather tangent to v: there, $\int_0^1 b^2/\omega$ is rather large.
2. The pieces rather tangent to ξ: there, $\int_0^1 b^2/\omega$ is rather small.

On the first kind of pieces, the geometric interpretation of the critical points at infinity holds. The curves are then close to a $\pm v_-$ orbit. Writing down, as in (D2), the equation function of the time s, satisfied by the form α_{x_0} in the transported frame along v, $((D\varphi^{\star}_{-s}\alpha)_{x_0}, (D\varphi^{\star}_{-s}\beta)_{x_0})$, we interpret (I5.4) as the transport equation of α along v. The condition $\int_0^1 b^2/\omega \to 0$ tells us, when $\omega \to +\infty$ (or $\varepsilon \to 0$) that these pieces run from one point to one of its conjugates; the error being of the order $o(\Delta t)$, where Δt is the time spent on such pieces.

For $\varepsilon \to 0$, the curve x_ε approximating the object (I6.1) on the deformation line forms a small bubble in section to v, i.e. if one projects a neighborhood of this piece rather tangent to v on a section to v, one finds:

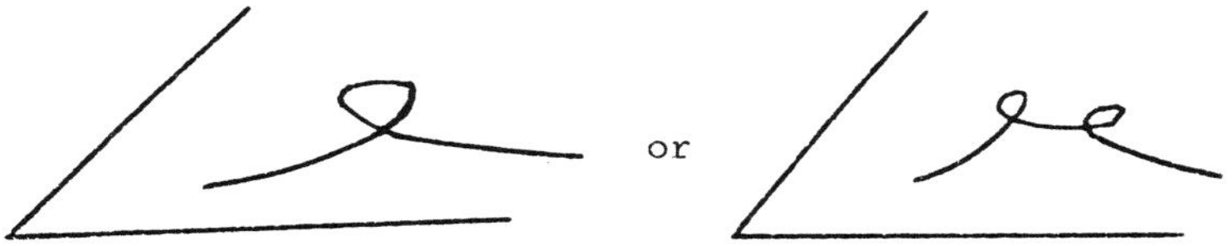

(I6.3)

or more bubbles.

These are thus points where the tangent vector to x_ε, when projected, completes rapidly an integer number of rotations, possibly growing when $\varepsilon \to 0$ through the following process:

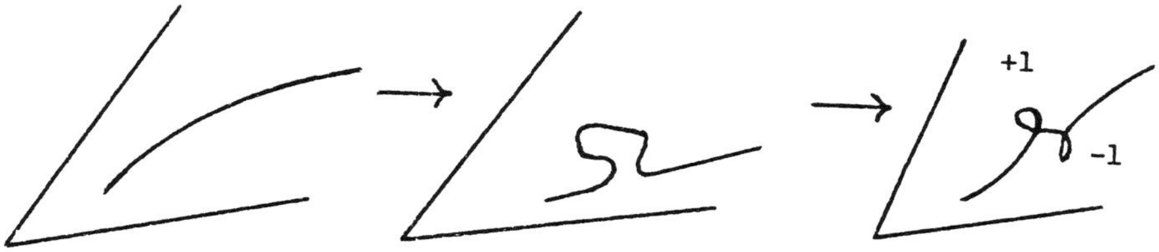

(I6.4)

no bubble

+1 and -1 bubble

However, the resulting movement is very particular: the bubble as deployed along v will go from one point to a conjugate of this point.

Therefore, generically, these bubbles build up at precise locations in M. We will see later on that these singularities have, further more, a very precise and restricted normal form. To see the phenomenon, we could draw v-orbits passing through each point x of M and distinguish on these orbits the coincidence points to x:

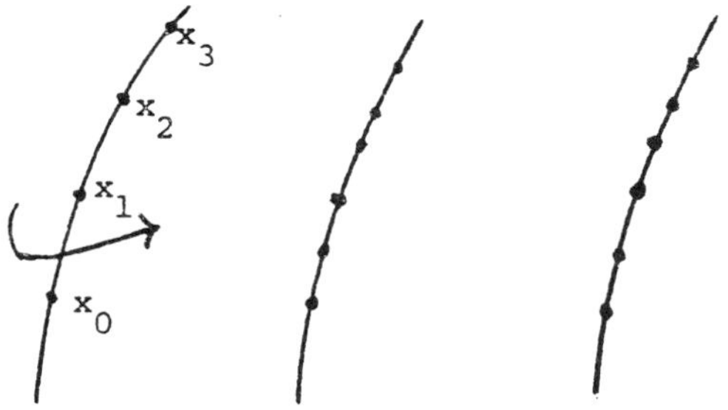

(I6.5)

We thus have a $\mathbb{Z}$-structure along M related to α along v. For some distinguished points, we have conjugate points:

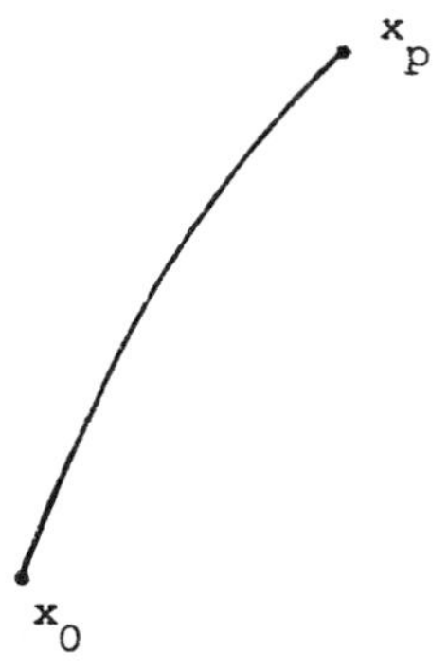

(I6.6)

$x_0 \in$ hypersurface of M.

If x_p corresponds to the time s_1 along the v-orbit starting at x_0, we have:

$$D\varphi^\star_{s_1} \alpha = \alpha \tag{I6.7}$$

and we may compute the second variation of α

$$\delta(D\varphi^\star_s \alpha - \alpha)(s). \tag{I6.8}$$

This gives rise, as we will see later on, to a quadratic form on tangent vectors to M at x_0, q_0; and an associated quadratic form on tangent vectors M at $x_{s_1} = x_p, q_1$.

Thus, these conjugate points come out with:

1 - a precise location;

2 - a precise normal form to the singularity;

3 - an integer (the rotation of α from x_0 to $x_{s_1} = x_0$);

4 - two quadratic forms q_0 and q_1;

5 - a way to approach them by curves which project on local sections on bubbles.

For the second kind of pieces - the ξ-pieces - these are pieces where the curve is tangent to the Reeb vector field; thus the curve is tangent to ξ (in this $\mathbb{Z}$-structure we introduced) until it hits a point admitting a conjugate point. Then, under certain conditions stated in Chapter 10, it jumps to the conjugate point.

The ξ-pieces come also with a quadratic form q_3 defined by the second variation of J along them with fixed ends. This quadratic form is related to a rotation of v along the ξ-piece. (I6.9)

The Reeb vector field ξ is, in the case of the cotangent unit sphere bundle of a Riemannian manifold Σ, such that its periodic orbits project on geodesics of Σ. In that case, there is no other conjugate point for a point x_0 than itself and the jumps describe a complete circle S^1 over a given point in Σ in $ST^\star\Sigma$.
In other simple, but more complicated cases, this is what happens:

Take the case of S^3 fibering over S^2 with the Hopf fibration

$$p : S^3 \to S^2. \qquad (I6.10)$$

Consider $\alpha = \lambda \alpha_0$, λ a positive function on S^3 and α_0 the standard contact form of S^3. Let v be the vector field of the fibers over S^2. In this case, the Reeb vector field ξ, when describing a fiber S^1 over a point x_0 of S^2, describes in the tangent plane to S^2 at x_0 the following:

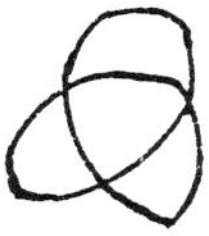

(I6.11)

i.e. two circles.

We thus have two choices of length on S^2, hence two notions of geodesics. Then, (I6.1) projects as:

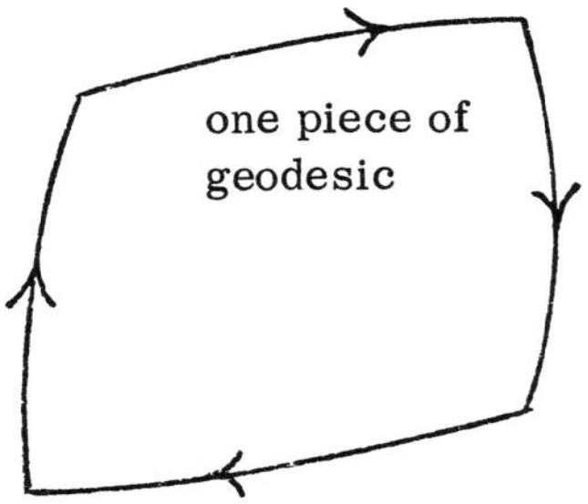

(I6.12)

The location of the corners is restricted and there is a Morse index related to q_0, q_1, q_3.

This is a general picture of what happens.

We reproduce here the theorem we announced in [O5]:

Assume:

(H_1) α turns well along v;

(H_2) v has a periodic orbit,

(H_3) for one vector field v_1, non-singular and collinear to v, we have: $\exists k_1 > 0$ such that $\|D\theta_s^1\| \le k_1$ $\forall s \in \mathbb{R}$, where θ_s^1 is the one-parameter group of v_1;

(H_4) $\exists k_2$ and $k_3 > 0$ such that $\forall i \in \mathbb{Z}$, we have:

$$k_2 d(x,y) \le d(f^i(x), f^i(y)) \le k_3 d(x,y) \quad \forall x, y \in M;$$

(H_5) $\exists k_4 > 0$ such that $|\mu_i(x) - \mu_i(y)| \le k_4 d(x,y)$ $\forall x, y \in M$;

(H_6) $\exists \rho > 0$ such that for any $x \in M$, the set $C_\rho(x) = \{f^i(x) / |\mu_i(x) - 1| < \rho;\ i \in \mathbb{Z}\}$ is finite.

Then, under these hypotheses, we have

Theorem: The critical points at infinity of the variational problem are continuous and closed curves made up with pieces $[x_{2i}, x_{2i+1}]$ tangent to ξ and pieces $[x_{2i+1}, x_{2i+2}]$ tangent to v. x_{2i+2} is conjugate to x_{2i+1}. If the Betti numbers of the loop space are unbounded, there are infinitely many of these curves. Furthermore, if n is the number of v-pieces of one of these curves, we have:

$$n \leq Ca \tag{I6.13}$$

where a is the length of the curve along ξ and C is a universal constant.

7. EXPANSION OF J NEAR INFINITY. THE INDEX OF A CRITICAL POINT AT INFINITY

7.1 The parameterization normal form

We are then left with these geometric curves made up of ξ-pieces and v-pieces. The v-pieces have been seen to run from a point to a conjugate.

On such a piece, as seen from Section 5 of this introduction onwards, the function b(t) is very particular.

Indeed, as stated in (I5.5), the vector:

$$\left.\begin{aligned} A_1^\star &= 1 - \frac{b^2}{2\omega a} + \frac{\int_0^1 b^2 dt}{2\omega a} \\ B_1^\star &= -\frac{b}{\omega} \\ C_1^\star &= -\frac{\dot{b}}{\omega a} \end{aligned}\right\} \tag{I7.1}$$

nearly satisfies the transport equation of the forms.

Furthermore, if we are looking at a v-piece between x_{2i} and x_{2i+1}, then near x_{2i},

$$\begin{bmatrix} A_1^\star \\ B_1^\star \\ C_1^\star \end{bmatrix}$$

is nearly $\begin{bmatrix} 1 \\ 0 \\ 0 \end{bmatrix}$.

Consequently b has in fact a first normal form on a critical point at infinity: Namely, we introduce the function φ_i on the v-piece between x_{2i} and x_{2i+1} satisfying:

$$\left.\begin{aligned} &\frac{\partial^2 \varphi_i}{\partial s^2} + \varphi_i + \bar{\mu}\frac{\partial \varphi_i}{\partial s} = 0 \\ &\varphi_i(0) = \varphi_i(s_i) = 1 \\ &\frac{\partial \varphi_i}{\partial s}(0) = \frac{\partial \varphi_i}{\partial s}(s_i) = 0. \end{aligned}\right\} \tag{I7.2}$$

Here s is the time parameter along the v-orbit from x_{2i} to x_{2i+1}. φ_i exists and is uniquely defined by (I7.2) as x_{2i} and x_{2i+1} are conjugate. We thus have:

$$\frac{b(t)}{\sqrt{\omega a}} \sim \pm \sqrt{1 - \varphi_i(s(t))} \; . \tag{I7.3}$$

We state this in:

Proposition O.9: Along a nearly tangent to v-piece between two conjugate points, $b(t)/\sqrt{\omega a}$ is equivalent to $\pm\sqrt{1 - \varphi_i(s(t))}$ where φ_i satisfies:

$$\left.\begin{aligned} &\frac{\partial^2 \varphi_i}{\partial s^2} + \varphi_i + \bar{\mu}\frac{\partial \varphi_i}{\partial s} = 0 \\ &\varphi_i(0) = \varphi_i(s_i) = 1 \\ &\frac{\partial \varphi_i}{\partial s}(0) = \frac{\partial \varphi_i}{\partial s}(s_i) = 0 \\ &s \in [0, s_i];\ \text{time on the } (\pm) \text{ v-orbit from } x_{2i} \text{ to} \\ &x_{2i+1} \text{ which are the conjugate points.} \end{aligned}\right\} \tag{I7.4}$$

Thus φ_i satisfies:

$$\left.\begin{aligned} &\frac{\partial^2 \varphi_i}{\partial s^2} + \varphi_i + \bar{\mu}\frac{\partial \varphi_i}{\partial s} = 0 \\ &\varphi_i(0) = 1; \quad \frac{\partial \varphi_i}{\partial s}(0) = 0. \end{aligned}\right\} \tag{I7.5}$$

φ_i is extremal only at the coincidence points of x_{2i} and the only possibility for b is to accomplish a piece of v-orbit from x_{2i} to x_{2i+1}, then come back from x_{2i+1} to x_{2i}, etc.

If we consider a deformation line of (I4.9) going to a critical point at infinity, these oscillations are in finite number (upperbounded). Otherwise, we leave an L^∞-neighborhood of this critical point at infinity and one constructs a deformation lemma to move all such curves away from infinity.

As we are dealing with an actual jump, this number is odd.

Thus, we are left, as a model, with only one jump and a definite sign for b on such a piece.

As we wish to present general ideas rather than justify all the technical details, we will assume for sake of simplicity that whenever a jumps occurs, a single oscillation is associated with it.

So that a critical point at infinity is this geometric curve, together with a parameter $\sqrt{\omega a} \to +\infty$, the v-pieces being described with $b(t)/\sqrt{\omega a} \sim \pm\sqrt{1 - \varphi_i(s(t))}$, where $\pm$ is fixed by the orientation along v of $[x_{2i}, x_{2i+1}]$.

7.2 The variations along a critical point at infinity inwards C_β

There are two kinds of variations along such a geometric curve with this limit parameterization we point out in Section 7.1.

The first kind, we will present here consists in opening up the oscillations in order to see if we are dealing with an actual critical point at infinity. This will be made clear later on.

These variations are inwards C_β.

We want to know if a sequence $(\varepsilon \to 0)$ of flow lines of (I4.9) does arrive at the limit object.

In order to discriminate between these two possibilities, we need a first expansion of J along inwards C_β-variation. An inward variation has to bring the length along ξ to be a strictly positive constant which is nearly a, a being the length along ξ of the limit object (the curve x).

We are thus led to introduce along a v-piece $[x_{2i}, x_{2i+1}]$ of x, which we will assume for sake of simplicity to be oriented by +v, the differential equation (see Chapters 1-2 of this book for details).

$$\left.\begin{aligned} &\dot{\widehat{\lambda + \bar{\mu}\eta}} - \eta = \frac{a}{\sqrt{\omega a}\sqrt{1 - \phi_i(s)}} \qquad s \in [0, s_i]. \\ &\dot{\eta} = -\lambda \end{aligned}\right\} \tag{I7.6}$$

In (I7.6), . is a derivation by $\frac{\partial}{\partial s} = v$; ω is a large positive parameter and a is the length along ξ of the curve as already stated.

Another way to see (I7.6) is to set:

$$\frac{\partial}{\partial t} = b\frac{\partial}{\partial s} = \sqrt{\omega a}\sqrt{1 - \varphi_i(s)}\ \frac{\partial}{\partial s} \tag{I7.7}$$

and we then have:

$$\left.\begin{aligned} &\frac{\partial}{\partial t}(\lambda + \bar{\mu}\eta) - b\eta = a \\ &\frac{\partial \eta}{\partial t} = -\lambda b \end{aligned}\right\} \quad z = \lambda\xi + \mu v + \eta w. \tag{I7.8}$$

The homogeneous equation:

$$\left.\begin{aligned} &\dot{\widehat{\lambda + \bar{\mu}\eta}} - \eta = 0 \qquad s \in [0, s_i] \\ &\dot{\eta} = -\lambda \end{aligned}\right\} \tag{I7.9}$$

has solutions satisfying:

$$(\lambda + \bar{\mu}\eta)(s_i) = (\lambda + \bar{\mu}\eta)(0) \tag{I7.9'}$$

as x_{2i} and x_{2i+1} are conjugate points. Indeed (I7.9) expresses the relations which have to be satisfied by a transported vector along a v-piece.

There is thus an indeterminacy in (I7.6) which we will discuss later on, when we will introduce the index of a critical point at infinity.

Notice that the parameterization introduced by (I7.8), with $b = \sqrt{\omega a}\sqrt{1-\varphi_i(s)}$ corresponds to the first normal form we pointed out in Proposition O.10.

In (I7.6) there is a problem:

Indeed, φ_i satisfies on $[x_{2i}, x_{2i+1}]$ parameterized by v:

$$\left.\begin{aligned} &\frac{\partial^2}{\partial s^2}\varphi_i + \varphi_i + \bar{\mu}\frac{\partial \varphi_i}{\partial s} = 0 \\ &\varphi_i(0) = \varphi_i(s_i) = 1 \\ &\frac{\partial \varphi_i}{\partial s}(0) = \frac{\partial \varphi_i}{\partial s}(1) = 0. \end{aligned}\right\} \tag{I7.10}$$

Thus, $1 - \varphi_i$ has a zero of second order at 0 and s_i. Near this point, we have:

$$1 - \varphi_i(s) \sim C_i s^2 \text{ at } 0; \; 1 - \varphi_i(s) \sim C_i'(s - s_i)^2 \tag{I7.11}$$

at s_i.

Thus

$$\int_0^{s_i} \frac{1}{\sqrt{1 - \varphi_i(s)}}\, ds \text{ diverges logarithmically at both ends.} \tag{I7.12}$$

This implies, by integration of the first equation in (I7.6), that $(\lambda + \bar{\mu}\eta, \eta)$ cannot possibly be L^∞, <u>a fortiori</u> L^∞-small.

We then have:

<u>Lemma O.2</u>: Consider a solution of (I7.6), $(\lambda, \eta)(s)$, and a solution of (I7.9), $(\lambda_1, \eta_1)(s)$ taking the same value at a point τ_i in $[0, s_i]$. Then:

$$|\eta(s) - \eta_1(s)| \le \frac{C}{\sqrt{\omega}}$$

$$\int_0^{s_i} |\lambda(s) - \lambda_1(s)|^2 ds \le \frac{C}{\omega}$$

$$\int_0^{s_i} |\dot{\eta}(s) - \dot{\eta}_1(s)|^2 ds \le \frac{C}{\omega}.$$

Furthermore:

$$\lambda + \bar{\mu}\eta(s) \text{ is equivalent to } -\frac{1}{\sqrt{C_i}}\sqrt{\frac{a}{\omega}} \log s \text{ near } s = 0$$

$\lambda + \bar{\mu}\,\eta(s)$ is equivalent to $-\dfrac{1}{\sqrt{C_i'}}\sqrt{\dfrac{a}{\omega}}\,\log(s_i - s)$ near $s = s_i$.

Lastly,

$$\Delta_i = \lim_{\substack{\varepsilon \to 0 \\ \varepsilon' \to 0}} \Big\{ \int_{\varepsilon}^{s_i - \varepsilon'} \frac{ds}{\sqrt{1 - \varphi_i(s)}} + \sqrt{\frac{\omega}{a}}\big[(\lambda + \bar{\mu}\,\eta)(\varepsilon) - (\lambda + \bar{\mu}\,\eta)(s_i - \varepsilon')\big]\Big\} \tag{I7.13}$$

exists and is independent of the solution of (I7.6) considered as well as on ω and a. This quantity is thus attached to $[x_{2i}, x_{2i+1}]$ and only to it.

A variation governed by (I7.6) comes now with two problems: the first one, which is not very serious, is due to the indeterminacy in it; one can add to such a variation any variation subject to (I7.9). In order to fix once and for all the variation we are looking at, we will impose:

$$\lambda(s_i/2) = \eta(s_i/2) = 0. \tag{I7.14}$$

The influence of the solutions of (I7.9) is analysed separately in section 7.3. The solution of (7.6)-(7.26) is denoted:

$$(\tilde{\lambda}, \tilde{\eta}). \tag{I7.15}$$

Notice that, by Lemma O.2, and the fact that $(\tilde{\lambda}_1, \tilde{\eta}_1) \equiv (0,0)$, where $(\tilde{\lambda}_1, \tilde{\eta}_1)$ is the solution of (I7.9) taking the same value (i.e. $(0,0)$) then $(\tilde{\lambda}, \tilde{\eta})$ at $\tau_i = s_i/2$, $(\tilde{\lambda}, \tilde{\eta})$ satisfies:

$$\left.\begin{aligned} &|\tilde{\eta}(s)| \le \frac{C}{\sqrt{\omega}}\,;\quad \int_0^{s_i} |\tilde{\lambda}(s)|^2 ds \le \frac{C}{\omega}\,;\quad \int_0^{s} |\dot{\tilde{\eta}}|^2 ds \le \frac{C}{\omega}\,; \\ &(\lambda + \bar{\mu}\,\eta)(s) \sim -\frac{1}{\sqrt{C_i}}\sqrt{\frac{a}{\omega}}\,\log s \text{ at } 0 \\ &\sim -\frac{1}{\sqrt{C_i'}}\sqrt{\frac{a}{\omega}}\,\log(s_i - s) \text{ near } s = s_i\,. \end{aligned}\right\} \tag{I7.16}$$

The second problem is more serious:

By (I7.28) such a variation cannot be made to be L^∞, <u>a fortiori</u> not L^∞ or

H^1 small.

Indeed $\tilde{\lambda} + \bar{\mu}\tilde{\eta}(s)$ diverges logarithmically at $s = 0$ and $s = s_i$. $\tilde{\eta}$ remains meanwhile bounded, and even, by (I7.17), going to zero with ω. Thus, at 0 and s_i, the variation is infinite along ξ. Nevertheless, we can try to extend it along the ξ-pieces, subject to the differential equation:

$$\left.\begin{aligned} &\dot{\overbrace{\lambda + \bar{\mu}\eta}} = \varphi(t) \\ & \qquad\qquad\qquad \text{here } \cdot = \xi = \text{derivation along } \xi. \\ &\dot{\eta} = \mu a \end{aligned}\right\} \qquad \text{(I7.17)}$$

We will state further on what are the conditions on φ near the points x_{2i+1}. For the moment, let us try to understand what is taking place on the v-pieces. The equation (I7.6) may be rewritten, in a more intrinsic form: setting:

$$z(s) = (\tilde{\lambda}\xi + \tilde{\eta}w) \qquad \text{(I7.18)}$$

we have:

$$[\frac{\partial}{\partial s} + v; z(s)] = \frac{a}{\sqrt{\omega a}} \frac{1}{\sqrt{1 - \varphi_i(s)}} \xi + \frac{a}{\sqrt{\omega a}} \theta v; \qquad \text{(I7.19)}$$

$$\theta \in C^{\infty}(\,]0, s_i[\,).$$

Indeed, (I7.6) and (I7.31) are equivalent as can be checked by applying α and β to (I7.31). As we are dealing with a variation along a v-piece and as the functional

$$J(x) = \int_0^1 \alpha_x(\dot{x})\,dt$$

(this is rephrasing of our functional which makes sense not only on C_β or $\mathcal{L}_\beta$, but even on unparameterized, however oriented, curves) is invariant under reparameterization, the θ-term is not relevant. What matters is the variation transverse to $\dot{x}$, i.e. to v along these pieces. Clearly:

$$z = \sqrt{\frac{a}{\omega}}\, z_0, \qquad \text{(I7.20)}$$

z_0 satisfying:

$$[\frac{\partial}{\partial s} + v, z_0] = \frac{1}{\sqrt{1 - \varphi_i(s)}} \xi + \theta v \qquad (I7.21)$$

$z_0(s_i/2) = 0$; z_0 has no component on v.

If we draw the variation, we have:

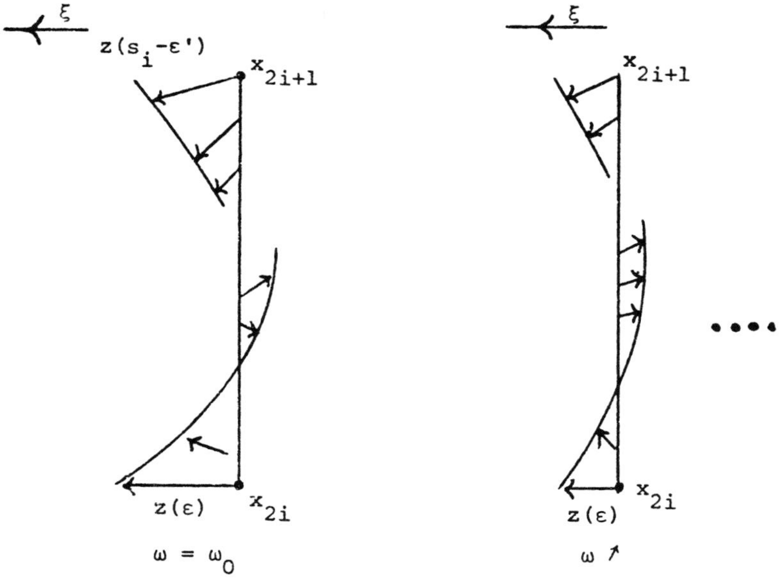

(I7.22)

<u>Remark</u>: The linking as shown in (I7.22) actually occurs.

There is thus, as $\omega \to +\infty$, more and more control on the variation z which is of the order of $C/\sqrt{\omega}$ on a fixed (when ω $+\infty$) compact set in $]0, s_i[$. Nevertheless, at the ends, 0 and s_i, we always have a logarithmic divergence along ξ. If we extend now to the ξ-pieces subject to (I7.17), we have:

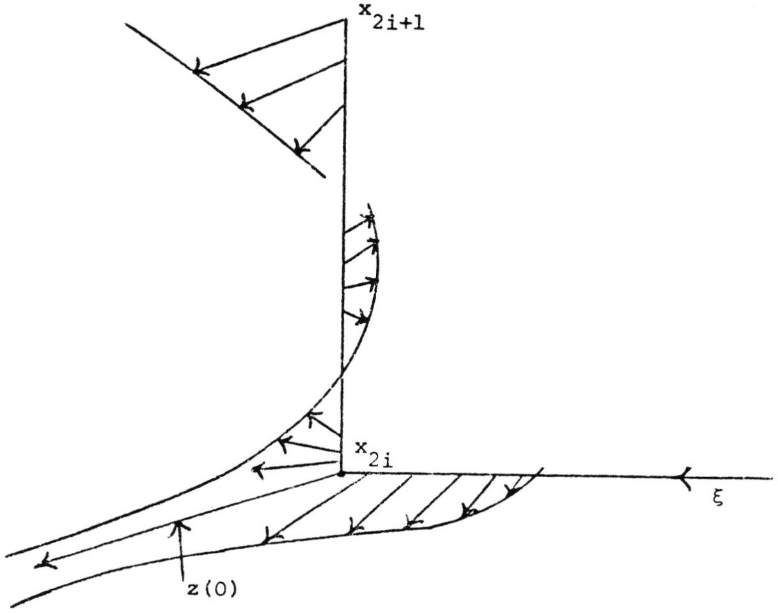

(I7.23)

Following a choice of φ in (I7.29) behaving as:

$$\varphi(t) \sim \sqrt{\frac{a}{\omega}}\,\frac{C_i''}{t - t_i} \quad \text{near} \quad t = t_i;$$

t_i corresponding to x_{2i} on the ξ-piece; (I7.24)

C_i'' bounded constant;

one can take care of the logarithmic divergence of $\tilde{\lambda} + \bar{\mu}\tilde{\eta}$ at x_{2i} (also x_{2i+1} on the other ξ-piece) and control in $\sqrt{(a/\omega)}$ the H_1-norm of the variation on each compact subset of the $]x_{2i-1}, x_{2i}[$.

This will be made precise later on.

The only problem is thus at the <u>corners</u>, i.e. at the points x_{2i}, x_{2i+1}.

There, in fact, it is natural that we find some problems, as $\dot{x}_\varepsilon$ changes very rapidly. On the other hand, the curves x_ε stay nevertheless in an L^∞-neighborhood of x.

We wish thus to cut out the variation on a neighborhood of the corners (a neighborhood which we can take to be smaller and smaller as ω goes to $+\infty$) in order to have an L^{∞} control on it (going to zero with ω); and then to make a first expansion of J.

Let us first derive what kind of first expansion we may expect, considering the variation as taking place in $\mathcal{L}_\beta$ with the functional

$$J(y) = \int_0^1 \alpha_y(\dot{y})\,dt$$

(which coincides with J on C_β and makes sense on unparameterized oriented curves of $\mathcal{L}_\beta$).

We have:

<u>Proposition O.10</u>: Consider x a curve of $\mathcal{L}_\beta$, unparametrized. Let

$J : \mathcal{L}_\beta \to \mathbb{R}$

$$y \to \int_0^1 \alpha_y(\dot{y})\,dt.$$

Consider the variation z defined by (I7.30), (I7.29), <u>cut out at the corners so that all the</u> (I7.28) <u>estimates hold.</u> Call z_1 this new variation. Then,

$$\partial J(x) \cdot z_1 = \sqrt{\frac{a}{\omega}}\,(\sum_i \Delta_i) + o(\frac{1}{\sqrt{\omega}}).$$

<u>Remark</u>: Such a curve x is thus a critical point at infinity (i.e. an end for a flow-line) if $\sum_i \Delta_i > 0$. If $\sum_i \Delta_i < 0$ x does not define a critical point at infinity. The case $\sum_i \Delta_i = 0$ would require further analysis.

<u>7.3 The variations tangent to the border-line. The index of a critical point at infinity</u>

We are now left with indeterminacies.

Namely, we are considering variations such that on the v-pieces, we have:

$$\left.\begin{aligned}&\dot{\widehat{\lambda + \bar{\mu}\eta}} - \eta = 0\\ &\dot{\eta} = -\lambda .\end{aligned}\right. \qquad \cdot = \frac{\partial}{\partial s} = \pm v; \ \mu \text{ arbitrary}, \quad s \in [0, s_i] \tag{I8.1}$$

On the ξ-pieces we have:

$$\left.\begin{aligned}&\dot{\widehat{\lambda + \bar{\mu}\eta}} = \varphi\\ &\dot{\eta} = \mu a\end{aligned}\right\} \tag{I8.2}$$

$\cdot = \frac{\partial}{\partial s} = \xi$; φL^{∞}-small.

(I8.1) defines, up to the μ-indeterminacy, the equation of a transported vector (by v) along the v-piece $[x_{2i}, x_{2i+1}]$. This can be easily checked by applying α and β to the transport equation:

$$\left[\frac{\partial}{\partial s} + v, \ \lambda\xi + \mu v + \eta w\right] = \psi(s)\, v, \quad \psi \text{ arbitrary.} \tag{I8.3}$$

Thus, calling:

$$z_i(s) \quad \text{the variation subject to (I8.1) on } [0, s_i], \tag{I8.4}$$

we have:

$$z_i(s_i) = D\varphi_{s_i}(z_i(0)) + \delta s_i v; \quad \delta s_i \in \mathbb{R} . \tag{I8.5}$$

Consequently, in order to compute $J(x + z)$, we may always see $x + z$ as being geometrically realized as follows:

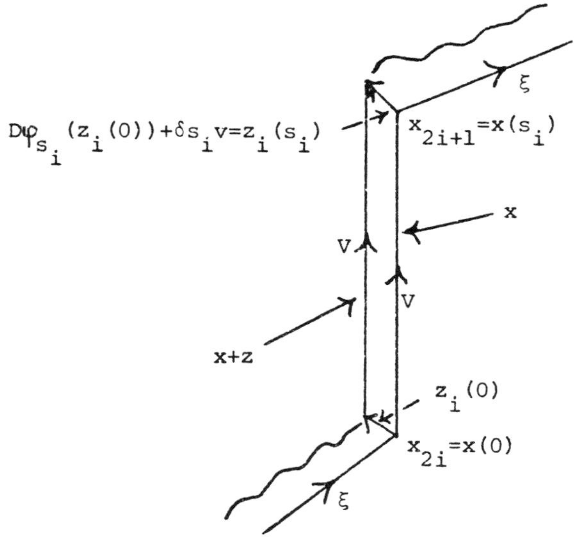

(I8.6)

i.e., along the v-pieces, we just push by v during a time $s_i + \delta s_i$ starting at $x_{2i} + z_i(0)$; along the ξ-pieces, we have as usual a tangential variation. In this way, $J(x+z) - J(x)$ comes only from the ξ-pieces and is thus equal to the variation of $J(x)$ along these pieces which is of second order (the first variation in zero. Indeed by (I8.1), (I8.2), this first variation is $\sum_i [(\lambda + \bar{\mu}\eta)(x_{2i+2}) - (\lambda + \bar{\mu}\eta)(x_{2i+1})] = 0$ as x_{2i} is conjugate to x_{2i+1}). The expansion of $J(x+z) - J(x)$ is:

$$
\begin{aligned}
J(x+z) - J(x) &= z \cdot (\partial J(x) \cdot z) \qquad\qquad (I8.7)\\
&= z \cdot \sum_i \{[(\lambda + \bar{\mu}\eta)(x_{2i+2}) - (\lambda + \bar{\mu}\eta)(x_{2i+1})] \\
&\quad - (\int_{x_{2i+1}}^{x_{2i+2}} b\eta)\} \\
&= z \cdot (\sum_i [(\lambda + \bar{\mu}\eta)(x_{2i}) - (\lambda + \bar{\mu}\eta)(x_{2i+1})] \\
&\quad - \int_{x_{2i+1}}^{x_{2i+2}} b\eta).
\end{aligned}
$$

We first compute:

$$z \cdot \left(\int_{x_{2i+1}}^{x_{2i+2}} b\eta\right) \tag{I8.8}$$

b is, as noted previously, equal to $d\alpha(\dot{x}, w) = \gamma(\dot{x})$. Thus, the first variation of b along z is:

$$\frac{d}{dt}\gamma(z) - d\gamma(\dot{x}, z) = \dot{\mu} + a\eta\tau - b\eta\bar{\mu}_{\xi}\,. \tag{I8.9}$$

On a ξ-piece, b is zero. Thus this variation is:

$$\dot{\mu} + a\eta\tau\,. \tag{I8.10}$$

Thus, using again the fact that b is zero on such a piece (thus $bz \cdot \eta = 0$ and $b\eta(x_{2i+2}) = b\eta(x_{2i+1}) = 0$), we derive:

$$z \cdot \left[-\int_{x_{2i+1}}^{x_{2i+2}} b\eta\right] = -\int_{x_{2i+1}}^{x_{2i+2}} (\dot{\mu} + a\eta\,\tau)\,\eta\,. \tag{I8.11}$$

We are thus left with:

$$z \cdot \left[(\lambda + \bar{\mu}\eta)(x_{2i}) - (\lambda + \bar{\mu}\eta)(x_{2i+1})\right] \tag{I8.12}$$

which has the following simple interpretation:

Consider the differential equations:

$$\left\{\begin{array}{l} \dot{\overbrace{\lambda + \bar{\mu}\eta}} = \eta \\ \dot{\eta} = -(\lambda + \bar{\mu}\eta) + \bar{\mu}\eta \cdot \text{ initial data given by } z_i(0) \\ \cdot = \dfrac{\partial}{\partial s} = \pm v \text{ on the } \pm v\text{-piece } \underline{\text{from}}\ \underline{x_{2i} + z_i(0)}\ \underline{\text{to}} \\ \underline{x_{2i+1} + z_i(s_i)} \text{ during time } [0, s_i + \delta s_i] \end{array}\right. \tag{I8.13}$$

$$\left\{\begin{array}{l} \dot{\overbrace{\lambda + \bar{\mu}\eta}} = \eta \\ \dot{\eta} = -(\lambda + \bar{\mu}\eta) + \bar{\mu}\eta \cdot \text{ initial data given by } z_i(0) \\ \cdot = \dfrac{\partial}{\partial s} = \pm v \text{ on the } \pm v\text{-piece } \underline{\text{from}}\ \underline{x_{2i}}\ \underline{\text{to}}\ \underline{x_{2i+1}} \\ \text{during the time } [0, s_i + \delta s_i]\,. \end{array}\right. \tag{I8.13'}$$

Then, with evident notation:

$$z \cdot \left[(\lambda + \bar{\mu}\eta)(x_{2i}) - (\lambda + \bar{\mu}\eta)(x_{2i+1})\right] \tag{I8.14}$$

$$= (\lambda + \bar{\mu}\eta)(x_{2i} + z_i(0)) - (\lambda + \bar{\mu}\eta))(x_{2i+1} + z_i(s_i))$$

$$- (\lambda + \bar{\mu}\eta)(x_{2i}) + (\lambda + \bar{\mu}\eta)(x_{2i+1})$$

at first order.

To give an intrinsic form to this expression, we come back to Proposition O. 4.

A transported vector $Z = \begin{bmatrix} A \\ B \\ C \end{bmatrix}(s)$ along v satisfies the differential equation:

$$\dot{Z} = \Gamma Z \text{ in the basis } (\xi, v, [\xi, v]). \tag{I8.15}$$

Let

$$\left.\begin{aligned} V(s) &= \begin{bmatrix} a_0(s) & b_0(s) & c_0(s) \\ a_1(s) & b_1(s) & c_1(s) \\ a_2(s) & b_2(s) & c_2(s) \end{bmatrix} \text{ be the resolvent matrix of (I8.15)} \\ V(0) &= Id . \end{aligned}\right\} \tag{I8.16}$$

Then:

$$\alpha_{x_s}(D\varphi_s(Z(0))) = [1, 0, 0]V(s)\begin{bmatrix} A(0) \\ B(0) \\ C(0) \end{bmatrix} = a_0(s)A(0) \tag{I8.17}$$

$$+ b_0(s)B(0) + c_0(s)C(0)$$

and

$$\alpha_{x_s}(D\varphi_s(Z(0))) - \alpha_{x_0}(Z(0)) = (a_0(s) - 1)A(0) \tag{I8.18}$$

$$+ b_0(s)B(0) + c_0(s)C(0).$$

Setting:

$$z(s) = \lambda\xi + \mu v + \eta w , \tag{I8.19}$$

we thus have:

$$(\lambda + \bar{\mu}\eta)(s) - (\lambda + \bar{\mu}\eta)(0) = (a_0(s) - 1)\lambda(0) \tag{I8.20}$$
$$+ b_0(s)\mu(0) + c_0(s)\eta(0).$$

In fact V depends also on the starting point of the transport differential equations, which we will denote y:

$$V(s) = V(s,y) . \tag{I8.21}$$

Thus

$$(\lambda + \bar{\mu}\eta)(s) - (\lambda + \bar{\mu}\eta)(0) = (a_0(s,y) - 1)\lambda(0) \tag{I8.22}$$
$$+ b_0(s,y)\mu(0) + c_0(s,y)\eta(0) .$$

With $y = x_{2i}$ and $s = s_i$, we have:

$$a_0(s_i, x_{2i}) = 1; \; b_0(s_i, x_{2i}) = c_0(s_i, x_{2i}) = 0 \tag{I8.23}$$

as x_{2i+1} is conjugate to x_{2i}.

Now, by (I8.5), the variation z we are considering, under (I8.1) and (I8.2) is defined by:

$$\left.\begin{array}{l} \lambda_i, \mu_i, \eta_i\text{: coordinates of } z_i(0) \text{ at } x_{2i} \\ \text{and } \delta s_i \text{ such that } z_i(s_i) = D\varphi_{s_i}(z_i(0)) + \delta s_i v \end{array}\right\} \tag{I8.24}$$

and we may view the variation along this v-piece as pushing along the transport vector of $z_i(0)$ along this v-piece during the time $s_i + \delta s_i$. The variation (I8.14), in view of (I8.22), (I8.23) and (I8.24) is thus:

$$z \cdot [(\lambda + \bar{\mu}\eta)(x_{2i}) - (\lambda + \bar{\mu}\eta)(x_{2i+1})] \tag{I8.25}$$
$$= -[\{\frac{\partial a_0}{\partial s}(s_i, x_{2i})\lambda_i + \frac{\partial b_0}{\partial s}(s_i, x_{2i})\mu_i + \frac{\partial c_0}{\partial s}(s_i, x_{2i})\eta_i]\delta s_i$$

$$+ \left(\frac{\partial a_0}{\partial y}\Big|_{(s_i, x_{2i})} \cdot z_i(0)\right)\lambda_i + \left(\frac{\partial b_0}{\partial y}\Big|_{(s_i, x_{2i})} \cdot z_i(0)\right)\mu_i$$

$$+ \left(\frac{\partial c_0}{\partial y}\Big|_{(s_i, x_{2i})} \cdot z_i(0)\right)\eta_i \}$$

We are now ready to define the index of a critical point at infinity. Consider:

$$\text{a } \pm v\text{-piece } [x_{2i}, x_{2i+1}]. \tag{I8.26}$$

Let

$V(s, y)$ be the resolvent matrix of the transport equation for the vectors starting at y along the v-orbit, in the basis $(\xi, v, [\xi, v])$, (I8.27)

i.e. $\dfrac{\partial V}{\partial s} = \Gamma V$; $\Gamma = \begin{bmatrix} 0 & 0 & -1 \\ 0 & 0 & \bar{\mu}_\xi(s, y) \\ 1 & 0 & \bar{\mu}(s, y) \end{bmatrix}$. Let

$$V(s, y) = \begin{bmatrix} a_0(s, y) & b_0(s, y) & c_0(s, y) \\ a_1(s, y) & b_1(s, y) & c_1(s, y) \\ a_2(s, y) & b_2(s, y) & c_2(s, y) \end{bmatrix} \tag{I8.28}$$

Let z be a variation of the critical point at infinity x satisfying:

$$\left.\begin{array}{l} \dot{\overbrace{\lambda + \bar{\mu}\eta}} - \eta = 0 \\ \dot{\eta} = -\lambda \\ \cdot = \dfrac{\partial}{\partial s} = \pm v; \ \mu \text{ arbitrary, } \ s \in [0, s_i] \text{ on the} \\ \text{piece } [x_{2i}, x_{2i+1}]; \end{array}\right\} \tag{I8.29}$$

$$\left.\begin{array}{l} \dot{\overbrace{\lambda + \bar{\mu}\eta}} = \varphi(s) \\ \dot{\eta} = \mu a \\ \cdot = \dfrac{\partial}{\partial s} = \xi \ \text{ on a } \xi\text{-piece; } \varphi \text{ is } L^\infty\text{-small.} \end{array}\right\} \tag{I8.30}$$

Let

$$z_i(0) = \lambda_i \xi + \mu_i v + \eta_i [\xi, v] \text{ be the value of this variation at } x_{2i}; \qquad (I8.31)$$

$$z_i(s_i) = \text{value of } z \text{ at } x_{2i+1} = D\varphi_{s_i}(z_i(0)) + \delta s_i v; \qquad (I8.32)$$

$$\hat{\mu}_i = a_1(s_i, x_{2i})\lambda_i + b_1(s_i, x_{2i})\mu_i + c_1(s_i, x_{2i})\eta_i ; \qquad (I8.33)$$

$$\hat{\eta}_i = a_2(s_i, x_{2i})\lambda_i + b_2(s_i, x_{2i})\mu_i + c_2(s_i, x_{2i})\eta_i ; \qquad (I8.34)$$

$$|z|^2_{H^1(x_{2i+1}, x_{2i+2})} = \int_{[x_{2i+1}, x_{2i+2}]} (\dot{\lambda}^2 + \dot{\mu}^2 + \dot{\eta}^2)\, ds . \qquad (I8.35)$$

We then have:

Proposition O. 11:

$$J(x+z) = a + \sum_i \Big[(\hat{\mu}_i + \delta s_i)\hat{\eta}_i - \mu_i \eta_i - \delta s_i \Big[\frac{\partial a_0}{\partial s}(s_i, x_{2i})\lambda_i$$

$$+ \frac{\partial b_0}{\partial s}(s_i, x_{2i})\mu_i + \frac{\partial c_0}{\partial s}(s_i, x_{2i})\eta_i \Big]$$

$$- \Big(\frac{\partial a_0}{\partial y}\Big|_{(s_i, x_{2i})} \cdot z_i(0)\Big)\lambda_i - \Big(\frac{\partial b_0}{\partial y}\Big|_{(s_i, x_{2i})} \cdot z_i(0)\Big)\mu_i$$

$$- \Big(\frac{\partial c_0}{\partial s}\Big|_{(s_i, x_{2i})} \cdot z_i(0)\Big)\eta_i + \frac{1}{a}\int_{[x_{2i}, x_{2i+2}]} (\dot{\eta}^2 - a^2\eta^2\tau)\, ds \Big]$$

$$+ o(|z|^2),$$

where η is an H^1-arbitrary function equal to η_i at x_{2i} and $\dot{\eta}_i$ at x_{2i+1}. This formula gives the index of the critical point at infinity. <u>Notice that it does not depend on φ.</u> Here

$$|z| = \sum_i |\delta s_i| + \sum_i [|\mu_i| + |\eta_i| + |\lambda_i|] + \sum_i |z|_{H^1(x_{2i+1}, x_{2i+2})}.$$

CONCLUSION

As one can clearly see throughout this description, the critical points at infinity, as a concept, are neither sequences, nor normal forms. They are geometrical objects together with a parameterization, involving quantities going to $+\infty$ and Morse indices to relate them to the difference of topology they produce in the level sets.

These are ends of orbits and, in this sense, are ordinary objects in dynamical systems which happen to have very nice representations when there is more structure in the problem, such as geometry and variational framework.

In fact, it is very much likely that such a conception is useful for solving the so-called "non-compactness"; out of it, one either loses the global variations (i.e. when one restricts to sequences) and therefore cannot derive general existence theorems; or one loses, developing only the topological aspects, the geometrical objects as such, which are seen in some other problems to be closely related to the critical points themselves.

REFERENCES

[O1] A. Weinstein, 'On the hypotheses of Rabinowitz' periodic orbit theorems', J. Diff. Equ. 33 (1979), 353-358.

[O2] S. Smale, 'Regular curves on Riemannian manifolds', Trans. Amer. Math. Soc. 87 (1958), 492-512.

[O3] D. Bennequin, 'Quelques remarques sur la rigidité symplectique', Seminaire Sud-Rhodanian de Geometrie III. Geometrie symplectique et de contact, 1-150.

[O4] A. Bahri, 'Un probleme variationnel sans compacité en geometrie de contact', Comptes. Rendus de l'Academie des Sciences Paris, t299, Serie I, no. 15 (1984).

[O5] P. H. Rabinowitz, 'Periodic solutions of Hamiltonian Systems' Comm. Pure and Applied Math. 31 (1978) p. 157-184.

[O6] C. Viterbo, 'A proof of the Weinstein conjecture in $\mathbb{R}^{2n}$' preprint. Summary by I. Ekeland in Proceedings of the NATO Conference on Hamiltonian Systems and their periodic orbits. NATO ASI Ser 25 Series C Vol. 209-Reidel.

[O7] H. Hofer, C. Viterbo, to appear.

[O8] A. Bahri, 'Pseudo Orbits of contact forms' Proceedings of the NATO Conference on Hamiltonian Systems and their periodic orbits NATO ASI Ser. 25 Series C Vol. 209-Reidel.

1 Geometric datas

α is a contact form on a three dimensional compact and orientable manifold M. ξ will denote its Reeb vector field, thus satisfying:

$$\alpha(\xi) \equiv 1 \; ; \; d\alpha(\xi, .) \equiv 0. \tag{1.1}$$

We assume the α-fiber bundle over M to be trivial; hence we can choose a non-vanishing vector field v in the kernel of α.

Once v is fixed, we get another vector field $\tilde{w}$, in the kernel of α, such that $d\alpha(v, \tilde{w}) \equiv 1$

$$\alpha(v) \equiv \alpha(\tilde{w}) \equiv 0 \; ; \; d\alpha(v, \tilde{w}) \equiv 1. \tag{1.2}$$

We have

$$\alpha \wedge d\alpha(\xi, v, \tilde{w}) \equiv 1. \tag{1.3}$$

We introduce the non-singular one-differential form $\beta = d\alpha(v, .)$ and we denote:

$$p = \beta \wedge d\beta\,(\xi, v, \tilde{w}) = d\beta(\xi, v) = -d\alpha(v, [\xi, v]). \tag{1.4}$$

p might vanish on M if β is not a contact form. We will assume at a certain point of this paper that this is not the case. However, for the moment, we introduce $\Sigma = \{x \in M | p(x) = 0\}$, which we may assume by a general position argument to be a hypersurface of M. In case Σ is not empty, we will need to study the trace of the plane-field β on Σ, i.e.

$$q = \beta \cap T\Sigma \; ; \; T\Sigma \text{ being the tangent space to } \Sigma. \tag{1.5}$$

Again, by a general position argument, q can be assumed to have nondegenerate singular points as well as non-degenerate periodic orbits.

An interesting question arises about the geometrical significance of β being a contact form. There is an example which shows this significance: let $N \overset{i}{\hookrightarrow} \mathbb{R}^4$ be a compact hypersurface in $\mathbb{R}^4$ and $\alpha = i^\star \alpha_0$, where $\alpha_0 = x_1 dx_2 - x_2 dx_1 + x_3 dx_4 - x_4 dx_3$.

We have the following proposition, whose proof is a straightforward computation:

<u>Proposition 1</u>: If N bounds a strictly convex open set in $\mathbb{R}^4$ and if v is a vector field in the kernel of α defining a Hopf fibration of N, then $\beta = d\alpha(v, .)$ is a contact form on N.

Hence the hypothesis β is a contact form means some convexity of (M, α) in the v-direction.

For the next three chapters, we will assume nothing on β but the fact that it lies in general position.

However, from the fourth chapter on, we will assume β is a contact form. In this situation, we normalize v by multiplication by a positive function λ so that:

$$\beta \wedge d\beta(\xi, v, \tilde{w}) \equiv 1. \tag{1.6}$$

As p is positive when β is a contact form (by transversality to α), the function λ is $\frac{1}{\sqrt{p}}$. We denote then:

$$\bar{\mu} = \alpha(w) \; ; \bar{\mu}_\xi = d\bar{\mu}(\xi) \; ; \bar{\mu}_{\xi\xi} = d\bar{\mu}_\xi(\xi) \tag{1.7}$$

where w is the Reeb vector field of β, thus satisfying:

$$\beta(w) \equiv 1 \; ; d\beta(w, \cdot) \equiv 0. \tag{1.8}$$

We then have the following proposition:

<u>Proposition 2</u>: $w = -[\xi, v] + \bar{\mu}\xi$; $[\xi, [\xi, v]] = -\tau v$; $\tau \in C^\infty(M, \mathbb{R})$; $\bar{\mu} = d\alpha(v, [\xi, v]])$; $\bar{\mu}_\xi = d\alpha([\xi, v], [v, [\xi, v]]))$: $\bar{\mu}_{\xi\xi} + \tau\bar{\mu} = -\tau_v$ where

$\tau_v = d\tau(v)$. All these relations hold under the hypothesis $\beta \wedge d\beta(\xi, v, w) \equiv 1$.

The proof of Proposition 2 is also a straightforward computation.

We will be dealing with submanifolds of $H^1(S^1;M)$, where $H^1(S^1;M)$ is the space of H^1 loops on M. If δ is a C^∞ one-differential on M, we can compute, along a curve $x(.)$ belonging to $H^1(S^1;M)$, the function $\delta_x(\dot{x})$ which belongs to $L^2(S^1)$. We will need the expression of the first variation of $\delta_x(\dot{x})$ along a tangent vector z to $H^1(S^1;M)$ at $x(.)$ which is given by the following formula; one can obtain it in local coordinates for instance.

Proposition 3: Let z be a tangent vector to $H^1(S^1;M)$ along the curve $x(.)$. The first variation of $\delta_x(\dot{x})$ with respect to z is $\frac{d}{dt}\delta(z) - d\delta(\dot{x}, z)$.

2 The spaces of variations $\mathcal{L}_\beta$ and C_β

We consider in this chapter the two following subsets of $H^1(S^1;M)$:

$$\mathcal{L}_\beta = \{x \in H^1(S^1;M) \mid \beta(\dot{x}) \equiv 0\} \tag{2.1}$$

$$C_\beta = \{x \in \mathcal{L}_\beta \mid \alpha_x(\dot{x}) \equiv \text{strictly positive constant}\}. \tag{2.2}$$

We want to know when $\mathcal{L}_\beta$ and C_β are submanifolds of $H^1(S^1;M)$; and more generally, if this is not the case, which type of singularities we can expect.

This will be carried out, for $\mathcal{L}_\beta$, through the study of its tangent space, which is given, by Proposition 3, by the following equation:

$x \in \mathcal{L}_\beta$; z is tangent to $\mathcal{L}_\beta$ along x if and only if (2.3)

$$\frac{d}{dt}\beta_x(z) = d\beta_x(\dot{x}, z)$$

while, for C_β, we get:

$x \in C_\beta$; z is tangent to C_β along x if and only if (2.4)

$$\begin{cases} \dfrac{d}{dt}\beta_x(z) = d\beta_x(\dot{x}, z) \\[2ex] \dfrac{d}{dt}\beta_x(z) - d\beta_x(\dot{x}, z) \equiv \text{constant.} \end{cases}$$

As $(\xi, v, \tilde{w})$ provides a basis to the tangent space to M, $\dot{x}$ can be decomposed on this basis, thus giving:

$$\dot{x} = a\xi + bv + c\tilde{w} \; ; \; a, b, c \in L^2(S^1) \tag{2.5}$$

$x \in \mathcal{L}_\beta$ if and only if $\beta_x(\dot{x}) = d\alpha_x(v, \dot{x}) = c \equiv 0$; while $x \in C_\beta$ if $c \equiv 0$ and if a is a positive constant.

Let z be a tangent vector to $\mathcal{L}_\beta$ (respectively C_β) at x. We decompose z in $z = \lambda\xi + \mu v + \eta\tilde{w}$; where $\lambda, \mu, \eta \in H^1(S^1;\mathbb{R})$.

Equations (2.3) and (2.4) then provide:

$z = \lambda \xi + \mu v + \eta \tilde{w}$ is tangent at x to $\mathcal{L}_\beta$ if and only if (2.6)

$\dot{\eta} = d\beta(\dot{x}, \tilde{w})\, \eta + (\mu a - \lambda b)\, p$

$z = \lambda \xi + \mu v + \eta \tilde{w}$ is tangent at x to $\mathcal{L}_\beta$ if and only if (2.7)

$\dot{\eta} = d\beta(\dot{x}, \tilde{w})\, \eta + (\mu a - \lambda b)\, p$; $\dot{\lambda} = b\eta$ + constant.

We then have the following theorem:

Theorem 1: If β is a contact form, $\mathcal{L}_\beta/M$ and C_β/M are submanifolds of $H^1(S^1;M)$. In the general case, when β is in a general position, the singularities of $\mathcal{L}_\beta/M$ are the contractible loops (with any parametrization) lying on the trajectories of q on Σ; while the singularities of C_β/M lie on the α-positively oriented periodic of q.

Proof of Theorem 1: We will discuss the general case.

From equation (2.6), we derive, given μ and λ, the function η:

$$\eta = e^{\int_0^t d\beta(\dot{x}, \tilde{w})\, ds} \left(\int_0^t e^{-\int_0^s d\beta(\dot{x}, \tilde{w})\, d\tau} (\mu a - \lambda b)\, p\, ds + C \right). \qquad (2.8)$$

Given μ and λ in $H^1(S^1, \mathbb{R})$, η has to be periodic. Hence:

$$C = e^{\int_0 d\beta(\dot{x}, \tilde{w})\, ds} \left(\int_0^1 e^{-\int_0^s d\beta(\dot{x}, \tilde{w})\, d\tau} (\mu a - \lambda b)\, p\, ds + C \right). \qquad (2.9)$$

Equation (2.9) allows us to compute C if $\int_0^1 d\beta(\dot{x}, \tilde{w})\, ds$ is not zero.

Hence, on those curves of $\mathcal{L}_\beta$, the tangent space is given by the choice of μ and λ in $H^1(S^1;\mathbb{R})$ and is thus $H^1(S^1;\mathbb{R}) \times H^1(S^1;\mathbb{R})$.

Otherwise, if $\int_0^1 d\beta(\dot{x}, \tilde{w})\, ds = 0$, we necessarily have:

$$\int_0^1 e^{-\int_0^s d\beta(\dot{x}, \tilde{w})\, d\tau} (\mu a - \lambda b)\, p\, ds = 0 \qquad (2.10)$$

while C is any constant.

Hence, if $p(x(\,.\,))$ is not identically zero, (2.10) provides a constraint on μ and λ which then belong to $H^1(S^1;\mathbb{R}) \times H^1(S^1;\mathbb{R})/\mathbb{R}$. But C is a free parameter; and the tangent space to $\mathcal{L}_\beta$ is thus $H^1(S^1;\mathbb{R}) \times H^1(S^1;\mathbb{R}) / \mathbb{R} \times \mathbb{R}$, providing no singularity.

However, if $p(x(.))$ is zero, hence if the curve lies on the trajectories of $q = B \cap T\Sigma$, then the constraint given by (2.10) disappears, providing a singularity.

It follows that singularities of $\mathcal{L}_\beta/M$ are drawn on the trajectories of q and must satisfy

$$\int_0^1 d\beta(\dot{x}, \tilde{w})\, dt = \int_c d\beta(\,.\,, \tilde{w}) = 0; \quad c \text{ being the curve.} \tag{2.11}$$

Assuming that β is in the general position, (2.11) implies we are dealing with contractible loops on the trajectories of q. Hence the result on $\mathcal{L}_\beta/M$.

The result about C_β/M follows from the analysis of the differential system:

$$\begin{pmatrix} \dot{\eta} \\ \dot{\lambda} \end{pmatrix} = a \begin{bmatrix} d\beta(\xi, \tilde{w}) & 0 \\ 0 & 0 \end{bmatrix} \begin{bmatrix} \eta \\ \lambda \end{bmatrix} + b \begin{bmatrix} 0 & -p \\ 0 & 0 \end{bmatrix} \begin{bmatrix} \eta \\ \lambda \end{bmatrix} + \begin{bmatrix} \mu ap \\ C \end{bmatrix} = [aA + bB] \begin{bmatrix} \eta \\ \lambda \end{bmatrix} + \begin{bmatrix} \mu ap \\ C \end{bmatrix} \tag{2.12}$$

Here again, we want periodic H^1 functions.

We know that a is a constant, while b is L^2.

Hence, the matrix differential equation:

$$\left| \begin{array}{l} \dot{R} = (aA + bB) R \\ R(0) = Id \end{array} \right. \tag{2.13}$$

has a unique solution, which is continuous and invertible. Its inverse G is continuous and satisfies:

$$\left| \begin{array}{l} \dot{G} = -G(aA + bB) \\ G(0) = Id. \end{array} \right. \tag{2.14}$$

The general solution of (2.12) can be written:

$$\begin{pmatrix} \eta \\ \lambda \end{pmatrix}(t) = z(t) = R(t)\ \left[\int_0^1 R(s)^{-1} \begin{bmatrix} \mu ap \\ C \end{bmatrix} ds + K\right] \tag{2.15}$$

where $K = \begin{bmatrix} K_1 \\ K_2 \end{bmatrix}$ is a constant.

As we want z to be periodic, we get:

$$K = R(1)\left[\int_0^1 R(s)^{-1} \begin{bmatrix} \mu ap \\ C \end{bmatrix} ds + K\right] \tag{2.16}$$

or:

$$[Id - R(1)]K = \int_0^1 R(s)^{-1} \begin{bmatrix} \mu ap \\ C \end{bmatrix} ds. \tag{2.17}$$

One case is easy to solve:

1. $\det[Id - R(1)] \neq 0$. Then $\begin{bmatrix} \mu ap \\ C \end{bmatrix}$ is arbitrary and K can be computed through (2.17).

The tangent space to C_β/M on such curves is thus $\mathbb{R} \times H^1(S^1, \mathbb{R})$.

We then have the following further case:

2. $R(1) = Id$.

Eq. (2.17) then becomes:

$$\int_0^1 R(s)^{-1} \begin{bmatrix} \mu ap \\ C \end{bmatrix} ds = 0; \text{ while } K \text{ is arbitrary.} \tag{2.18}$$

Eq. (2.18) gives in general two constraints on C and μ.

We have to check that these two constraints do not degenerate and are independent.

Let us denote:

$$R(s)^{-1} = \begin{bmatrix} \hat{A}(s) & \hat{B}(s) \\ \hat{C}(s) & \hat{D}(s) \end{bmatrix} \text{ where } \hat{A}, \hat{B}, \hat{C}, \hat{D} \text{ are continuous functions.} \tag{2.19}$$

Eq. (2.18) then amounts to:

$$a[\int_0^1 \hat{A}(s)\mu p\,ds] + C\int_0^1 \hat{B}(s)\,ds = 0$$
$$a[\int_0^1 \hat{C}(s)\mu p\,ds] + C\int_0^1 \hat{D}(s)\,ds = 0. \qquad (2.20)$$

If these two constraints degenerate, we have:

$$(\int_0^1 \hat{D}(s)\,ds)\,\hat{A}p = (\int_0^1 \hat{B}(s)\,ds)\,\hat{C}p. \qquad (2.21)$$

We are assuming that $p(x(.))$ is not identically zero. Otherwise, such a curve would lie on an α-positively oriented periodic orbit of q.

By the obvious S^1 invariance of the expression of the tangent space, we may also assume that $p(x(0))$ is not zero.

As $\hat{A}(0) = 1$, while $\hat{C}(0) = 0$, we deduce from (2.21) that $\int_0^1 D(s)\,ds = 0$, which implies, if the constraints degenerate, that $\int_0^1 \hat{B}(s)\,ds = 0$ and $\hat{C}p = \lambda\,\hat{A}p$. But $C(0) = 0$ while $\hat{A}(0) = 1$; hence $\lambda = 0$ and we get:

$$\hat{C}p \equiv 0; \quad \int_0^1 \hat{D}(s)\,ds = \int_0^1 \hat{B}(s)\,ds = 0. \qquad (2.22)$$

On the other hand, $R(s)^{-1} = G$ satisfies (2.14); hence:

$$\dot{\hat{D}} = -bp\hat{C}. \qquad (2.23)$$

Thus, by (2.22), $\hat{D}(s) = 1$; which contradicts $\int_0^1 \hat{D}(s)\,ds = 0$.

It follows that the two constraints do not degenerate; the tangent space to C_β/M on such curves is $[H^1(S^1;\mathbb{R}) \times \mathbb{R}/\mathbb{R} \times \mathbb{R}] \times \mathbb{R} \times \mathbb{R} \simeq H^1(S^1;\mathbb{R}) \times \mathbb{R}$. Hence there is again no singularity in this case.

The remaining case to study is:

3. $\dim \mathrm{Ker}(\mathrm{Id} - R(1)) = 1$.

Let $E_1 = \mathrm{Im}(\mathrm{Id} - R(1))$, E_2 be a supplement in $\mathbb{R}^2$ of E_1 and P the projection on E_2.

Eq. (2.17) can only be satisfied if $\int_0^1 R(s)^{-1}\begin{bmatrix}\mu ap\\ C\end{bmatrix}ds$ belongs to E_1; thus if:

$$P(\int_0^1 R(s)^{-1}\begin{bmatrix}\mu ap\\ C\end{bmatrix}ds) = 0 \qquad (2.24)$$

which is equivalent to:

$$\int_0^1 PR(s)^{-1}\begin{bmatrix}\mu ap\\ C\end{bmatrix}ds = 0. \qquad (2.25)$$

These constraints can be seen not to degenerate, by the same technique as in case 2. Under (2.25), (2.17) has a unique solution in K up to translations in Ker(Id - R(1)), whose dimension is one.

The tangent space is then $[\mathbb{R} \times H^1(S^1;\mathbb{R})/\mathbb{R}] \times \mathbb{R} \simeq \mathbb{R} \times H^1(S^1;\mathbb{R})$ and the proof of theorem 1 is thereby complete. ■

Remark 1: We point out here that α-positively oriented periodic orbits q are not likely to exist. Indeed, these periodic orbits are due to elliptic singularities of q on Σ; they appear then as limit cycles. But, on such a singularity, $dp(\xi) = dp(v) = 0$. Hence, there is a curve $dp(v) = 0$ passing through this point. This curve is likely to intersect the limit cycle which is tangent to $q = (-dp(v), dp(\xi))$. But, in this case, dp(v) changes sign on the periodic orbit, forbidding α-positiveness (or negativeness).

Remark 2: It should be noted for the remainder of this book that, when β is a contact form, we normalize v such that $\beta \wedge d\beta = \alpha \wedge d\alpha$ (this is equivalent to (1.6)). Then $\tilde{w}$ can be replaced by w in a given basis of the tangent space to M and the equation of the tangent space to C_β becomes

$$\begin{cases} \dot{\overbrace{\lambda + \bar{\mu}\eta}} = b\eta + \text{Cte} \\ \dot{\eta} = \mu a - \lambda b \end{cases},$$

where $\bar{\mu}$ has been defined in Proposition 2.

Further information on the topology of the spaces of variations $\mathcal{L}_\beta$ and C_β is provided later on in the book.

3 The functional

We assume from now on that β is a contact form.

We are dealing with a normalized vector field v, such that $\beta \wedge d\beta(\xi, v, w) = 1$; or else $\beta \wedge d\beta = \alpha \wedge d\alpha$. w is the Reeb vector field of β; and $w = -[\xi, v] + \bar{\mu}\xi$ ($\bar{\mu} = \alpha(w)$). As well, we have $[\xi, [\xi, v]] = -\tau v$ and $\bar{\mu}_{\xi\xi} + \tau v = -\tau_v$ (see Proposition 2).

On C_β, we consider the functional:

$$J(x) = \alpha_x(\dot{x}) = \int_0^1 \alpha_x(\dot{x})\, dt. \tag{3.1}$$

We first compute the first variation of J. The gradient of J with respect to some fixed metric will be denoted ∂J:

Proposition 4: J is a C^2 functional on C_β whose critical points are periodic orbits to ξ. If z is a tangent vector to C_β along the curve x(.), then:

$$\partial J(x) \,.\, z = -\int_0^1 d\alpha_x(\dot{x}, z)\, dt.$$

Proof: The first variation of $\alpha_x(\dot{x})$ is $\frac{d}{dt}\alpha(z) - d\alpha(\dot{x}, z)$ (see Proposition 3). Hence, the first variation of J(x) is

$$\partial J(x) \,.\, z = \int_0^1 \left(\frac{d}{dt}\alpha(z) - d\alpha(\dot{x}, z)\right) dt = -\int_0^1 d\alpha(\dot{x}, z)\, dt.$$

Hence the formula in Proposition 4.

Now if $z = \lambda\xi + \mu v + \eta w$ and $\dot{x} = a\xi + bv$, we have:

$$\partial J(x) \,.\, z = -\int_0^1 b\eta\, dt. \tag{3.2}$$

Eq. (3.2) shows that J is C^1, as b is L^2 and η is H^1.

To see that J is C^2, we notice that:

$$b = d\alpha_x(\dot{x}, w) = \gamma_x(\dot{x})$$

where

$$\gamma_x = d\alpha_x(\,.\,, w) \text{ is a } C^\infty\text{-one-differential form.} \tag{3.3}$$

The first variation of $\gamma_x(\dot{x})$ along z is:

$$\frac{d}{dt}\gamma(z) - d\gamma(\dot{x}, z).$$

Hence, this first variation is also:

$$\frac{d}{dt}\gamma(z) - d\gamma(\dot{x}, z) = \dot{\mu} - d\gamma(a\xi + bv, \lambda\,\xi + \mu v + \eta w) =$$

$$= \dot{\mu} - (\mu a - \lambda b)\, d\gamma(\xi, v) - a\eta\; d\gamma(\xi, w) - b\eta\; d\gamma(v, w) \tag{3.4}$$

which is again an L^2-function.

As $\partial J(x)\,.\,z = -\int_0^1 b\eta\, dt = \int_0^1 \gamma_x(\dot{x})\,\eta\, dt$ and as the first variation of $\gamma_x(\dot{x})$ is L^2, $J(x)$ is at least C^2 on C_β.

It remains to show that the critical points of J are periodic orbits of ξ. Indeed, $\partial J(x)\,.\,z = 0$ is equivalent to $\int_0^1 b\eta\, dt = 0$ for any variation η such that:

$$\begin{cases} \dfrac{d\alpha(z)}{dt} - d\alpha(\dot{x}, z) = Cte \\[2ex] \dfrac{d\alpha(z)}{dt} - d\beta(\dot{x}, z) = 0 \end{cases} \iff \begin{cases} \dot{\overline{\lambda + \bar{\mu}\eta}} = b\eta + C \\[2ex] \eta = \mu a - \gamma b \end{cases} \quad \lambda,\ \mu,\ \eta \text{ periodic of period one,} \tag{3.5}$$

(see Remark 2).

In (3.5), by the periodicity of $\lambda + \bar{\mu}\eta$, C has to be equal to $-\int_0^1 b\eta\, dt$. Hence, x is critical to J if and only if (3.5) has no periodic solution with a non-vanishing C. This implies $b \equiv 0$ by similar arguments to the ones used in previous chapters to study the tangent space to C_β.

There is also a heuristic way to see this fact, if we assume b to be C^1. Then, if b is not zero, we can compute $\lambda + \bar{\mu}\eta = \int_0^t b\eta\, ds - t\int_0^1 b\eta\, ds$ from the first equation of (3.5) choosing η in H^2 such that $\int_0^1 b\eta\, ds$ is not zero. μ is then easily derived as an $H^1(S^1, \mathbb{R})$ function equal to $\dfrac{\dot{\eta}+\lambda b}{a}$ from the second equation of (3.5). Hence the result with this heuristic argument (which can be made rigorous).

But if $b \equiv 0$, $\dot{x} = a\xi$, with a being a strictly positive constant.

Hence x is a periodic orbit to ξ. ■

The result given by Proposition 4 extends to the more general case where β is not a contact form, but remains, however, in a general position:

Proposition 5: Assume β is in a general position. Then the critical points of J on C_β of finite Morse index are periodic orbits to ξ.

The proof is somewhat technical and hence will be published elsewhere as it is not our purpose here to deal with this situation.

4 The Palais-Smale condition and the lack of compactness

As we will see later on in this book, there are intuitive reasons which imply that J does not satisfy the Palais-Smale condition or else any weaker compactness criterion. This will become clear after we study the topology of C_β and the differences of topology induced by J, which are by far too drastic to be due only to the ξ periodic orbits.

For the moment, we can get a first understanding of the problem: along (pseudo)-gradient lines, one knows that $J(x)$ remains bounded; which provides a bound on the ξ-component of $\dot{x} = a\xi + bv$.

But we have no control on the v-component and it might happen that this v-component becomes infinite on a gradient line.

On the other hand, if $\partial J(x)$ goes to zero, we get:

$$\partial J(x) \cdot z = -\int_0^1 b\eta \, dt \to 0 \quad (z = \lambda\xi + \mu v + \eta w \text{ satisfying } (18)). \tag{4.1}$$

Hence

$$\left| \int_0^1 b\eta \, dt \right| \leq \varepsilon \, |z|_{H^1}. \tag{4.2}$$

The H^1-norm of z contains an L^2-norm of $\dot{\eta}$ and other terms also. Hence (4.2) will give a very weak information on b and anyway, no bound whatsoever (there are b-terms in $|z|_{H^1}$); hence the failure of the Palais-Smale condition.

In this situation, we construct a special deformation lemma, which will compactify the situation, by introducing a certain 'viscosity term' which allows us to control the b-component. The deformation depends on a certain parameter $\varepsilon > 0$ and is induced by a vector field Z_ε such that the scalar product $\partial J(x) \cdot Z_\varepsilon(x)$ is positive. However, Z_ε can be zero while $\partial J(x)$ is not zero; which will produce the differential equation giving rise to critical points at infinity. Let

$$I(x) = \int_0^1 b^2 dt \qquad (\dot{x} = a\xi + bv). \tag{4.3}$$

I is a C^2 functional on C_β on which we wish to have control along deformation lines of the level sets of J.

For this purpose, we introduce the following vector field on C_β:

$$Z = |\partial J|\,\partial I + |\partial I|\,\partial J, \tag{4.4}$$

∂J and ∂I are the gradients of I and J and $|\partial I|$, $|\partial J|$ are their norms with respect to the H^1-metric of C_β (inherited from the H^1-metric on $H^1(S^1, M)$). We note (,) the H^1 scalar product on the tangent space to C_β.

We then have the following:

Lemma 1: Z is a locally lipschitz vector field on C_β. $(Z, \partial J)$ is positive as well as $(Z, \partial I)$. If one of these quantities is zero, then Z is zero.

Proof: The only delicate point is the fact that Z is lipschitz. This is evident, as I and J are C^2, outside those points x of C_β where $\partial J(x)$ or $\partial I(x)$ vanishes. But, in those points, $|\partial J|$ and $|\partial I|$ are locally lipschitz: hence the result. ■

Let $\varepsilon > 0$ be given. We consider:

$$\begin{cases} \omega_\varepsilon : \overline{\mathbb{R}} \to [0,1] \\ \omega_\varepsilon(x) = 1 \text{ if } x \geq \varepsilon \\ \omega_\varepsilon(x) = 0 \text{ if } x \leq \varepsilon/2 \\ \omega_\varepsilon \in C^\infty \end{cases} \tag{4.5}$$

$$\begin{cases} l_\varepsilon : C_\beta \to [0,1];\ l_\varepsilon \in C^\infty \\ l_\varepsilon = 0 \text{ on an } \varepsilon/2\text{-neighbourhood of the critical points of } J \\ l_\varepsilon = 1 \text{ outside an } \varepsilon\text{-neighbourhood of these points.} \end{cases} \tag{4.6}$$

$$\begin{cases} \varphi : C_\beta - \{\text{critical points of } J\} \to \overline{\mathbb{R}} \\ \varphi(x) = \dfrac{|\partial J|(x)}{|\partial I|(x)} \, I(x) \quad \text{if } |\partial I|(x) \text{ is non zero} \\ \varphi(x) = +\infty \quad \text{if } |\partial I|(x) = 0. \end{cases} \tag{4.7}$$

We introduce then the following vector field on C_β:

$$Z_\varepsilon(x) = l_\varepsilon(x)\,(\omega_\varepsilon(\varphi(x))\,\partial J(x) + Z(x)). \tag{4.8}$$

This formula of Z_ε is a priori defined only on $C_\beta / \{\text{critical points of } J\}$ as φ is only defined on this set. But it clearly extends to all of C_β by setting $Z_\varepsilon = 0$ on a critical point of J. In fact, Z_ε is zero on an $\varepsilon/2$-neighbourhood of these critical points, by the definition of l_ε.

We have the following proposition:

Proposition 6: Z_ε is locally lipschitz on C_β.

Proof: The proof is reduced to the fact that $\omega_\varepsilon(\varphi)$ is locally lipschitz on the set where l_ε is non zero.

If x is such that $l_\varepsilon(x)$ is non zero, then $|\partial J|(x)$ is non zero and $I(x)$ is bounded away from zero, implying that $I(y)$ is bounded away from zero on a neighbourhood of x. $\omega_\varepsilon(\varphi(y))$ is then C^1 on this neighbourhood. This completes the proof of Proposition 6. ∎

We now consider the differential equation:

$$\begin{cases} \dfrac{\partial x}{\partial s} = -Z_\varepsilon(x) \\ x(0) \text{ given.} \end{cases} \tag{4.9}$$

Compactifying Deformation Lemma:

1. On an integral curve on (4.8) $x(s)$, $J(x(s))$ decreases and $I(x(s))$ remains bounded by a constant depending of ε, $J(x(0))$ and $I(x(0))$, for all time $s \geq 0$ such that $J(x(s))$ is positive.

2. Let $0 < a_0 < a_1$. Suppose $J^{a_0} = \{x \in C_\beta \,|\, J(x) \leq a_0\}$ is not retract by deformation of $J^{a_1} = \{x \in C_\beta \,|\, J(x) \leq a_1\}$.

Then there exists $\varepsilon_0 > 0$ such that for any $0 < \varepsilon < \varepsilon_0$, there is a point x_ε in C_β with:

$$a_0 \leq J(x_\varepsilon) \leq a_1 \text{ and either } \partial J(x_\varepsilon) = 0 \text{ or } Z(x_\varepsilon) = 0; \varphi(x_\varepsilon) \leq \varepsilon. \tag{4.10}$$

Proof: Let $f(x) = I(x(s))$, $g(s) = J(x(s))$

$$g'(s) = -(\partial J, Z) = -l_\varepsilon(x(s)) \quad (\omega_\varepsilon |\partial J|^2 + (Z, \partial J)).$$

We know that $(Z, \partial J)$ is positive. Hence:

$$\begin{aligned} g(s) - g(0) &= -\int_0^s l_\varepsilon(x(\tau))[\omega_\varepsilon |\partial J|^2 + (z, \partial J)](x(\tau))d\tau \\ &\leq -\int_0^1 l_\varepsilon \omega_\varepsilon |\partial J|^2 d\tau. \end{aligned} \tag{4.11}$$

Hence, for positive s and as long as $g(s)$ remains positive, we have:

$$\int_0^s \omega_\varepsilon l_\varepsilon(x(\tau)) \; |\partial J|^2 d \leq g(0) = J(x(0)). \tag{4.12}$$

We first notice that $g'(s)$ is negative; hence $J(x(s))$ decreases.

On the other hand, $f'(s) = -(\partial I, Z_\varepsilon) = -l_\varepsilon(\omega_\varepsilon(\partial I, \partial J) + (Z, \partial I))$. As $(Z, \partial I)$ is positive, we have:

$$f'(s) \leq -l_\varepsilon \omega_\varepsilon(\partial I, \partial J) \leq \omega_\varepsilon l_\varepsilon |\partial I| \, |\partial J| = l_\varepsilon \omega_\varepsilon(\varphi) \frac{|\partial I|}{|\partial J|} \cdot \frac{1}{I}(I|\partial J|^2). \tag{4.13}$$

Hence

$$f'(s) \leq l_\varepsilon \frac{\omega_\varepsilon(\varphi)}{\varphi} I |\partial J|^2. \tag{4.14}$$

But, by the very definition of ω_ε, we have:

$$\omega_\varepsilon(\varphi) \leq 2/\varepsilon\varphi. \tag{4.15}$$

Hence:

$$f'(s) \leq (2l_\varepsilon/\varepsilon)\, \omega_\varepsilon(\varphi) \, |\partial J|^2 I = (2l_\varepsilon/\varepsilon \omega_\varepsilon(\varphi) \, |\partial J|^2) f(s). \tag{4.16}$$

Hence, using (4.12),

$$f(s) \leq f(0)\, e^{2/\varepsilon \int_0^1 \omega_\varepsilon(\varphi)\, |\partial J|^2 l_\varepsilon d\tau} \leq f(0)\, e^{2/\varepsilon\, g(0)} . \tag{4.17}$$

The first statement of the lemma is then proven.

The proof of the second statement requires the following two lemmas whose proof is straightforward.

<u>Lemma 2</u>: Let (x_n) be a sequence in C_β such that $0 < a_0 \leq J(x_n) \leq a_1$ and $(I(x_n))$ is bounded. If $\partial J(x_n)$ goes to zero, there is a sub-sequence converging weakly to x in C_β with $0 < a_0 \leq J(x) \leq a_1$ and $\partial J(x) = 0$.

<u>Lemma 3</u>: Let (x_m) be a sequence in C_β such that $0 < a_0 \leq J(x_m) \leq a_1$ and $(I(x_m))$ is bounded. If $(\partial J(x_m), Z(x_m)) \to 0$ and if (x_m) converges weakly to x in C_β, with $\partial J(x)$ non zero, there is a strongly convergent sub-sequence to x.

<u>Remark</u>: If $(\partial J(x_m), Z(x_m)) \to 0$, then $|Z(x_m)| \to 0$; and if (x_m) converges weakly to x such that $\partial J(x)$ is non zero, with $J(x_m)$ and $I(x_m)$ bounded, then $|\partial J(x_m)|$ is bounded away from zero. Therefore $\partial I(x_m) \to 0$; $I(x_m)$ and $J(x_m)$ are bounded. It is then easy to prove Lemma 3.

<u>Proof of the second statement</u>: Arguing by contradiction, we may assume there is no critical point of J in the set $\{x \,|\, a_0 \leq J(x) \leq a_1\}$. Hence, for $0 < \varepsilon \leq \varepsilon_0$, l_ε is equal to 1 on this set.

Now let x_0 be such that $a_0 \leq J(x_0) \leq a_1$.

We denote by $x(s, x_0)$ the solution of (4.8) having x_0 as initial data. The situation divides into two cases.

<u>1st case</u>: $\forall s \geq 0,\ J(x(s, x_0)) > a_0 > 0$.

Then $I(x(s, x_0))$ and $J(x(s, x_0))$ are uniformly bounded on $[0, +\infty[$ and the solution of (4.8) exists for all positive s.

Remembering that $g(s) = J(x(s, x_0))$, we have

$$g'(s) = -l_\varepsilon(x(s, x_0))\,(\omega_\varepsilon |\partial J|^2 + (Z, \partial J)). \tag{4.18}$$

Hence, as $l_\varepsilon(x(s, s_0)) = 1$,

$$g'(s) = -\omega_\varepsilon |\partial J|^2 - (Z, \partial J). \tag{4.19}$$

From the boundedness of $g(s)$, we deduce

$$\int_0^{+\infty} [\omega_\varepsilon |\partial J|^2 + (Z, \partial J)]\, d\tau < +\infty. \tag{4.20}$$

As $(Z, \partial J)$ is positive, (4.20) yields the existence of a sequence (x_n) such that:

$$a_0 \leq J(x_n) \leq a_1 \tag{4.21}$$

$$\omega_\varepsilon(\varphi)(x_n) |\partial J|^2(x_n) + (\partial J, Z)(x_n) \to 0 \tag{4.22}$$

$$I(x_n) \leq I(x_0) e^{2/\varepsilon\, a_1} \quad (\text{see } (56)). \tag{4.23}$$

As $(J(x_n))$ and $(I(x_n))$ are bounded, (x_n) is H^1-bounded and we can extract from (x_n) a weakly convergent sub-sequence to x belonging to C_β, with $a_0 \leq J(x) \leq a_1$. We will call this sub-sequence (x_n) again.

Our hypothesis is that x is not critical.

Hence (x_n) is a sequence such that $a_0 \leq J(x_n) \leq a_1$; $(I(x_n))$ is bounded and, by (4.22), $(\partial J, Z)(x_n)$ goes to zero. Applying Lemma 3, we derive that (x_n) converges in fact strongly to x; (4.22) then implies:

$$\omega_\varepsilon(\varphi)(x) |\partial J|^2 + (\partial J, Z)(x) = 0. \tag{4.24}$$

Hence, as $\partial J(x)$ is non zero,

$$Z(x) = 0 \text{ and } \varphi \leq \varepsilon \tag{4.25}$$

which proves the second statement of the lemma is this case.

<u>2nd case:</u> For any x_0 belonging to $\{x | a_0 \leq J(x) \leq a_1\}$, there exists a positive s such that $J(x(s, x_0)) = a_0$.

Then let $s(x_0)$ be the s first used such that $J(x(s(x_0), x_0)) = a_0$. Let

$$z_0 = x(s(x_0), x_0). \tag{4.26}$$

If $(\partial J(z_0), Z(z_0))$ is not zero, the function $s(.)$ is continuous at x_0.

Hence, if

$$(\partial J(x), Z_\varepsilon(x)) > 0 \quad \forall\, x \text{ such that } a_0 \le J(x) \le a_1 \tag{4.27}$$

then $s(.)$ is globally continuous and the map:

$$J^{a_1} \times [0,1] \to J^{a_1}$$

$$(x_0, t) \to \begin{cases} x(ts(x_0), x_0) & \text{if } J(x_0) \ge a_0 \\ x_0 & \text{if } J(x_0) \le a_0 \end{cases}$$

defines a retraction by deformation of J^{a_1} on J^{a_0}.

This is excluded from our hypothesis; and (4.27) is consequently impossible; which yields the existence of z_0 such that:

$$J(z_0) = a_0; \ (\partial J(z_0), Z_\varepsilon(z_0)) = 0 \text{ i.e. } l_\varepsilon(\omega_\varepsilon(\varphi)\,|\partial J|^2 + (Z, \partial J))(z_0) = 0. \tag{4.28}$$

But $l_\varepsilon(z_0)$ is equal to 1 $(J(z_0) = a_0)$ and $|\partial J|(z_0)$ is non zero. Hence:

$$(\partial J(z_0), Z_\varepsilon(z_0)) = 0 \text{ and } \omega_\varepsilon(\varphi)(z_0) = 0 \tag{4.29}$$

which implies

$$Z_\varepsilon(z_0); \ \varphi(z_0) \le \varepsilon. \tag{4.30}$$

This ends the proof of the compactifying deformation lemma. ■

5 The pendulum equation in the (α, ξ, v) dynamics and in the variational problem

5.1 Geometric aspects

In this chapter of the book, we explore some properties of the dynamics of α along v. The two main features of these dynamics are, on one hand, a differential equation which governs the transport of α along v; and, on the other hand, a property of rotation of α in a transported frame along a v-orbit.

These two aspects interfere heavily with the equation of the critical points at infinity that we will write down in section 5.2.

For the moment, we consider at a point x_0, belonging to M, a vector tangent to M which we denote by u_0. Let $\phi_s^\star = D\phi_s$ be the differential of the one-parameter group ϕ_s generated by v.

We wish to compute, in the $(\xi, v, [\xi, v])$ basis the coordinates of

$$u_s = \phi_s^\star(u_0)\,;\quad u_0 = a_0\xi + b_0 v + c_0[\xi, v]\,. \tag{5.1}$$

We will denote these coordinates $A(s)$, $B(s)$, $C(s)$, so that:

$$u_s = A(s)\,\xi + B(s)\,v + C(s)\,[\xi, v]\,. \tag{5.2}$$

Let:

$$Z(s) = \begin{bmatrix} A(s) \\ B(s) \\ C(s) \end{bmatrix} \tag{5.3}$$

and

$$\Gamma = \begin{bmatrix} 0 & 0 & -1 \\ 0 & 0 & \bar{\mu}_\xi \\ 1 & 0 & \bar{\mu} \end{bmatrix}\,. \tag{5.4}$$

We have the following proposition:

Proposition 7: Z satisfies the differential equation:

$$\dot{Z} = \Gamma Z; \quad Z(0) = \begin{bmatrix} a_0 \\ b_0 \\ c_0 \end{bmatrix} . \tag{5.5}$$

We defer for the moment the proof of the proposition and we derive the differential equation satisfied by a transported form rather than a transported vector.

For this purpose, we introduce the dual basis to $(\xi, v, [\xi, v])$ which is:

$$(\xi^\star, v^\star, [\xi, v]^\star) = (\alpha, \gamma, -\beta); \text{ where} \tag{5.6}$$

$$\gamma = d\alpha(\,.\,, w). \tag{5.7}$$

Indeed we have:

$$\begin{cases} \alpha(\xi) = 1; \quad \alpha(v) = \alpha([\xi, v]) = 0 \\ \gamma(\xi) = 0; \quad \gamma(v) = d\alpha(v, w) = 1; \quad \gamma([\xi, v]) = d\alpha([\xi, v], w) \\ \qquad\qquad\qquad\qquad\qquad\qquad = d\alpha([\xi, v], -[\xi, v] + \bar{\mu}\xi) = 0 \\ -\beta(\xi) = 0; \quad -\beta(v) = 0; \quad -\beta([\xi, v]) = -d\alpha(v, [\xi, v]) = 1 . \end{cases} \tag{5.8}$$

Let:

$$u_0^\star = a_0^\star \alpha_{x_0} + b^\star \gamma_{x_0} + c^\star(-\beta_{x_0}) \tag{5.9}$$

be an initial data in the cotangent space to M at x_0, and let:

$$u_s^\star = u_o^\star(D\phi_s^{-1}(\,.\,)) = u_0^\star(\phi_s^{\star -1}(\,.\,)) = u_0^\star(\phi_{-s}^\star(\,.\,)) \tag{5.10}$$

$u_s^\star$ is the form $u_0^\star$ transported by v.

$u_s^\star$ can be expressed in the basis $(\alpha_{x_s}, \gamma_{x_s}, -\beta_{x_s})$:

$$u_s^\star = A^\star(s)\,\alpha_{x_s} + B^\star(s)\,\gamma_{x_s} - C^\star(s)\,\beta_{x_s} \tag{5.11}$$

where $x_s = \phi_s(x_0)$.

We then have the following proposition, which is similar to the previous one:

Proposition 8: $Z^\star(s) = \begin{bmatrix} A^\star(s) \\ B^\star(s) \\ C^\star(s) \end{bmatrix}$ satisfies the differential equation:

$$\begin{cases} \dot{Z}^\star(s) = -{}^t\Gamma Z^\star(s) \\ Z^\star(0) = Z_0^\star = \begin{bmatrix} a_0^\star \\ b_0^\star \\ c_0^\star \end{bmatrix} \end{cases} . \quad (5.12)$$

Remark 3: (81) gives rise between $A^\star$ and $C^\star$ to the pendulum equation in its full expression.

Proof of Propositions 7 and 8: The proof of Proposition 8 follows easily from Proposition 7. Indeed, let $V(s)$ be the matrix of $\phi_s^\star$ in the basis $(\xi, v, [\xi, v])$. $V(s)$ is the resolvent matrix to the differential equation (5.12) (by the very definition of this differential equation) and thus satisfies:

$$\dot{V} = \Gamma V; \quad V(0) = Id. \quad (5.13)$$

As $u_s^\star = u_0^\star(\phi_{-s}^\star(.)) = u_0^\star(D\phi_s^{-1}(.))$, we have:

$$Z^\star(s) = \begin{bmatrix} A^\star(s) \\ B^\star(s) \\ C^\star(s) \end{bmatrix} = {}^tV(s)^{-1} \begin{bmatrix} a_0^\star \\ b_0^\star \\ c_0^\star \end{bmatrix} = {}^tV(s)^{-1} Z_0^\star . \quad (5.14)$$

Hence:

$$\begin{cases} \dot{Z}^\star(s) = ({}^tV(s)^{-1})^\cdot Z_0^\star = -{}^t\Gamma\, {}^tV(s)^{-1} Z_0^\star = -{}^t\Gamma Z^\star \\ Z^\star(0) = Z_0^\star \end{cases} \quad (5.15)$$

where we derived from (5.13) the fact that $({}^tV(s)^{-1})^\cdot = -{}^t\Gamma\, {}^tV(s)$.

The proof of Proposition 7 goes as follows:

u_s being transported by v satisfies:

$$[\frac{\partial}{\partial s} + v, u_s] = [\frac{\partial}{\partial s} + v, A(s)\,\xi + B(s)\,v + C(s)\,[\xi, v]] = 0 \tag{5.16}$$

which yields:

$$\mathring{A}(s)\,\xi + \dot{B}(s)\,v + \dot{C}(s)\,[\xi, v] + A(s)\,[v, \xi] + C(s)\,[v, [\xi, v]] = 0. \tag{5.17}$$

Hence:

$$\dot{A}(s) + \dot{C}(s)\,\alpha([v, [\xi, v]] = 0 \tag{5.18}$$

by applying α to (5.17); hence:

$$\dot{A}(s) + C(s) = 0 \qquad (\text{as } \alpha([v, [\xi, v]]) = 1). \tag{5.19}$$

Now

$$-\dot{C}(s) + A(s) + C(s)\,\bar{\mu} = 0 \tag{5.20}$$

by applying $d\alpha(v, .)$ to (5.17). (Notice that $d\alpha(v, [\xi, v]) = -1$ and $d\alpha(v, [v, [\xi, v]]) = \bar{\mu}$; see Proposition 2.)

Also

$$\dot{B}(s) + C(s)\,\bar{\mu}_\xi = 0. \tag{5.21}$$

Indeed, $d\alpha([\xi, v], [v, [\xi, v]]) = \bar{\mu}_\xi$ (see Proposition 2 again).

Now (5.19), (5.30) and (5.21) yield (5.12) immediately.

This completes the proof of Propositions 7 and 8. ■

We exhibit now a property of rotation of α in its transport along v. This goes as follows:

Let x_0 be a point of M and let $\phi_s(x_0)$ be the integral curve of v passing through x_0.

Let $e_1(x_0)$ and $e_2(x_0)$ be two tangent vectors to M at x_0 such that:

$$(\alpha \wedge d\alpha)_{x_0}(e_1(x_0), e_2(x_0), v(x_0)) < 0. \tag{5.22}$$

We denote:

$$e_1(s) = D\phi_s(e_1(x_0))\,;\quad e_2(s) = D\phi_s(e_2(x_0)). \tag{5.23}$$

At $x(s) = \phi_s(x_0)$, $(e_1(s), e_2(s, v(x(s)))$ is again a basis of the tangent space at $x(s)$ and we have:

$$(\alpha \wedge d\alpha)_{x(s)}(e_1(s), e_2(s), v(x(s))) < 0. \tag{5.24}$$

As $\alpha(v) = 0$, there is a well-defined trace of α in the plane $(e_1(s), e_2(s))$ which is given by:

$$\alpha(e_2(s))\, e_1(s) - \alpha(e_1(s))\, e_2(s) = u(s). \tag{5.25}$$

We follow this trace when s changes.

The following proposition expresses, by a property of this trace, the fact that α is a contact form.

<u>Proposition 9</u>: The trace of α in the plane $(e_1(s), e_2(s))$ rotates from e_1 to e_2 when s grows.

<u>Proof</u>: We express the fact that $\alpha \wedge d\alpha(e_1(s), e_2(s), v)$ is negative. Hence:

$$\alpha(e_1)\, d\alpha(e_2, v) - \alpha(e_2)\, d\alpha(e_1, v) < 0. \tag{5.26}$$

Let

$$\alpha(e_2) = \rho \cos \tilde{\varphi} \qquad\qquad \alpha(e_1) = -\rho \sin \tilde{\varphi} \tag{5.27}$$

so that $u = \rho[\cos \tilde{\varphi}\, e_1 + \sin \tilde{\varphi}\, e_2]$.

We fix a local coordinate chart in a neighbourhood of $x_s = \phi_s(x_0)$ where v is read as a constant vector; derivation along $\frac{\partial}{\partial s}$ can be seen as well as v-derivation. $e_1(s)$ and $e_2(s)$ are also constant as they are v-transported. We may extend them as constant vector fields on $\mathbb{R}^3$, denoted e_1 and e_2. In this chart α becomes a new contact form but we keep the same notation.

We compute:

$$\alpha(e_1)\, d\alpha(e_2, v) - \alpha(e_2)\, d\alpha(e_1, v) \tag{5.28}$$

$d\alpha(e_1, v) = e_1 \cdot \alpha(v) - v \cdot \alpha(e_2) - \alpha([\frac{\partial}{\partial s}, e_2])$, but $\alpha(v) = 0$, $[\frac{\partial}{\partial s}, e_2] = 0$ as e_2 is constant when s changes, and $v.\, \alpha(e_2) = \frac{\partial}{\partial s}\alpha(e_2)$

Hence:

$$d\alpha(e_2, v) = -\frac{\partial}{\partial s}\alpha(e_2) = -\frac{\partial}{\partial s}(\rho \cos \tilde{\varphi}) \tag{5.29}$$

Similarly,

$$d\alpha(e_1, v) = -\frac{\partial}{\partial s}\alpha(e_1) = \frac{\partial}{\partial s}(\rho \sin \tilde{\varphi}) \tag{5.30}$$

Hence,

$$\begin{aligned} &\alpha(e_1)\, d\alpha(e_2, v) - \alpha(e_2)\, d\alpha(e_1, v) \\ &\qquad = \rho \sin \tilde{\varphi} \frac{\partial}{\partial s}(\rho \cos \tilde{\varphi}) - \rho \cos \tilde{\varphi} \frac{\partial}{\partial s}(\rho \sin \tilde{\varphi}) \\ &\qquad = -\rho^2 \dot{\tilde{\varphi}} < 0. \end{aligned} \tag{5.31}$$

Hence $\dot{\tilde{\varphi}}$ is positive and the proposition is thereby proven. ■

We conclude this subsection with the following definitions.

Let

$$\tilde{\varphi}(s) \tag{5.32}$$

be the angle of the trace of α in $(e_1(s), e_2(s))$ with the e_1-direction, as previously noted.

Definition 1: A coincidence point to x_0 along the v-orbit passing through x_0 is a point $x_{s_0} = \phi_{s_0}(x_0)$ such that the α-direction in any frame of type $(e_1(s), e_2(s))$ has done k complete revolutions $(k \in \mathbb{Z})$, between 0 and s_0; hence $\tilde{\varphi}(s_0) - \tilde{\varphi}(0) = k\pi$; $k \in \mathbb{Z}$.

An oriented coincidence point is a coincidence point such that $k = 2p$; $p \in \mathbb{Z}$.

Definition 2: A conjugate point to x_0 is a point x_0 such that

$$\phi^{\star}_{s_0}(\alpha_{x_0}) = \alpha_{x_{s_0}} \tag{5.33}$$

As can be easily seen, coincidence points are those points such that:

$$\phi^{\star}_{s}\alpha_{x_0} = \lambda\alpha_{x_s} \; ; \; \lambda \in \mathbb{R}^{\star}. \tag{5.34}$$

We introduce also the following hypothesis which we will assume to hold at some point in this paper:

Hypothesis (A_1): α turns well along v, i.e. any point x_0 in M has a coincidence point distinct from itself.

Under this hypothesis, we define:

Definition 3: $\bar{s} : M \to \mathbb{R}$ is the function which associates to $x \in M$ the first time $\bar{s}$ on the v-orbit x_s through x such that $x_{\bar{s}}$ is a coincidence point of x. $\bar{s}$ is $C^{\infty}(M, \mathbb{R})$. We denote then $\psi(x) = x_{\bar{s}(x)}$.

5.2 The equation of critical points at infinity

In this section, we study these sequences, given by the compactifying lemma, which satisfy:

$$Z(x_\varepsilon) = 0 \; ; \; \varphi(x_\varepsilon) \leq \varepsilon \; ; \; a_0 \leq J(x_\varepsilon) \leq a_1. \tag{5.35}$$

We will discuss later on the existence of such sequences.

For the moment, we are interested in making explicit the equation satisfied by these sequences.

For this purpose, we introduce:

$$\dot{x}_\varepsilon = a\xi + bv \; ; \; \omega = \frac{|\partial I|(x_\varepsilon)}{|\partial J|(x_\varepsilon)}. \tag{5.36}$$

We will drop for the sake of simplicity the subscripts ε in the variables we will use. a is a constant. b is an $L^2(S^1)$-function.

We will assume:

$$\int_0^1 b^2 dt \to +\infty \qquad \text{when } \varepsilon \to 0. \tag{5.37}$$

This is indeed the interesting case, when there is no compactness.

We then have the following proposition which gives the equation satisfied by b.

Proposition 10: Under (5.37), (5.35) is equivalent to:

$$\begin{cases} \ddot{b} + b(-\frac{\omega a}{2} + \frac{b^2}{2} - \int_0^1 \frac{b^2}{2}) + a^2 b\tau - ab^2\bar{\mu}_\xi + b\dot{b}\bar{\mu} = 0 \\ b(0) = b(1) \; ; \; \dot{b}(0) = \dot{b}(1) \\ \int_0^1 \frac{b^2}{\omega} \to 0. \end{cases} \tag{5.38}$$

Proof: (5.35) is equivalent to:

$$\omega\, \partial J(x) + \partial I(x) = 0; \; \int_0^1 b^2 \frac{dt}{\omega} \to 0; \; J(x) \to \bar{a} \; ; \; a_0 \le \bar{a} \le a_1 \tag{5.39}$$

(at least on a sub-sequence; we dropped as we previously said the εs).

Let z be a tangent vector to C_β along the curve $x, z = \lambda\xi + \mu v + \eta w$. We compute $(\partial J(x), z)$ and $(\partial I(x), z)$. We remember that $\gamma(.) = d\alpha(., w)$ (see (5.7)). I(x) is equal to $\int_0^1 \gamma_x(\dot{x})^2 dt$. Hence, the first variation of I(x) along z is given, through Proposition 3, by the following formula:

$$(\partial I(x), z) = 2\int_0^1 b(\dot{\mu} - (\mu a - \lambda b)\, d\gamma(\xi, v) - a\eta\, d\gamma(\xi, w) - b\eta\, d\gamma(v, w))\, dt. \tag{5.40}$$

Now,

$$d\gamma(\xi, v) = \xi\cdot\gamma(v) - v\cdot\gamma(\xi) - \gamma([\xi, v]) = -d\alpha([\xi, v], w)$$

$$= -d\alpha([\xi, v], -[\xi, v] + \bar{\mu}\xi = 0.$$

By similar computations, we get:

$$d\gamma(\xi, w) = -\tau \; ; \; d\gamma(v, w) = \bar{\mu}_\xi. \tag{5.41}$$

Hence, $(\partial I(x), z)$ is:

$$(\partial I(x), z) = 2 \int_0^1 b(\dot{\mu} + a\eta\tau - b\eta\bar{\mu}_\xi)\, dt. \tag{5.42}$$

On the other hand, (4.19) gives:

$$(\partial J(x), z) = -\int_0^1 b\eta\, dt. \tag{5.43}$$

Hence, (5.39) reads:

$$\begin{cases} -\omega \int_0^1 b\eta\, dt + 2\int_0^1 b(\dot{\mu} + a\eta\tau - b\eta\bar{\mu}_\xi)\, dt = 0 \\ \int_0^1 b^2 dt/\omega \to 0 \\ J(x) \to \bar{a}\ ;\ a_1 \geq \bar{a} \geq a_0 > 0. \end{cases} \tag{5.44}$$

The functions λ, μ, η, defining z, satisfy (3.5):

$$\begin{cases} \dot{\overbrace{\lambda + \bar{\mu}\eta}} = b\eta + C \\ \dot{\eta} = \mu a - \lambda b. \end{cases} \tag{3.5}$$

As we already noticed, by the periodicity of $\lambda + \bar{\mu}\eta$, C is equal to $-\int_0^1 b\eta dt$. Hence, the equation satisfied by a point x of the sequence we study is (5.44) under (3.5).

To avoid lengthy computations on systems of differential equations, we will assume that b is C^2 and derive then the equation it satisfies. From this equation that b has, anyway, even if not C^2, to satisfy in the distributional sense, it will appear clearly that our assumption was correct; thus justifying the computation.

Now, if b is C^2, μ is well-defined in H^1 with the second equation of (3.5) provided η is H^2, as equal to:

$$\mu = \frac{\dot{\eta} + \lambda b}{a}. \tag{5.45}$$

The first equation of (3.5) allows us to compute λ as a function of:

$$\lambda = \int_0^t b\eta \, ds - t \int_0^1 b\eta \, ds - \bar{\mu}\eta + C_1 \quad (C_1 \text{ arbitrary}). \tag{5.46}$$

We thus obtain for the equation satisfied by x:

$$-\frac{\omega}{2}\int_0^1 b\eta \, dt + \int_0^1 b(\overset{\bullet}{\overbrace{\frac{\dot{\eta} + \lambda b}{a}}}) + a\eta\tau - b\eta\bar{\mu}_\xi) \, dt = 0 \tag{5.47}$$

where η is $H^2(S^1)$, which is equivalent to:

$$-\frac{\omega}{2}\int_0^1 b\eta \, dt + b(1)\frac{\dot{\eta}+\lambda b}{a}(1) - b(0)\frac{\dot{\eta}+\lambda b}{a}(0) - \int_0^1 \dot{b}\left(\frac{\dot{\eta}+\lambda b}{a}\right) dt \tag{5.48}$$

$$+ \int_0^1 (ab\tau - b^2\bar{\mu}_\xi)\,\eta \, dt = 0$$

hence

$$\int_0^1 (ab\tau - b^2\bar{\mu}_\xi - \frac{\omega b}{2})\,\eta \, dt + \frac{\int_0^1 \ddot{b}\,\eta}{a} dt + \left[\frac{b^2(0)}{2a} - \frac{b^2(1)}{2a}\right]\lambda(0) \tag{5.49}$$

$$-\frac{\dot{b}(1)\,\eta(1)}{a} + \frac{\dot{b}(0)\,\eta(0)}{a} + \int_0^1 \frac{b^2}{2a}\dot{\lambda} \, dt + b(1)\frac{\dot{\eta}+\lambda b}{a}(1) - b(0)\frac{\dot{\eta}+\lambda b}{a}(0) = 0$$

which yields

$$\int_0^1 \left(\frac{\ddot{b}}{a} + ab\tau - b^2\bar{\mu}_\xi - \frac{\omega}{2}b\right)\eta \, dt + \int_0^1 \frac{b^2}{2a}\,(\overset{\bullet}{\overbrace{\lambda + \bar{\mu}\eta}} - b\eta) \, dt \tag{5.50}$$

$$+ \int_0^1 \frac{b^2}{2a}\,(b\eta - \overset{\bullet}{\overbrace{\bar{\mu}\eta}}) + \frac{\dot{b}(0)\,\eta(0)}{a} - \frac{\dot{b}(1)\,\eta(1)}{a} + \lambda(0)\left[\frac{b^2}{2a}(0) - \frac{b^2}{2a}(1)\right]$$

$$+ b(1)\frac{\dot{\eta}+\lambda b}{a}(1) - b(0)\frac{\dot{\eta}+\lambda b}{a}(0) = 0.$$

As $\overset{\bullet}{\overbrace{\lambda + \bar{\mu}\eta}} = b\eta - \int_0^1 b\eta$, (5.50) yields:

$$\int_0^1 \left[\frac{\ddot{b}}{a} + \frac{b^3}{2a} - \frac{b}{2a}\int_0^1 b^2 + ab\tau - b^2\bar{\mu}_\xi - \frac{\omega}{2}b + \bar{\mu}\frac{b\dot{b}}{a}\right]\eta \, dt \tag{5.51}$$

$$+ \frac{b^2}{2a}(1) - \frac{b^2}{2a}(0) \quad \bar{\mu}(0) \quad \eta(0) + \frac{\dot{b}(0)\,\eta}{a}(0) - \frac{\dot{b}(1)\,\eta}{a}(1)$$

$$+ b(1)\frac{\dot{\eta}+\lambda b}{a}(1) - b(0)\frac{\dot{\eta}+\lambda b}{a}(0) + \lambda(0)\left[\frac{b^2}{2a}(0) - \frac{b^2}{2a}(1)\right] = 0.$$

By (5.46) $\lambda(1) = \lambda(0)$ is arbitrary (C_1 is arbitrary) and η is arbitrary in $H^2(S^1)$.

We first take η to be zero and λ an arbitrary constant. We get

$$\frac{b^2}{a}(0) = \frac{b^2}{a}(1). \tag{5.52}$$

We then take $\eta(0) = \eta(1) = \dot{\eta}(1) = 0$; $\eta \in H^2(S^1)$. We then get:

$$\ddot{b} + b(-\frac{\omega}{2}a + \frac{b^2}{2} - \int_0^1 \frac{b^2}{2}) + a^2 b\tau - ab^2\bar{\mu}_\xi + b\dot{b}\bar{\mu} = 0. \tag{5.53}$$

Finally, taking $\eta(0) = \eta(1)$ and $\dot{\eta}(0) = \dot{\eta}(1)$ to be two arbitrary constants we get

$$b(0) = b(1) \; ; \; \dot{b}(0) = \dot{b}(1). \tag{5.54}$$

Hence the proposition. ∎

5.3 Geometric interpretation of the equation of critical points at infinity

We write down the equation satisfied by the critical points at infinity in a matrix form:

Let

$$\begin{cases} A_1^\star = 1 - \dfrac{b^2}{\omega a} + \displaystyle\int_0^1 \frac{b^2 dt}{\omega a} \\ B_1^\star = -\dfrac{2b}{\omega} \\ C_1^\star = -\dfrac{2\dot{b}}{\omega a} \end{cases} \tag{5.55}$$

Then we have:

$$\begin{cases} \dot{A}_1^{\star} = -\dfrac{2b\dot{b}}{\omega a} = -bC_1^{\star} \\ \dot{B}_1^{\star} = -\dfrac{2\dot{b}}{\omega} \\ \dot{C}_1^{\star} = -\dfrac{2\ddot{b}}{\omega a} = bA_1^{\star} - \dfrac{2ab\tau}{\omega} - b\bar{\mu}_{\xi}B_1^{\star} - b\bar{\mu}\, C_1^{\star} \end{cases} \tag{5.56}$$

or else

$$\dot{Z}_1^{\star} = \begin{bmatrix} \dot{A}_1^{\star} \\ \dot{B}_1^{\star} \\ \dot{C}_1^{\star} \end{bmatrix} = b\begin{bmatrix} 0 & 0 & -1 \\ 0 & 0 & 0 \\ 1 & -\bar{\mu}_{\xi} & -\bar{\mu} \end{bmatrix} \begin{bmatrix} A_1^{\star} \\ B_1^{\star} \\ C_1^{\star} \end{bmatrix} + \begin{bmatrix} 0 \\ -\dfrac{2\dot{b}}{\omega} \\ -\dfrac{2ab}{\omega} \end{bmatrix} \tag{5.7}$$

$$= b\begin{bmatrix} 0 & 0 & -1 \\ 0 & 0 & 0 \\ 1 & -\bar{\mu}_{\xi} & -\bar{\mu} \end{bmatrix} Z_1^{\star} + \begin{bmatrix} 0 \\ -\dfrac{2\dot{b}}{\omega} \\ \dfrac{2ab}{\omega} \end{bmatrix}$$

The matrix $\begin{bmatrix} 0 & 0 & -1 \\ 0 & \underline{0} & \underline{0} \\ 1 & -\bar{\mu}_{\xi} & -\bar{\mu} \end{bmatrix}$ is ${}^{t}\Gamma$ where Γ is defined in (5.4). Hence (5.57) is also:

$$\dot{Z}_1^{\star} = -b\,{}^{t}\Gamma\, Z_1^{\star} + \begin{bmatrix} 0 \\ -\dfrac{2\dot{b}}{\omega} \\ -\dfrac{2ab\tau}{\omega} \end{bmatrix} . \tag{5.58}$$

Eq. (5.58) and (5.12), which gives the dynamics of α along v, have to be brought together.

Indeed these equations are very close:

In (5.12), defining the critical points at infinity, $\dot{Z}^{\star}(s)$ is the derivative of $Z^{\star}(s)$ with respect to either s or to v, as s represents the time on the v-orbit.

In equation (5.58) $\dot{Z}_1^{\star}$ is the derivative of $Z_1^{\star}$ either with respect to time t along x(t) or with respect to $\dot{x} = a\xi + bv$.

Let us rewrite (5. 58) with respect to:

$$\frac{\partial}{\partial s} = \frac{1}{b}\frac{\partial}{\partial t} = \frac{a}{b}\xi + v. \qquad (5.59)$$

We then find:

$$\frac{\partial Z_1^\star}{\partial s} = -{}^t\Gamma\, Z_1^\star + \begin{bmatrix} 0 \\ -\frac{2\dot{b}}{\omega b} \\ -\frac{2a\tau}{\omega} \end{bmatrix}. \qquad (5.60)$$

Now, when b is very large, $\frac{a}{b}$ is small; hence $\frac{\partial}{\partial s}$ looks like v. As ω goes to infinity, $\frac{a\tau}{\omega}$ goes to zero.

Hence (5. 60) is very close to the equation governing the dynamics of α along v, provided b is very large and $\frac{\dot{b}}{\omega b}$ is small.

Under these conditions, which will amount later on to the fact that b is large, (5. 60) acquires a geometric significance: it is very close to the transport equations of forms along v.

This feature is fundamental for the rest of this work.

6 Topological aspects of the problem and the statement of the result

6.1 The topology of $\mathcal{L}_\beta$ and C_β

The topology of a space of curves whose tangent direction belongs to the kernel of a one-differential form has been studied by S. Smale [2] and W. Boothby [3].

Their main results deal with the case when this one-differential form, we call β in our context, is a contact form.

In this context, S. Smale [2] and W. Boothby [3] prove the following theorem:

Theorem [2], [3]: Let β be a contact form. Then the injection of $\mathcal{L}_\beta$ in $H^1(S^1;M)$ is a weak homotopy equivalence.

In [2], S. Smale proves that $\mathcal{L}_\beta$ fibers over M, the fibration being given through the natural projection of $\mathcal{L}_\beta$ over M.

Using the Darboux reduction of a contact form in local coordinates, he exhibits, given two close enough points on M, a path, which depends continuously on these two points, between them, whose tangent direction belongs to the kernel of β; this allows us to prove the theorem as, with this result, the situation is the same as that of $\Omega(M)$ with its projection over M.

We give here a slight extension of this result which deals with a case where β is not a contact form.

We assume β to be in a general position. We remember that

$$p = \beta \wedge d\beta(\xi, v, \tilde{w}). \tag{1.4}$$

We introduce in the first section of this paper the hypersurface $\Sigma = \{x \in M \mid p(x) = 0\}$. In this context, we have the following proposition:

Proposition 11: Let x_0 be a point of M. Assume either $p(x_0)$ is not zero or β is transverse to Σ at x_0.

Then there exists a neighbourhood U of x_0 in M such that for any x_1 in U, there exists a continuous piecewise differentiable path starting at x_0 and ending at x_1, whose tangent direction is in the kernel of β. This path depends continuously on x_1 in U and is constant equal to x_0 if x_1 is x_0.

We delay for the moment the proof of Proposition 2.

This proposition allows us to prove, just as in the case of the path space and the closed curve space on a manifold M, that the maps:

$$\begin{cases} p : \mathcal{L}_\beta \to M \\ \qquad x \to x(0) \\ p_{x_0} : \hat{\mathcal{L}}_\beta^{x_0} \to M \qquad \hat{\mathcal{L}}_\beta^{x_0} = \{H^1\text{-paths on } M \text{ starting at } x_0\} \\ \qquad x \to x(1) \end{cases} \tag{6.1}$$

are Serre-fibrations.

Hence, the following theorem:

<u>Theorem 2</u>: 1). Assume β is a non-singular one-differential form in a general position. Then $\pi_0(\mathcal{L}_\beta) = \pi_0(\Omega(M))$ (where $\Omega(M)$ is the space of closed continuous curves on M).

2). Assume furthermore that β is transverse to Σ; then the injection of $\mathcal{L}_\beta$ in $H^1(S^1;M)$ is a weak homotopy equivalence.

<u>Proof of theorem 2</u>: As noted, Proposition 11 implies, if β is transverse to Σ, that the maps $p : \mathcal{L}_\beta \to M$ and $p_{x_0} : \hat{\mathcal{L}}_\beta^{x_0} \to M$ are Serre-fibrations.
Statement 2). of Theorem 2 easily follows.
For statement 1)., we notice that, with β in a general position, β will be transverse to Σ outside a finite number of stationary points of the trace q (see (1.5)) of β on Σ.

As the dimension of M is three, any two-dimensional deformation in M, or else any one-dimensional deformation of curves, may be assumed to avoid these stationary points, by a general position argument.

If $\hat{M}$ is the complement in M of these stationary points, we thus have:

$$\pi_0(\mathcal{L}_\beta(M)) = \pi_0(\mathcal{L}_\beta(\hat{M}));\ \pi_0(H^1(S^2,\hat{M})) = \pi_0(H^1(S^1,M)).$$

But, on $\hat{M}$, β is transverse to Σ. Hence, the injection of $\mathcal{L}_\beta(\hat{M})$ in $H^1(S^1;\hat{M})$ induces a weak homotopy equivalence. Therefore, $\pi_0(\mathcal{L}_\beta(\hat{M})) = \pi_0(H^1(S^1;\hat{M}))$; which completes the proof of Theorem 2.

Now, the results on the topology of $\mathcal{L}_\beta$ given in Theorem 2 imply results on the topology of C_β, in case β is a contact form and the hypothesis A_1 (that α turns well along v) holds. They are given in the following theorem:

<u>Theorem 3:</u> Assume β is a contact form and (A_1) holds. Then the injection of C_β in $H^1(S^1,M)$ is a weak homotopy equivalence.

<u>Remark 4:</u> We wish to remark that the following result should be true: "Assume (A_1) holds. Then the injection of C_β in $\mathcal{L}_\beta$ is a weak homotopy equivalence." However, if (A_1) does not hold, it is likely that this result is no longer valid. This is an interesting situation, which we will examine later on in this paper, when we will choose v in the kernel of a codimension one foliation transverse to α.

<u>Remark 5:</u> We wish also to point out that the study of $p : \mathcal{L}_\beta \to M$ in a neighbourhood of a singular point of q is likely to give a complete understanding of the topology of $\mathcal{L}_\beta$, as soon as β is a non singular one-differential form in general position.

The intuition about the results of Theorem 3 can be seen to come, once more, from a situation studied by S. Smale [2].

Indeed, S. Smale proves in [2] the following result:

<u>Theorem [2]:</u> Let N be a compact surface without boundary. Let $\mathrm{Imm}(S^1,N)$ be the space of immersions of S^1 in N and $\Omega(ST^\star N)$ be the space of continuous closed curves in the cotangent S^1-bundle over N.

Then the inclusion $\mathrm{Imm}(S^1,N) \to \Omega(ST^\star N)$ is a homotopy equivalence.

This result is related to Theorem 3 as follows:

Consider α_0 the standard contact form on $ST^\star N$ and let α be a constant form obtained by multiplying α_0 by a C^∞ positive function λ. Let v be the vector field defining the natural fibration of $ST^{\star}N$ over N.

It is not difficult to see that, in this context, the form α_0, hence the form α, makes one (and only one) complete revolution along the fiber (in the sense of rotation introduced in Section 5.1; see Definitions 1, 2 and 3).

Suppose now $\beta = d\alpha(v,.)$ is also a contact form, which means, from the Hamiltonian point of view, the convexity of the kinetic energy in the cotangent variable. As $\beta(v)$ is zero, β is also tangent to the fibers. Being transverse to α, β makes also one (and only one) complete revolution along the fiber.

Let x be an immersion of S^1 in N; $x(t_0)$ be a fixed point on the curve and $\dot{x}(t_0)$ be its tangent vector.

Let $p : ST^{\star}N \rightarrow N$ be the natural projection.

As β turns once along the fiber, ξ, which belongs to the kernel of β, turns also once along this fiber. This implies the existence and the unicity of a point $(x(t_0), \theta)$, projecting on $x(t_0)$, in $ST^{\star}N$ such that:

$$dp_{(x(t_0),\theta)}(\xi_{(x(t_0),\theta)}) = \mu\,\dot{x}(t_0);\ \mu > 0, \tag{6.2}$$

where $dp_{(x(t_0),\theta)}$ is the differential of p at $(x(t_0),\theta)$ and $\xi_{(x(t_0),\theta)}$ is the value of ξ at $(x(t_0),\theta)$.

When t_0 runs in S^1, this device allows the lifting of the immersion x in a closed curve $y(.) = (x(.), \theta(.))$ belonging to $\Omega(ST^{\star}N)$. This curve y is defined by the following equation:

$$\begin{cases} p(y(t)) = x(t) \\ dp(\dot{y}(t)) = \dot{x}(t);\ dp_{y(t)}(\xi_{y(t)}) = \mu(t)\ \dot{x}(t);\ \mu > 0. \end{cases} \tag{6.3}$$

Eq. (6.3) yields

$$\dot{y}(t) = \frac{1}{\mu(t)}\xi_{y(t)} + \lambda(t)\,v_{y(t)}.$$

Hence:

$$\dot{y}(t) = a_1\xi_{y(t)} + b_1 v;\quad a_1 > 0. \tag{6.4}$$

As a_1 is continuous and positive, one can reparametrize the curve $y(.)$; hence obtaining $z(.)$ such that:

$$\dot{z}(t) = a\xi_{z(t)} + bv_{z(t)}; \quad a = \text{Cste.} > 0. \tag{6.5}$$

We defined in this way a map h:

$$\text{Imm}(S^1, N) \xrightarrow{h} C_\beta. \tag{6.6}$$

Conversely, if y is a given curve in C_β, $p(y)$ belongs to $\text{Imm}(S^1, N)$. Indeed:

$$\dot{y} = a\xi + bv; \; a = \text{positive constant.} \tag{6.7}$$

Hence

$$dp(\dot{y}) = \dot{\widehat{p(y)}} = a\, dp(\xi). \tag{6.8}$$

As $dp(\xi)$ is not zero, $p(y)$ belongs to $\text{Imm}(S^1, N)$.

We thus have another map:

$$\hat{h} : C_\beta \to \text{Imm}(S^1, N). \tag{6.9}$$

One can check that $\hat{h} \circ h(x)$ is a reparametrization of $x(.)$ while $h \circ \hat{h}(y)$ is a reparametrization of y.

This, together with some approximation argument appropriate for dealing with the H^1-topology, proves that C_β has the same topology as $\text{Imm}(S^1, N)$; hence as $\Omega(ST^\star N)$ by Smale's theorem.

The same argument works if M is a finite covering of $ST^\star N$; $\text{Imm}(S^1, N)$ and $\Omega(ST^\star N)$ are in this context to be replaced by $\text{Imm}_M(S^1, N)$ and $\Omega_M(ST^\star N)$ where these indexed sets denote the results of curves which lift to M as closed curves.

The general proof of Theorem 3 will be published elsewhere as it differs from the main stream of this book.

We give now the proof of Proposition 11:

In the case when $p(x_0)$ is not zero, $p = \beta \wedge d\beta\, (\xi, v, \tilde{w})$, then β is locally a contact form and Proposition 11 follows then from the results of Smale [2] or Boothby [3]. If $p(x_0) = 0$ and β is transverse to Σ at x_0, we have the following reduction lemma:

Lemma 4: Assume β is transverse to Σ at x_0. Then there is a local system of coordinates (x, y, z) where β reads:

$$\beta = \lambda (xy\, dy + dz) \ ; \ \lambda \text{ is a strictly positive function.} \tag{6.10}$$

With the help of Lemma 4, the arguments of [3], where $\beta = x\,dy + dz$, adapt to the situation given by (6.10) with minor technical modifications.

We now give the proof of Lemma 4:

Consider the local hypersurfaces $p = \varepsilon$, where ε runs in $[-\varepsilon_0, \varepsilon_0]$; $\varepsilon_0 > 0$. The trace q of β on $\Sigma = \{x/p(x) = 0\}$ extends in a neighbourhood of x_0 in $\hat{q}$ which is the trace of β on the hypersurfaces $p = \varepsilon$.

As $\hat{q}$ is not singular and $dp(\hat{q}) = 0$, we may choose local coordinates where $p = \varepsilon$ reads as $Y = \varepsilon$ while $\hat{q}$ is $\frac{\partial}{\partial X}$.

In these coordinates, β reads:

$$\beta = l\, dY + m\, dZ \ ; \ m(x_0) \neq 0 \ \text{(as } \beta \text{ is transverse to } dp = dY). \tag{6.11}$$

We assume, without loss of generality, that $(\xi, v, \tilde{w})$ and $(\frac{\partial}{\partial X}, \frac{\partial}{\partial Y}, \frac{\partial}{\partial Z})$ have the same orientation. This yields the existence of a strictly positive function such that:

$$\beta \wedge d\beta\, (\xi, v, \tilde{w}) = p = Y = \left(\frac{l}{m}\right)_X. \tag{6.12}$$

We now make the following change in the X-coordinate, leaving Y and Z fixed:

$$X' = \int_0^X \frac{1}{\mu(\tau, Y, Z)}\, d\tau\ . \tag{6.13}$$

In these new coordinates (X', Y, Z), (6.12) yields:

$$Y = \left(\frac{l}{m}\right)_{X'} \tag{6.14}$$

hence

$$\frac{l}{m} = YX' + \varphi(Y, Z) \qquad (6.15)$$

and β becomes:

$$\beta = m(X'Y\,dY + dZ + \varphi(Y, Z)\,dY). \qquad (6.16)$$

Consider now:

$$\tilde{w} = dZ + \varphi(Y, Z)\,dY, \qquad (6.17)$$

$\tilde{w}$ is a non-singular foliation.

Hence, in a small neighbourhood of x_0, $\tilde{w} = h\,df$, with f non-singular and h strictly positive:

$$h\,df = dZ + \varphi(Y, Z)\,dY. \qquad (6.18)$$

We define then the following change of coordinates:

$$\begin{cases} x = \dfrac{X'}{h} \\ y = Y \\ z = f(X', Y, Z) \end{cases} . \qquad (6.19)$$

In these new coordinates, the reduction of Lemma 4 is completed. ■

6.2 The change of topology induced by J

We first introduce the following definitions; d is a distance on M.

Definition 4: Let v_1 be a non-singular vector field collinear to v ($v_1 = \lambda v$; $\lambda > 0$) and let ϕ_s^1 be the one-parameter group generated by v_1. v is said to be ξ-conservative if there exist v_1 and a constant K such that

$$\| D\phi_s^1(\xi) \| \leq K \ \forall s \in \mathbb{R}. \qquad (6.20)$$

Here $\| \ \|$ stands for any norm of differentiable maps from M to M.

Definition 5: v is said to be α-non-resonant if there exists a non-singular one-differential form θ, transverse to α and tangent to v, such that:

$$\text{there exists } \varepsilon_4 > 0 \text{ such that for any } x \text{ in } M \text{ and } s_1 \text{ in } \mathbb{R} \text{ satisfying } d(\phi_{s_1}(x), x) < \varepsilon_4, \ (\phi^{\star}_{s_1}\alpha)_x \text{ is transverse to } \theta_x. \quad (6.21)$$

Definition 6: C_β is said to be branched on v if there exists $C > 0$ such that for any $\varepsilon > 0$, there exists x in C_β, with $\dot{x} = a\xi + bv$, $0 < a < \varepsilon$ and $\left| \int_0^1 b(t)\,dt \right| \geq C$.

Here are some comments about these definitions:

First, if v is ξ-conservative, there is no hyperbolic invariant set to v. Indeed, on such a set, (6.20) cannot hold. Hence, v might be called "elliptic" in this situation.

Second, we consider a point x in M which is recurrent to v. Let f_x be the local Poincaré map on a section transverse to v at x.

As α rotates monotonically along v from x to $f_x(x)$, we can define:

$$\tilde{\varphi}(x) = \text{total rotation of } \alpha \text{ from } x \text{ to } f_x(x) \quad (6.22)$$

and

$$n_x = \left[\frac{\tilde{\varphi}(x)}{\pi} \right]. \quad (6.23)$$

If v is α-non-resonant, n_x is constant on the connected components of the v-recurrent set.

Third, C_β is branched on v if there exist curves of C_β close to recurrent orbits of v.

Notice that if v has a periodic orbit and if (A_1) holds, C_β is branched on v.

Finally, we notice that in the case when v is the fiber vector field of an S^1-fibration over a surface N, then v is ξ-conservative, α-non-resonant and C_β is branched on v.

With these definitions, we have the following theorem:

Theorem 4: 1). Assume (A_1) holds and either v is ξ-conservative and C_β is not branched on v or v is ξ-conservative, α-non-resonant and C_β is branched on v. Then J induces a change of topology in C_β.

2). Assume (A_1) holds and v is ξ-conservative, α-non-resonant and C_β is branched on v.

Then for $\varepsilon > 0$ small enough, $\pi_0(J^\varepsilon)$ is infinite, where $J^\varepsilon = \{x \in C_\beta \mid J(x) \leq \varepsilon\}$.

The proof of statement 1). will be given elsewhere.

For the second statement, we consider x in C_β such that:

$$\dot{x} = a\xi + bv \qquad 0 < a < \varepsilon \qquad \left|\int_0^1 b(t)\,dt\right| \geq C > 0 \tag{6.24}$$

where C and ε are given by Definition 6. Let

$$u_x(t) = -\int_0^t \frac{b(\tau)}{\lambda(x(\tau))}\,d\tau \tag{6.25}$$

where λ is given by:

$$v_1 = \lambda v; \quad \lambda > 0. \tag{6.26}$$

The existence of v_1, hence of λ, is due to the ξ-conservativity of v. Let

$$y(t) = \phi^1_{u_x(t)}(x(t)) \tag{6.27}$$

where ϕ^1_s is the one-parameter group generated by v_1. We have:

$$y(0) = \phi^1_0(x(0)) = x(0) \tag{6.28}$$

$$y(1) = \phi^1_{u(1)}(x(1)) = \phi^1_{u_x(1)}(x(0)). \tag{6.29}$$

y(1) is thus obtained by pushing y(0) along v_1 during the time $u_x(1) = -\int_0^1 \frac{b(\tau)}{\lambda(x(\tau))}\,d\tau$. Equivalently, y(1) is obtained by pushing y(0) along v during the time $\tilde{u}_x(1) = -\int_0^1 b(\tau)\,d\tau$. By (6.24), we have:

$$|\tilde{u}_x(1)| \geq C > 0. \tag{6.30}$$

We have furthermore:

$$\dot{y}(t) = D\phi^1_{u(t)}(\dot{x}) + \dot{u}(t)v_1 = D\phi^1_{u_x(t)}(a\xi+\frac{b}{\lambda}v_1) - \frac{b}{\lambda}v_1 = a\,D\phi^1_{u_x(t)}(\xi) \qquad (6.31)$$

v being ξ-conservative, (6.20) yields:

$$\|\dot{y}(t)\| \leq Ka \leq K\varepsilon. \qquad (6.32)$$

Therefore:

$$d(y(1), y(0)) \leq K\varepsilon \text{ or else } d(\phi_{\tilde{u}_x(1)}(x(0)), x(0)) \leq K\varepsilon, \qquad (6.33)$$

as $y(0) = x(0)$ while $y(1) = \phi^1_{u_x(1)}(x(0)) = \phi_{\tilde{u}_x(1)}(x(0))$.

Now let $\varepsilon > 0$ be such that:

$$0 < \varepsilon < \frac{\varepsilon_4}{K} \qquad (6.34)$$

where ε_4 is given by Definition 5 (the α-non-resonance of v).

Let, in x belonging to M, z_x be the orthogonal vector to v in the kernel of θ_x. Orthogonality can be taken with respect to any scalar product. As θ is transverse to α, z is never collinear to $[\xi, v]$. In fact, $(v, z, [\xi, v])$ is a basis of the tangent space to M. Let:

$$e_1(0) = z_{x(0)}; \; e_2(0) = [\xi, v]_{x(0)}. \qquad (6.35)$$

We consider

$$e_1(x) = D\phi_s(z_{x(0)}) = D\phi_s(e_1(0)); \; e_2(s) = D\phi_s([\xi, v]_{x(0)}) = D\phi_s(e_2(0)). \qquad (6.36)$$

In this transported frame, when s runs from 0 to $\tilde{u}(1)$, α rotates.

Let $\tilde{\varphi}(s, x(0))$ be the angle of the trace of α in $(e_1(s), e_2(s))$ with $e_1(s)$. We consider:

$$\delta_x = \frac{\tilde{\varphi}(u_x(1), x(0) - \tilde{\varphi}(0, x(0))}{\pi} \qquad (6.37)$$

$$\begin{cases} n_x = [\delta_x] \text{ if } \delta_x \text{ is positive} \\ n_x = -[-\delta_x] \text{ if } \delta_x \text{ is negative.} \end{cases} \tag{6.38}$$

Under (6.34), we have:

$$\text{for any } x_1, x_2 \text{ belonging to the same component of } J^{\varepsilon}, \quad |n_{x_2} - n_{x_1}| < 2. \tag{6.39}$$

Indeed, if $|n_{x_2} - n_{x_1}| \geq 2$, δ_x takes at least one integer value on a path in J^{ε} from x_1 to x_2. This value will be achieved on a curve x_3.

But, then, at $\phi_{\tilde{u}_x(1)}(x_3(0))$, the trace of α and e_1 coincide.

Now, $e_1(\tilde{u}_{x_3}(1))$ is the trace of the form $\theta_{x_3(0)}(D\phi_{-\tilde{u}_{x_3}(1)}(\cdot))$. Hence:

$$(\phi^{\star}_{\tilde{u}_{x_3}(1)}\alpha)_{x_3(0)} = \lambda\theta_{x_3(0)}. \tag{6.40}$$

Inequality (6.33) holds for $x(0) = x_3(0)$, as this inequality was proved with x in J^{ε} (the fact that $|\int_0^1 b(t)\,dt| \geq C$ was not relevant to the inequality). Hence,

$$d(\phi_{\tilde{u}_{x_3}(1)}(x_3(0)), x_3(0)) \leq K\varepsilon < \varepsilon_4. \tag{6.41}$$

(6.40) and (6.41) contradict then the α-non-resonance of v and (6.39) holds.

Consider now $0 < \varepsilon_5 < \frac{\varepsilon_4}{K}$.

We want to show that $\pi_0(J^{\varepsilon_5})$ is infinite:

As (A_1) holds, α turns well along v. Hence, with a suitable uniform positive constant C, the following holds:

$$\tilde{\varphi}(s, x(0)) - \tilde{\varphi}(0, x(0)) \geq C_1\pi|s|. \tag{6.42}$$

Let $x_1, \ldots, x_2$ be curves in J^{ε_5} and $n_{x_1}, \ldots, n_{x_r}$ be the corresponding integers defined through (6.38).

Let

$$N_r = \sup_{i=1,\dots,r} |n_{x_i}| + 2 . \tag{6.43}$$

Consider $q \in N$ such that:

$$q\, C_1\, C > N_r . \tag{6.44}$$

Let x be a curve, given through Definition 6, such that:

$$0 < a < \frac{\varepsilon_5}{q} \quad \left|\int_0^1 b(t)\, dt\right| \geq C. \tag{6.45}$$

Let x_{r+1} be the q-th iterate of x.

If $\dot{x}_{r+1} = a_{r+1} \xi + b_{r+1} v$, we have:

$$a_{r+1} = qa < \varepsilon_5 \; ; \; |\tilde{u}_{x_{r+1}}(1)| = \left|\int_0^1 b_{r+1}(t)\, dt\right| = q\left|\int_0^1 b(t)\, dt\right| \geq qC. \tag{6.46}$$

Hence, x_{r+1} belongs to J^{ε_5}.

But now, by (6.42) and (6.43), we have:

$$\frac{|\tilde{\varphi}(\tilde{u}_{x_{r+1}}(1), x_{r+1}(0)) - \tilde{\varphi}(0, x_{r+1}(0))|}{\pi} = |\delta_{x_{r+1}}| \tag{6.47}$$

$$\geq C_1 |\tilde{u}_{x_{r+1}}(1)| \geq q\, C\, C_1 > N_r .$$

Hence:

$$N_r \leq |n_{x_{r+1}}| . \tag{6.48}$$

This implies that $\pi_0(J^{\varepsilon_5})$ is infinite.

7 Statement of the results

Our results are concerned with understanding the ends of the flow-lines of (4.9) when ε goes to zero.

We prove, under suitable hypotheses which we will discuss later on, that there is an end, a "critical point at infinity", to these flow-lines whose closure is non-compact.

The starting point of our analysis relies on the similarity we pointed out in Chapter 5 between the equation of the critical points at infinity and the transport equation of the form α along v.

Chapter 8 of the book is devoted to the full analysis of the similarity. We sum up in what follows this analysis, which holds under the sole hypothesis (A_1), i.e. that α turns well along v.

In Chapter 9, we introduce new hypotheses (A_2) and (A_3), $(\mathfrak{a}_2(x))$. Under these hypotheses, we prove local convergence theorems; which globalize in Theorem 5 of this chapter, which we will state later on.

In Chapter 10, we display a geometric localization of these limit curves or "ends". We explain also why this phenomenon takes place. This section is rather qualitative. The proof of the results has been deferred to a later publication, to avoid us being involved again in very technical details, after the lengthy Chapter 8.

In Chapter 11, we derive local expansions of the functional J near these "ends". This allows us to discriminate between actual and "fake" critical points at infinity.

Indeed, the analysis of Chapters 8 and 9 shows us the possible ends. A first expansion of J, as we derive it in Chapter 11, allows us to see the actual ends. We will see that a certain quantity defined of the limit object has to be positive in order to be an end of flow-lines of (4.9) when ε goes to zero.

Then, whenever it is an actual end, we define a Morse index related to this critical point at infinity which allows us to compute, as in the case when we are dealing with a usual critical point in Morse theory, the difference topology due to this singularity at infinity.

In Chapter 12, we completely shift the hypotheses. We drop (A_1) and aim at studying exactly the opposite situation: namely when α does not turn along v. We show that this situation occurs when v belongs to the kernel of a foliation transverse to α. We compute the transport equations in this case, which are quite explicit; and we show the features that the same variational problem has in this framework.

Roughly speaking, in this opposite situation, critical points at infinity are not allowed, as these are related to the existence of conjugate points, which implies (A_1) holds, at least somewhere in M.

However, new problems in the Palais-Smale condition as well as in the difference of topology, arise, which we point out in a later publication.

Chapters 10 to 12 may be read directly after the following summary of the analysis completed in Chapters 8 and 9.

In the following, we summarize the technical results of Chapters 8 and 9 and we state the related convergence theorems.

7.1 Analysis of the phenomenon

We start now with a sequence of lemmas, namely Lemmas 5, 6, 7, 8 of Chapter 8, which aim:

[1] at defining "concentration" intervals: these are intervals where the curve x_ε is rather tangent to ξ. Time intervals $I_1(s)$ where this concentration occurs are defined, together with their main properties (evolution of $\frac{b^2}{\omega}$, $\frac{|\dot{b}|}{\omega}$, end points $t^+(s)$ and $t^-(s)$ related to each interval). The total time spent in these intervals is shown to go to 1 as ε goes to zero (ω goes to $+\infty$).

Thus, the curve x_ε becomes more and more tangent to ξ globally on the total time interval $[0,1]$, in the Lebesgue measure sense (Lemma 5).

We state here these four lemmas:

Lemma 5: Let $\varepsilon_0 > 0$ be given. Let

$$A = \{t \in S^1 \mid \frac{b^2}{\omega}(t) \geq \varepsilon_0 \text{ or } \frac{|\dot{b}|}{\omega}(t) \geq \varepsilon_0\}.$$

Then

$$\mu(A) \xrightarrow[\omega \to +\infty]{} 0 \quad (\mu \text{ is the Lebesgue measure of } A).$$

Lemma 6: Let $\alpha_1 > 0$ and $s \in [0,1]$ be given such that:

$$\frac{|\dot{b}|}{\omega}(s) \geq \sqrt{\alpha_1} > 0, \quad \frac{b^2}{\omega}(s) \leq \frac{\varphi(\alpha_1)}{2M},$$

where φ is a continuous and increasing function from $\mathbb{R}^+$ to $\mathbb{R}^{+\star}$ such that $\varphi(0) = 0$ and M is a positive constant larger than 1. Let $I(s)$ be the maximal time interval containing s such that:

$$\forall t \in I(s) \qquad \frac{b^2}{\omega}(t) \leq \frac{\varphi(\alpha_1)}{M}.$$

We then have:

$$\forall t \in I(s) \quad (1 - \varepsilon(\omega, M)) \, |\dot{b}(s)| \leq |\dot{b}(t)| \leq (1+\varepsilon(\omega, M)) \, |\dot{b}(s)|$$

where $\varepsilon(\omega, M) \to 0$ when M and $\omega \to +\infty$ (α_1 fixed).

Lemma 7 (convexity lemma): There exists a constant $\alpha_0 > 0$ such that the function b^2 is convex in every point s such that:

$$\frac{|\dot{b}|}{\omega}(s) < \alpha_0, \quad \frac{b^2}{2\omega}(s) < \alpha_0 \qquad (\omega \geq \omega_0).$$

Lemma 8: Let $\overline{M} > 1$ and ω_1 be chosen such that $\varepsilon(\omega, \overline{M}) \leq \frac{1}{2}$ for $\omega \geq \omega_1$. Let $\varepsilon_1 > 0$ be small enough such that:

$$\varepsilon_1 < \frac{\alpha_0}{2}; \quad \sqrt{\varphi^{-1}(4\overline{M}\varepsilon_1)} < \alpha_0.$$

Let $\alpha_1 = \varphi^{-1}(4\overline{M}\varepsilon_1)$. Let $0 < \varepsilon_2 < \inf(\alpha_0, \frac{\sqrt{\alpha_1}}{3})$ be given. Let $s \in [0,1]$ be such that:

$$\frac{b^2(s)}{\omega} \leq \varepsilon_1 = \frac{\varphi(\alpha_1)}{\overline{M}} \quad \text{and} \quad \frac{|\dot{b}(s)|}{\omega} < \varepsilon_2 .$$

Let $I_1(s)$ be the maximal time interval containing s such that $\forall\ t \in I_1(s)$,

$$\frac{b^2(t)}{\omega} \leq 2\varepsilon_1 .$$

Under these hypotheses, the function b^2 is convex over all $I_1(s)$. Moreover if $I_1(s) = [t^-(s), t^+(s)]$, we have:

$$\frac{b^2}{\omega}(t^+) = 2\varepsilon_1; \quad \frac{|\dot{b}|}{\omega}(t^+) < \sqrt{\alpha_1}; \quad \frac{b^2}{\omega}(t^-) = 2\varepsilon_1; \quad \frac{|\dot{b}|}{\omega}(t^-) < \sqrt{\alpha_1};$$

$$\frac{b^2}{\omega}(t) \leq 2\varepsilon_1; \quad \frac{|\dot{b}|}{\omega}(t) < \sqrt{\alpha_1}, \quad \forall\ t \in I_1(s).$$

Finally, there exists a unique point $t_1(s)$ in $I_1(s)$ such that:

$$\text{either } b(t_1(s)) = 0 \quad \text{or} \quad \dot{b}(t_1(s)) = 0.$$

After these four lemmas, the curve x_ε on the $I_1(s)$ intervals can be of two types depending on the sign of $b(t^+(s)) . b(t^-(s))$

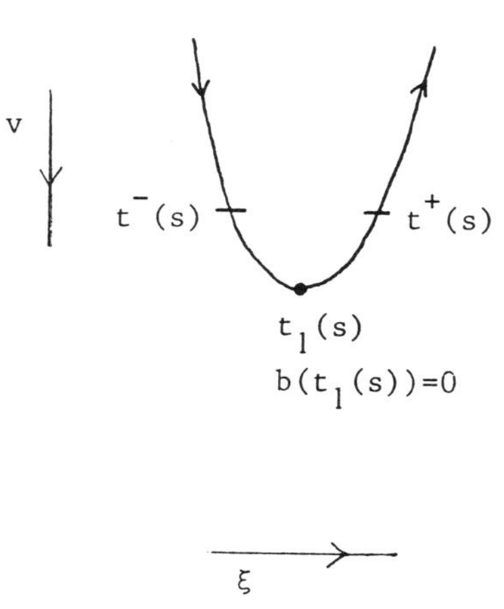

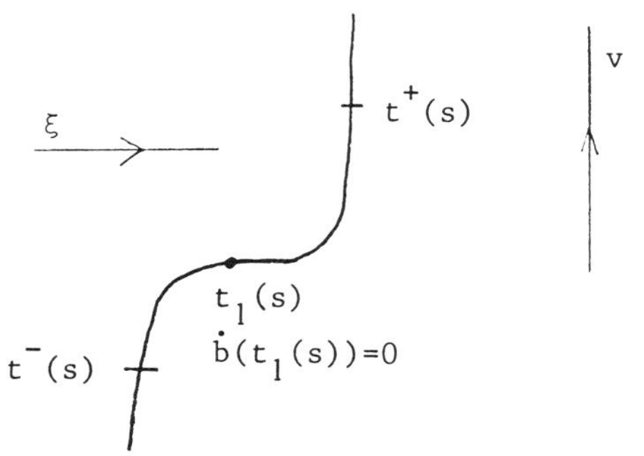

2 After having defined these concentration intervals $I_1(s)$, we restrict ourselves to one of them, which we call $[t^-(s), t^+(s)]$. We start a process of <u>straightening up</u> the curve on such an interval. This is done in two steps. As a first step, <u>we push the curve</u> x_ε <u>on</u> $[t^-(s), t^+(s)]$ <u>along</u> v to straighten it up during the time:

$$u(t) = -\left[\int_{t_1(s)}^{t} b(x)\,dx + \bar{c}(t_1(s)) \right] \tag{7.1}$$

$t_1(s)$ is defined in Lemma 8. $\bar{c}$ is an appropriate constant subject to

$$|\bar{c}(t_1(s))| \leq C\sqrt{\alpha_1} \tag{7.2}$$

α_1 is involved in Lemmas 5, 6, 7, 8.

ϕ being the one-parameter group generated by v, the new curve we have defined this way is called y, the "incomplete straightened-up" curve on $I_1(s)$ and has the following equation:

$$y(t) = \phi(u(t), x_\varepsilon(t)), \quad t \in (t^-(s), t^+(s)). \tag{7.3}$$

To obtain the "complete straightened-up" curve on $I_1(s)$, we add to y two v-pieces: the first one starting at $x(t^-(s))$ and tangent to v during the time $\int_{t_1}^{t^-(s)} b(x)\,dx + \bar{c}(t_1(s))$, the second one starting at $x(t^+(s))$ and tangent to v during the time $\int_{t_1}^{t^+(s)} b(x)\,dx + \bar{c}(t_1(s))$.

This operation is illustrated in the following drawings:

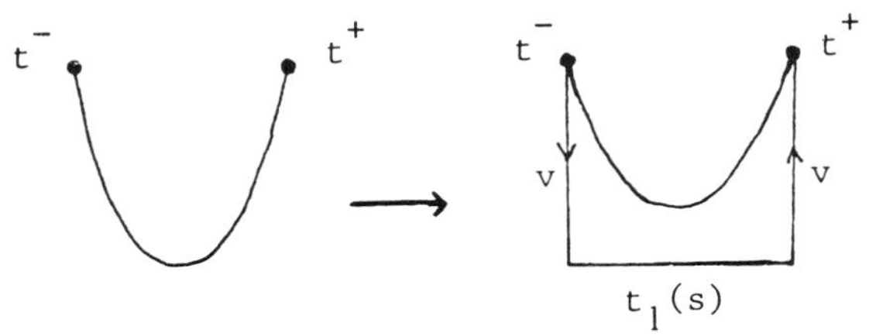

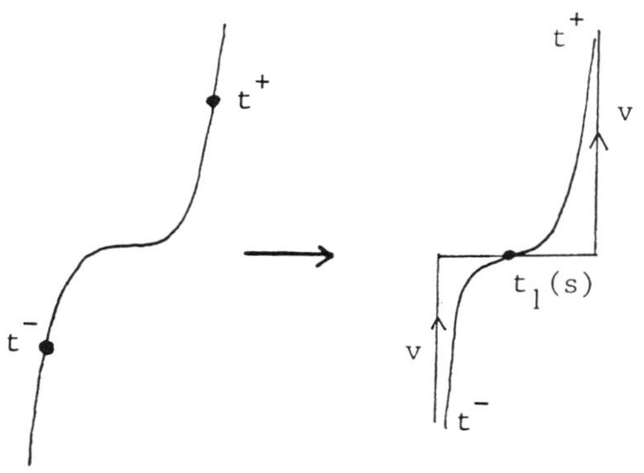

which we make more precise in the first case:

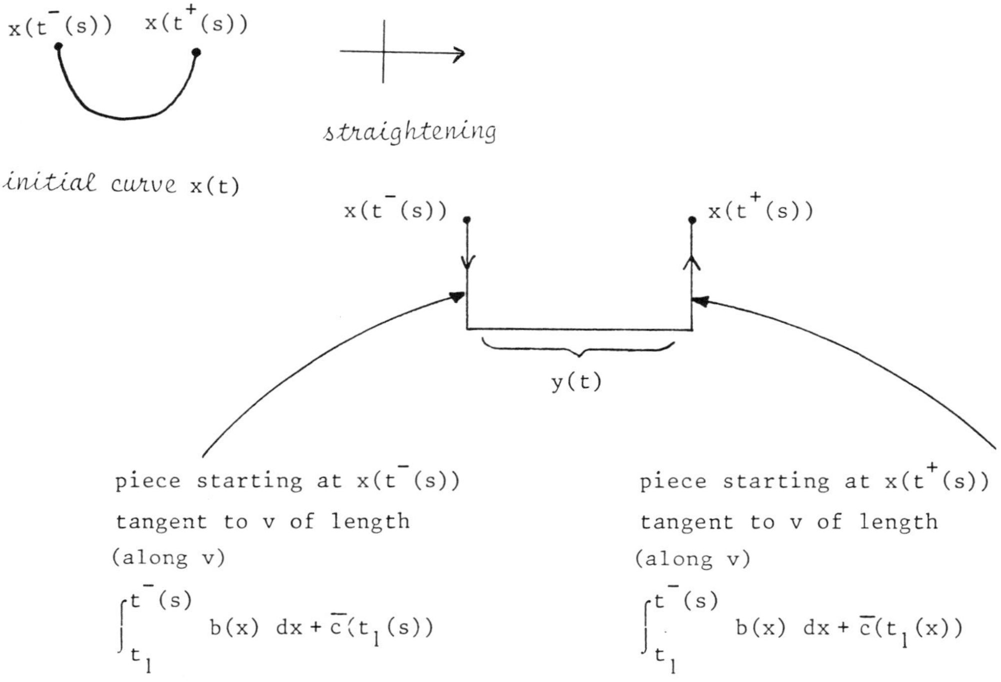

complete straightened curve

In a second step, we then prove that the curve y is on one hand close to x_ε at order $0(\sqrt{\alpha_1})$ (α_1 involved in Lemmas 5, 6, 7, 8) and, on the other hand, close to a piece of ξ-orbit z, defined by the equation:

$$\begin{cases} \dot{z} = a\ \xi(z(t)) \\ z(t^-(s)) = y(t^-(s)) \end{cases} \qquad t \in [t^-(s), t^+(s)]. \tag{7.4}$$

This is stated in:

Lemma 9: Let ε_1 be defined as in Lemma 8. There exist γ, $K_1, K_2, K_3 > 0$ depending only on a_0 ($0 < a_0 \leq a$) and on the geometry such that if $0 < \varepsilon_1 < \gamma$, and $\omega \geq \omega_2$, we have:

$$u_s = \sup_{t \in I_1(s)} |u(t)| \leq K_1 \sqrt{\alpha_1}$$

(α_1 as in Lemma 8) ;

$$\Delta_s = \sup_{t \in I_1(s)} \|\dot{y}(t) - a\xi(y(t))\| \leq K_2 \sqrt{\alpha_1} ;$$

$$\delta_s = \sup_{t \in I_1(s)} d(y(t), z(t)) \leq K_3 \sqrt{\alpha_1}\, \mu(I_1(s))$$

$$= K_3 \sqrt{\alpha_1} (t^+(s) - t^-(s))$$

This completes the analysis on the concentration intervals where the curve x_ε is thus:

a) rather tangent to ξ

b) well approximated by a piece of ξ-orbit, plus two small pieces (at $t^-(s)$ and $t^+(s)$) of v-orbits

c) the estimates are uniform on all type $I_1(s)$-intervals.

Next, we turn to analyse what happens between two concentration intervals. Our aim is to prove:

1). that the curve on such pieces is "almost tangent to v"

2). that between two limit points in such pieces, the form α has completed k ($k \in \mathbb{Z}$) revolutions and has come back on itself as a form.

This is illustrated in the following drawing:

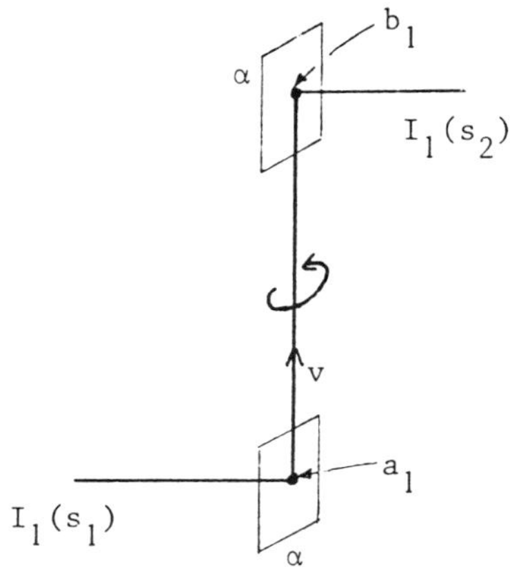

In the transport equation defined in Chapter 5, α has turned along the v-piece and come back on itself as a form.

However, this aim will be reached under further assumptions than (A_1). We do not feel that all these assumptions are necessary. We will discuss this later on.

To complete the analysis, we first distinguish two types of intervals: for $\varepsilon_1 > 0$ given, let $I_1(s_1)$ and $I_1(s_2)$ be two consecutive concentration intervals given through Lemma 8. Let

$$\begin{cases} t^+(s_1) < t_1 < t_2 < \dots < t_k < t^-(s_2) \\ \dot{b}(t_i) = 0 \end{cases} \tag{7.5}$$

$t_1, \dots, t_k$ are the zeros of $\dot{b}$ on the interval $[t^+(s_1), t^-(s_2)]$.
Between t_i and t_{i+1}, b takes at most once the value zero, as the t_i's cover all zeros of b.

We then have:

the type (I) intervals, where b never takes the value zero in $[t_i, t_{i+1}]$. (7.6)

The type (I) intervals where b never takes the value zero in $[t_i, t_{i+1}]$.
We draw the curve between these two times as follows:

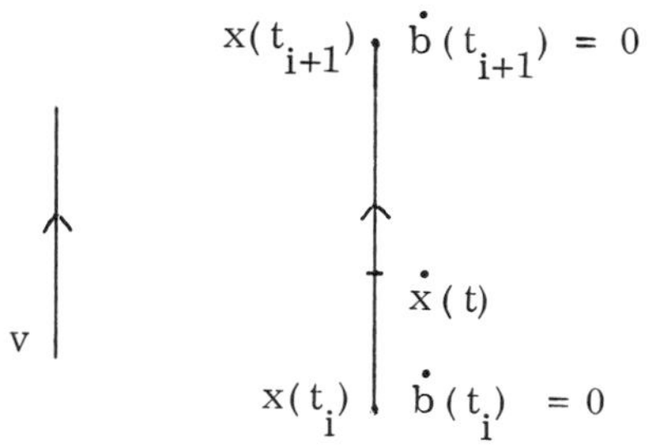

the type (II) intervals, where b is zero somewhere between t_i and t_{i+1}. (7.7)

The type (II) intervals, where b is zero somewhere between t_i *and* t_{i+1}.
We draw the curve then as follows:

(if $b(t_i) > 0$)

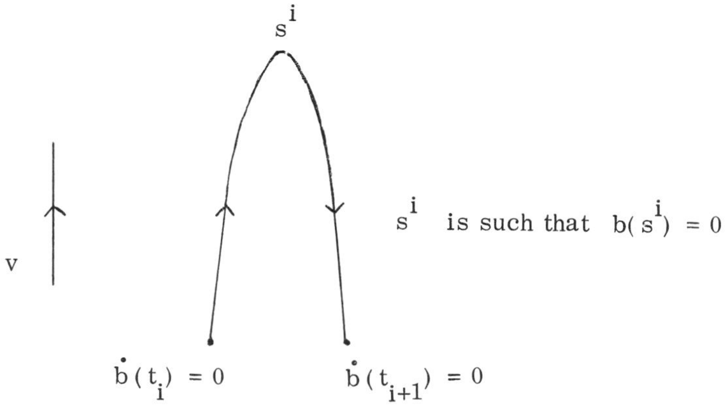

There are also two intervals, <u>the initial and final one</u>, we analyse separately.

<u>On the type</u> (I) -<u>intervals</u>:

In a similar way to that used on the concentration intervals, we introduce two new pieces of curves:

$$\begin{cases} \bar{z}(\,.\,) \text{ defined by} \\ \dot{\bar{z}}(t) = b(t)\ v(\bar{z}(t))\,;\ \bar{z}(t_i) = x(t_i) \\ t \in [t_i, t_{i+1}]. \end{cases} \tag{7.8}$$

and

$$\bar{y}(t) = \theta_{\bar{u}(t)}(x_\varepsilon(t)) \tag{7.9}$$

where

$$\theta_s \text{ is the one-parameter group generated by } \xi \tag{7.10}$$

$$\bar{u}(t) = -\int_{t_i}^{t} a(\tau)\,d\tau = -a(t - t_i) \qquad (a(\tau) = \text{Cte} = a). \tag{7.11}$$

$\bar{z}$ is tangent to v and $\bar{y}$ is straightened up from x_ε by pushing along ξ.

We then have the following lemma (Lemma 10) which holds in a more general setting (Lemma 10'). These lemmas provide us with the necessary estimates on the distance between the curve $x_\varepsilon(t)$ (denoted $x(t)$ for sake of simplicity) and $\bar{z}(t)$, on these intervals:

Lemma 10: Assume $\left|\int_{t_i}^{t_{i+1}} b(t)\,dt\right| \leq M_3$; where M_3 is a given constant. There exist $K_4 > 0$ and ω_4 such that for $\omega \geq \omega_4$, we have:

$$\sup_{t\in[t_i, t_{i+1}]} d(x(t), \bar{y}(t)) \leq K_4(t_{i+1} - t_i)$$

$$\sup_{t\in[t_i, t_{i+1}]} d(\bar{y}(t), \bar{z}(t)) \leq K_4(t_{i+1} - t_i)^2$$

$$\sup_{t\in[t_i, t_{i+1}]} d(x(t), \bar{z}(t)) \leq K_4[(t_{i+1} - t_i) + (t_{i+1} - t_i)^2],$$

ω_4 is uniform on all intervals of type (I).

Lemma 10': Let $[t^1, t^2]$ be a time interval such that b is never 0 in $]t^1, t^2[$. Assume $\left|\int_{t^1}^{t^2} b(x)\,dx\right| \leq C$, where C is a given constant.

There exist then constants $\varepsilon_0(C) > 0$ and $\bar{K}$ such that if $|t^2 - t^1| < \varepsilon_0(C)$, one has:

$$\sup_{t\in[t^1,t^2]} d(x(t),\bar{y}(t)) \le \bar{K}|t^2 - t^1|$$

$$\sup_{t\in[t^1,t^2]} d(\bar{y}(t),\bar{z}(t)) \le \bar{K}|t^2 - t^1|^2$$

$$\sup_{t\in[t^1,t^2]} d(x(t),\bar{z}(t)) \le \bar{K}[\,|t^2 - t^1| + |t^2 - t^1|^2].$$

Here $\bar{y}$ and $\bar{z}$ are defined by (8.101) and (8.104) where we replace the interval $[t_i, t_{i+1}]$ by $[t^1, t^2]$.

Finally, the condition $\omega \le \omega_4$ of Lemma 10 is here to be replaced by $|t^2 - t^1| < \varepsilon_0(C)$. Lemma 10' is useful for the remainder of this book.
Thus the curve on these pieces is "almost tangent to v" and the first part of the program has been completed.
The second part now has to be completed and it is a lot more delicate.

We first need to control the transport equation of the form α on such a piece and to compare it to the equation giving rise to the critical points at infinity which we found.

For this, we introduce a hypothesis:

Hypothesis (H): a type (I) interval $[t_i, t_{i+1}]$ is said to satisfy the (H)-hypothesis if

$$0 < \delta_0 \le \left|1 - \frac{b^2(t_i)}{2\omega a} + \frac{\int_0^1 b^2}{2\omega a}\right| \le \beta_0; \ \delta_0 \text{ and } \beta_0 \text{ fixed}$$

constants. (7.12)

We consider the solution of the transport equation:

$$\begin{cases} \dot{Z}^{\star} = -{}^{t}\Gamma(x_s)\, Z^{\star} \\ Z^{\star}(0) = Z_0^{\star} = \begin{bmatrix} 1 - \dfrac{b^2(t_i)}{2\omega a} + \dfrac{\int_0^1 b^2}{2\omega a} \\ 0 \\ 0 \end{bmatrix} \end{cases} \tag{7.13}$$

where x_s is the v-orbit through $x(t_i)$.
$Z_1^{\star}$ has been introduced in (5.58)

$$Z_1^{\star}(t) = \begin{bmatrix} 1 - \dfrac{b^2}{2\omega a}(t) + \dfrac{\int_0^1 b^2}{2\omega a} \\ -\dfrac{b(t)}{\omega} \\ \dfrac{\dot{b}(t)}{\omega a} \end{bmatrix} . \tag{7.14}$$

Finally, $\bar{s}$ has been introduced in Definition 3, and we have on the v-orbit x_s starting at $x(t_i)$:

$$\psi(x(t_i)) = x_{\bar{s}(x(t_i))} . \tag{7.15}$$

We assume $b(t_i)$ is positive (the other case is similar).
The following lemma (Lemma 11) shows that $x(t_{i+1})$ is close to $x_{\bar{s}(x(t_i))}$ and $Z_1^{\star}(t_{i+1})$ is close to $Z^{\star}(x_{\bar{s}(x(t_i))})$, at the order $0(t_{i+1} - t_i + \frac{1}{\sqrt{\omega}})$. We point out here that Lemma 11 relies on hypothesis (A_1).

Lemma 11: Assume that hypothesis (H) is satisfied, there exist $\delta_1 > 0$, $\overline{M}_3 > 0$, $K > 0$ and $\overline{\omega}_5 > 0$ such that, if $\omega \geq \overline{\omega}_5$, we have:

$$0 < \delta_1 \leq \int_{t_i}^{t_{i+1}} b(x)\, dx \leq \overline{M}_3$$

$$d(x(t_{i+1}), x_{\bar{s}(x(t_i))}) = d(x(t_{i+1}), \psi(x(t_i)) \leq K(t_{i+1} - t_i + \frac{1}{\sqrt{\omega}})$$

$$\|Z_1^\star(t_{i+1}) - Z^\star(x_{\bar{s}(x(t_i))})\| = \|Z_1^\star(t_{i+1}) - Z^\star(\psi(x(t_i))\|$$

$$\le K[t_{i+1} - t_i + \frac{1}{\sqrt{\omega}}].$$

Here $\left\|\begin{bmatrix} a \\ b \\ c \end{bmatrix}\right\| = |a| + |b| + |c|$. By $Z^\star(x_{\bar{s}(x(t_i))})$ we mean $Z^\star(\bar{s}(x(t_i)))$ although this is a misuse of notation.

The following drawing sums up the content of Lemma 11:

<u>If</u> t_i <u>satisfies (H)</u>

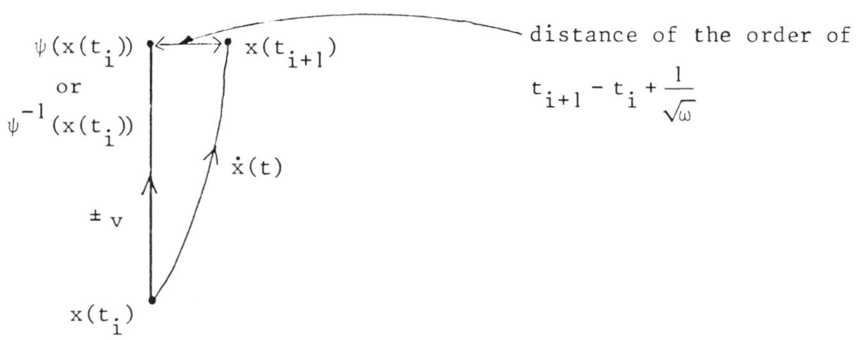

Difference of the transported equations of the order of $t_{i+1} - t_i + \frac{1}{\sqrt{\omega}}$

<u>On the type</u> (II) <u>intervals</u>, the situation is different.

We introduce the point s^i in $[t_i, t_{i+1}]$ where:

$$b(s^i) = 0. \tag{7.16}$$

We also assume (H) is satisfied at t_i.

$Z_1^\star$ is defined as in (7.14).

We then derive that $x(t_i)$ and $x(t_{i+1})$, as well as $Z_1^\star(t_i)$ and $Z_1^\star(t_{i+1})$, are close at the order of $(t_{i+1} - t_i + \frac{1}{\sqrt{\omega}})$.

This is summed up in:

<u>Lemma 14</u>: Assume that in t_i hypothesis (H) is satisfied. There exist δ_3,

K_8, M_4, ω_6 positive, M_4, δ_3, K_8 being geometric constants such that, if $\omega \geq \omega_6$, we have:

$$\delta_3\varepsilon_2 \leq \left|\int_{t_i}^{s^i} b(t)\,dt\right| \leq M_4; \quad \delta_3\varepsilon_2 \leq \left|\int_{s^i}^{t_{i+1}} b(t)\,dt\right| \leq M_4.$$

$$d(x(t_i), x(t_{i+1})) \leq \overline{K}_8(t_{i+1} - t_i + \frac{1}{\sqrt{\omega}})$$

$$\|Z_1^\star(t_{i+1}) - Z_1^\star(t_i)\| \leq \overline{K}_8(t_{i+1} - t_i + \frac{1}{\sqrt{\omega}}),$$

ω_6 is uniform on all type (II) intervals.

We illustrate this lemma by the following drawing:

<u>if</u> t_i <u>satisfies (H)</u>

$\delta\varepsilon_2 \leq$ length along $v \leq M_4$

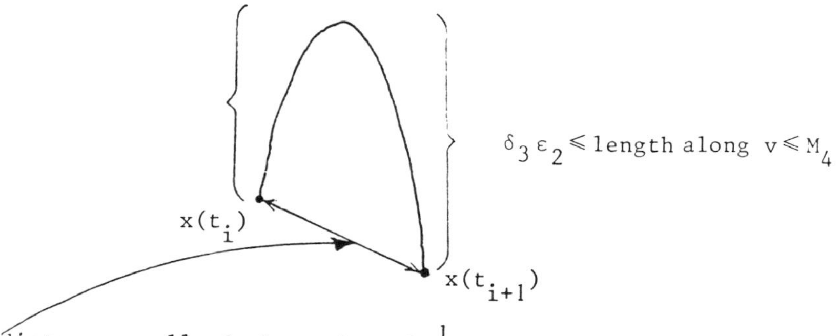

Difference of the transport equations small, of the order of $\frac{1}{\sqrt{\omega}} + t_{i+1} - t_i$

<u>On the initial and final intervals,</u> we define distinguished points $x_{\tilde{s}_2^+}$ for the initial one, $x_{\tilde{s}_2^-}$ for the final one.

Let us consider the case of an initial interval $[t^+(s_1), t_1]$.

We are thus coming from a concentration interval $I_1(s_1) = [t^-(s_1), t^+(s_1)]$ and

$\dot{x}_\varepsilon$ is shifting from being rather tangent to ξ to being rather tangent to v. Lemma 8 provides us with a unique point $t_1(s_1)$ in $[t^-(s_1), t^+(s_1)]$ where $b\dot{b}$ vanishes. Lemma 9 shows that the curve $y(.) = \phi(u(t), x_\varepsilon(.))$ is close to the curve $z(.)$, starting at $y(t_-(s_1))$ and tangent to $a\xi$ on $[t^-(s_1), t^+(s_1)]$ at the order $\sqrt{\alpha_1}(t^+(s_1) - t^-(s_1))$.

Here, $u(t) = -[\int_{t_1(s_1)}^{t} b(x)\,dx + \bar{c}(t_1(s_1))]$.

The choice of $\bar{c}(t_1(s_1))$ was left open, under the condition:

$$|\bar{c}(t_1(s_1))| \leq C\sqrt{\alpha_1}\,. \tag{7.17}$$

This was summed up in:

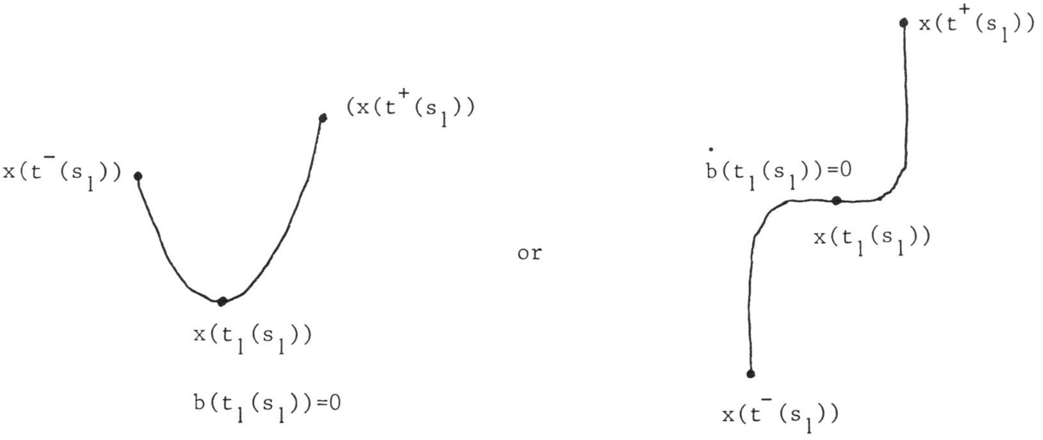

In order to define $x_{\tilde{s}_2^+}$, we consider two v-orbits x_s and x'_s:

x'_s is the v-orbit starting at $x(t_1(s_1))$; (7.18)

x_s is the v-orbit starting at $x(t^+(s_1))$. (7.19)

We consider the transport equations on these v-orbits:

$$\left\{\begin{array}{l}\dot{Z}'^{\star} = -{}^t\Gamma(x'_s)\,Z'^{\star}\\[2ex] Z'^{\star}(0) = Z'^{\star}_0 = \end{array}\right. \begin{bmatrix} 1 - \dfrac{b^2(t_1(s_1))}{2\omega a} + \dfrac{\int_0^1 b^2}{2\omega a} \\ 0 \\ \dfrac{\dot{b}(t_1(s_1))}{\omega a} \end{bmatrix} = \begin{bmatrix} a^{\star\prime}(0) \\ 0 \\ c^{\star\prime}(0) \end{bmatrix} \qquad (7.20)$$

$$\left\{\begin{array}{l}\dot{Z}^{\star} = -{}^t\Gamma(x_s)\,Z^{\star}\\[2ex] Z^{\star}(0) = Z^{\star}_0 = \end{array}\right. \begin{bmatrix} 1 - \dfrac{b^2({}^+(s_1))}{2\omega a} + \dfrac{\int_0^1 b^2}{2\omega a} \\ 0 \\ \dfrac{\dot{b}(t^+(s_1))}{\omega a} \end{bmatrix} = \begin{bmatrix} a^{\star}(0) \\ 0 \\ c^{\star}(0) \end{bmatrix} \qquad (7.21)$$

The initial conditions $Z'^{\star}_0$ and $Z^{\star}_0$ satisfy:

$$|c^{\star\prime}(0)| < \sqrt{\alpha_1}\ ; \quad |c^{\star}(0) < \sqrt{\alpha_1} \qquad (7.22)$$

$$|a\,'(0)| \geq \tfrac{1}{2}\ ; \quad |a^{\star}(0) \geq \tfrac{1}{2}. \qquad (7.23)$$

Thus, the forms initial data are close to α, provided α_1 is small. Due to (A_1) then, and to the property of rotations of α along v expressed in Proposition 9, there are two unique points on x'_s and x_s respectively, near $x(t_1(s_1))$ and $x(t^+(s_1))$ respectively where $Z'^{\star}$ and $Z^{\star}$ are collinear to $\begin{bmatrix}1\\0\\0\end{bmatrix}$, i.e. the transported vector is collinear to α.

We denote these points by $x_{\tilde{s}'_2}$ and $x_{\tilde{s}^+_2}$ respectively. This equation is thus:

$$\left\{\begin{array}{ll} c^{\star\prime}(s'_2) = 0 & c^{\star}(s^+_2) = 0 \\ x'_{\tilde{s}'_2} \text{ near } x(t_1(s_1))\ ; & x_{\tilde{s}^+_2} \text{ near } x(t^+(s_1)) \end{array}\right. . \qquad (7.24)$$

The existence of such points (and their uniqueness) is provided by the following abstract lemma (Lemma 13) on a v-orbit x_s and the transport equation $Z^{\star}(s)$.

Lemma 13: Let $s_1 \in \mathbb{R}$; M and m given positive such that:

$$|a^\star| + |c^\star|(s_1) \leq M, \quad |a^\star|(s_1) \geq m > 0.$$

There exist then $\chi(M,m) > 0$, $\theta(M,m) > 0$, $\mu(M,m) > 0$, $K_6(M,m) > 0$, $K_7(M,m) > 0$ such that if $|c^\star(s_1)| < \chi(M,m)$, the function $c^\star(s)$ vanishes once and only once in the time interval $]s_1 - \theta(M,m), s_1 + \theta(M,m)[$.
Furthermore, we have

$$\begin{cases} c^\star(s_1 - \theta(M,m))\, c^\star(s_1 + \theta(M,m)) < 0 \\ |c^\star(s_1 \pm \theta(M,m))| \geq \mu(M,m) \\ |a^\star|(s) + |c^\star(s)| \leq M+1 \text{ over }]s_1 - \theta(M,m), s_1 + \theta(M,m)[\end{cases}$$

$$\begin{cases} \|Z^\star(s_2) - Z^\star(s_1)\| \leq K_6 |c^\star(s)| \\ |c^\star(s)| = |c^\star(s) - c^\star(s_2)| \geq K_7 |s - s_2| \end{cases}$$

$\forall\, s \in]s_1 - \theta(M,m), s_1 + \theta(M,m)[$.
As already stated, Lemma 13 means that each time α_{x_s} coincides with a transported form (as a field of planes), then, in a small neighbourhood, there is no other coincidence point and α_{x_s} diverges from this transported form backwards and forwards before and after the coincidence time. This is merely Proposition 9.

Now, if a transported form is close to α somewhere, then by (A_1), α coincides with this transported form in a neighbourhood of the considered point. This sums Lemma 13; which implies the existence of $x_{\tilde{s}_2^+}$ and $x'_{\tilde{s}'_2}$.
We set $\bar{c}(t_1(s_1)) = -\tilde{s}'_2$.
We then have the following lemmas about these points: Lemma 15 is concerned with initial intervals and expresses that $x_{\tilde{s}_2^+}$ is close to $y(t^+(s_1))$ at the order $\frac{1}{\sqrt{\omega}} + \sqrt{\alpha_1}(t^+(s_1) - t_1(s_1))$, that $Z^\star(s_2^+) = \begin{bmatrix} a^\star(s_2^+) \\ 0 \\ 0 \end{bmatrix}$,
$|a^\star(\tilde{s}_2^+) - a'^\star(\tilde{s}'_2)| = 0\left(\frac{1}{\sqrt{\omega}} + \sqrt{\alpha_1}(t^+(s_1) - t_1(s_1))\right)$ and lastly that
$|a^\star(\tilde{s}_2^+) - 1| = 0\left(\frac{\int_0^1 b^2}{\omega} + \varepsilon_1 + \sqrt{\alpha_1}\right) = o(1)$.

Lemma 15' is concerned with the final intervals, where the same properties hold with a distinguished point $x_{\tilde{s}_2^-}$ instead of $x_{\tilde{s}_2^+}$. Such a final interval will be called $[t_k, t^-(s_2)]$.

We state these lemmas:

Lemma 15: There exist ω_8 and $\bar{\bar{\varepsilon}}_1$; ω_8 and $\bar{\bar{\varepsilon}}_1$ uniform on all concentration intervals such that:

$$d(x_{\tilde{s}_2^+}, y(t^+(s_1))) < K_{12}\left[\frac{1}{\sqrt{\omega}} + \sqrt{\alpha_1(t^+(s_1) - t_1(s_1))}\right];$$

$$z^\star(\tilde{s}_2^+) = \begin{bmatrix} a^\star(\tilde{s}_2^+) \\ 0 \\ 0 \end{bmatrix};$$

$$|a^\star(\tilde{s}_2^+) - a'^\star(\tilde{s}_2')| < K_{12}\left[\frac{1}{\sqrt{\omega}} + \sqrt{\alpha_1(t^+(s_1) - t_1(s_1))}\right]$$

$$\left\|z^\star(\tilde{s}_2^+) - \begin{bmatrix} 1 \\ 0 \\ 0 \end{bmatrix}\right\| = |a^\star(\tilde{s}_2^+) - 1| < K_{12}\left[\frac{\int_0^1 b^2}{\omega} + \varepsilon_1 + \sqrt{\alpha_1}\right],$$

for $\omega \geq \omega_8$; $0 < \varepsilon_1 < \bar{\bar{\varepsilon}}_1$.

Here K_{12} is a uniform geometric constant.

Lemma 15': There exist ω_8, $\bar{\bar{\varepsilon}}_1$, uniform on all intervals of that type such that:

$$d(x_{\tilde{s}_2^-}, y(t^-(s_1))) < K_{12}\left[\frac{1}{\sqrt{\omega}} + \sqrt{\alpha_1(t_1(s_1) - t^-(s_1))}\right]$$

$$z^\star(\tilde{s}_2^-) = \begin{bmatrix} a^\star(\tilde{s}_2^-) \\ 0 \\ 0 \end{bmatrix};$$

$$|a^\star(\tilde{s}_2^-) - a'^\star(\tilde{s}_2')| < K_{12}\left[\frac{1}{\sqrt{\alpha}} + \sqrt{\alpha_1(t_1(s_1) - t^-(s_1))}\right]$$

$$\left\|z^\star(\tilde{s}_2^-) - \begin{bmatrix} 1 \\ 0 \\ 0 \end{bmatrix}\right\| = |a^\star(\tilde{s}_2^-) - 1| < K_{12}\left(\int_0^1 \frac{b^2}{\omega} + \varepsilon_1 + \sqrt{\alpha_1}\right),$$

for $\omega \geq \omega_8$; $0 < \varepsilon_1 < \bar{\bar{\varepsilon}}_1$; K_{12} is a geometric constant.

Next, we assume for the sake of simplicity (the other cases are similar) that:

$$b(t^+(s_1)) > 0. \tag{7.25}$$

$Z^\star(s)$ has been defined in (7.21).

$Z_1^\star(t)$ is as usual $\begin{bmatrix} 1 - \frac{b^2}{2\omega a}(t) + \frac{\int_0^1 b^2}{2\omega a} \\ \frac{b(t)}{\omega} \\ \frac{\dot{b}(t)}{\omega a} \end{bmatrix}$.

$\bar{s}$ is defined by:

$$x_{\bar{s}} = \psi(x_{\tilde{s}_2^+}). \tag{7.26}$$

We then show that the piece of curve x_ε on $[t^+(s_1), t_1]$ is rather tangent to v on a consistent and limited piece of v-orbit x_s while $x_\varepsilon(t_1)$ is close to $\psi(x_{\tilde{s}_2^+})$ and $Z_1^\star(t_1)$ is close to $Z^\star(\bar{s})$ at the order $\frac{1}{\sqrt{\omega}} + |t_1 - t^+(s_1)|$.

This is stated in Lemma 16, under assumption (7.25) and the general situation, in all possible cases, for initial as well as final intervals, is summarized in the drawings which follow this lemma:

<u>Lemma 16</u>: Under (7.25), there exist ω_9, M_4, K_{13}, $\bar{\bar{\varepsilon}}_1$, $\gamma_1 > 0$ uniform such that: if the interval we consider is $[t^+(s_1), t_1]$ (i.e. $\dot{b}(t_1) = 0$; t_1 the first zero of $\dot{b}$ after $t^+(s_1)$), we have:

$$\gamma_1 \le \left| \int_{t^+(s_1)}^{t_1} b(x)\,dx \right| \le M_4; \quad Z^\star(\bar{s}) = \begin{bmatrix} a^\star(\bar{s}) \\ 0 \\ 0 \end{bmatrix}$$

$$d(\psi(x_{\tilde{s}_2^+}), x(t_1)) = d(x_{\bar{s}}, x(t_1)) < K_{13}\left[\frac{1}{\sqrt{\omega}} + |t_1 - t^+(s_1)|\right]$$

$$\|Z^\star(\bar{s}) - Z_1^\star(t_1)\| = \|Z^\star(\psi(x_{\tilde{s}_2^+})) - Z_1^\star(t_1)\| < K_{13}\left[\frac{1}{\sqrt{\omega}} + |t_1 - t^+(s_1)|\right],$$

for $\omega \ge \omega_9$ and $\varepsilon_1 < \bar{\bar{\varepsilon}}_1$.

a) If $[t^+(s_1), t_1]$; $b(t^+(s_1)) > 0$.

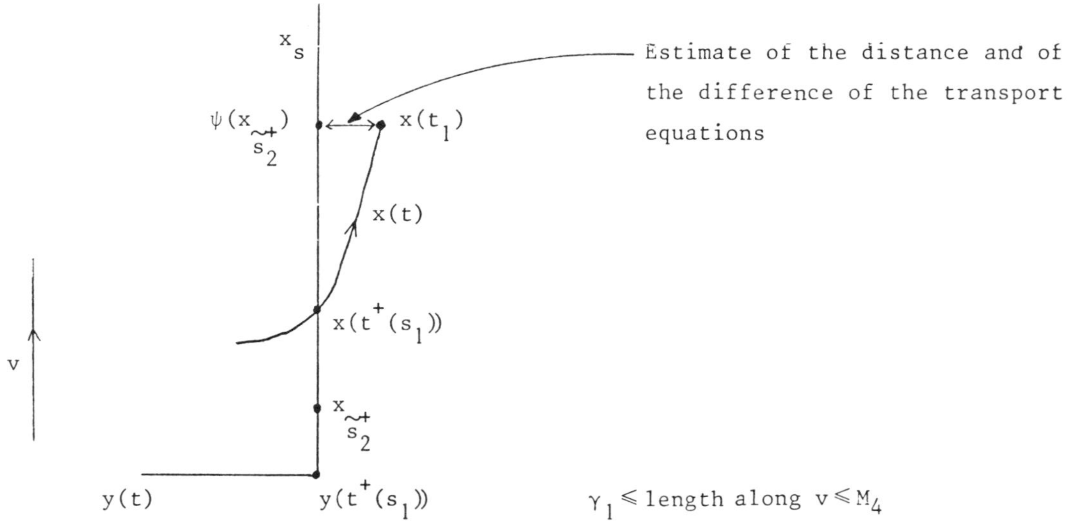

b) If $[t^+(s_1), t_1]$; $b(t^-(s_2)) < 0$.

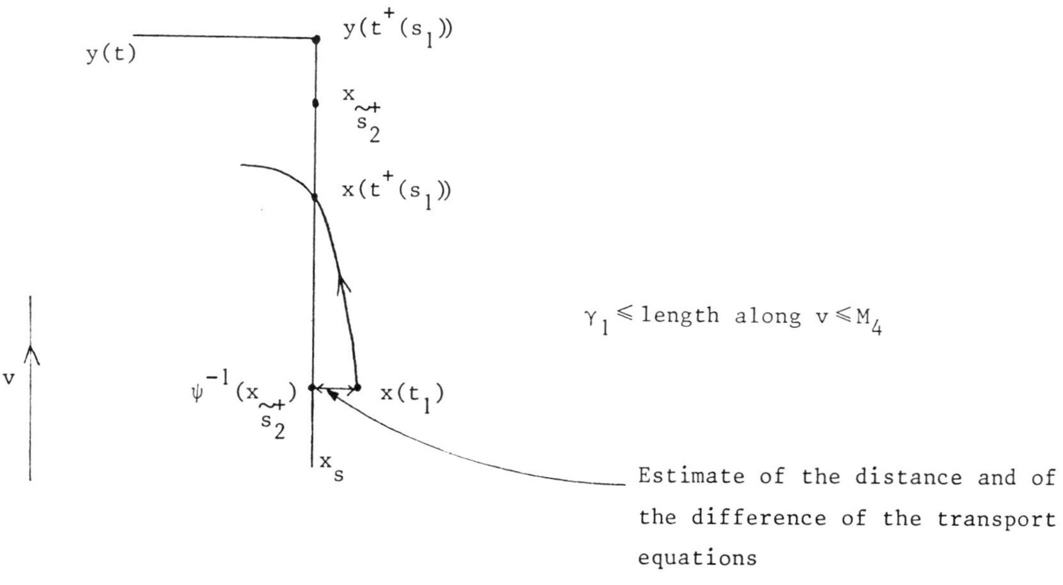

Conclusion of 7.1:

In Chapter 8, we will thus be completing all the local analysis which relies on three key points:

1) the concentration intervals; and the straightening up process;
2) the non-concentration intervals, where the curve is rather tangent to v. We need then the (H)-hypothesis in order to be able to analyse an interval of type (I), (II) or initial or final;
3) the property of rotation of α and the (A_1)-hypothesis which allows us to show that on the type I, as well as on the initial and final type intervals, x_ε runs from a point to the image by $\psi^{\pm 1}$ of this point, nearly. This also uses the (H)-hypothesis.

Of course, in each of these approximations, something is lost: on the concentration intervals, the approximation is at order $\sqrt{\alpha_1}$, where α_1 can be chosen to be quite small with ω going to $+\infty$.
On the other intervals, say $[t_i, t_{i+1}]$, it is at the order $\frac{1}{\sqrt{\omega}} + t_{i+1} - t_i$.

Do all these losses cumulate in a consistent loss; or are there hypotheses allowing a global convergence process? This is the question we look at in Chapter 9, and which we describe now.

Before ending 7.1, we state the following useful lemma (Lemma 18) which says that the total number of type I, type II, initial and final intervals is $o(\frac{1}{\sqrt{\omega}})$. We feel that this lemma could be improved a lot using the fact that x_ε lies on the same flow-line of (4.9) with ε going to zero (or ω going to $+\infty$).

Lemma 18: Let ε_1 be given small enough so all the preceding lemmas are valid for $\omega \geq \omega_{10}$.
Let j_1 be the total number of type (I) and (II) intervals satisfying the hypothesis (H) with given δ_0 and β_0 (i.e., at t_i, (H) is satisfied with δ_0 and β_0).
Let j_2 be the total number of initial and final intervals.

Then $\frac{j_1+j_2}{\sqrt{\omega}} \to 0$ when $\omega \to +\infty$.

7.2 The convergence theorem

After having completed the local analysis in Chapter 8, we turn to semi-local results.

We first need to control the oscillations between two concentration intervals $I_1(s_1)$ and $I_1(s_2)$, along v.

Assume the (H)-hypothesis is satisfied on the t_i's separating two given concentration intervals.

To avoid any confusion, we denote $x_{\tilde{s}_2^+}$ the point built on $I_1(s_1)$ by Lemma 15 and $x_{\tilde{s}_2^-}$ the point built on $I_1(s_2)$ by Lemma 15'. Hence, to each concentration interval, $I_1(s_i)$, we associate two points $(x_{\tilde{s}_2^+})_i$ and $(x_{\tilde{s}_2^-})_i$. When two intervals $I_1(s_i)$ and $I_1(s_{i+1})$ are consecutive and we must estimate quantities involving $(x_{\tilde{s}_2^+})_i$ and $(x_{\tilde{s}_2^-})_{i+1}$, we will denote these $x_{\tilde{s}_2^+}$ and $\bar{x}_{\tilde{s}_2^-}$ to avoid useless repetition of indices.

Then we have the following drawing where $[t_3, t_4]$ and $[t_5, t_6]$ are type (II) intervals.

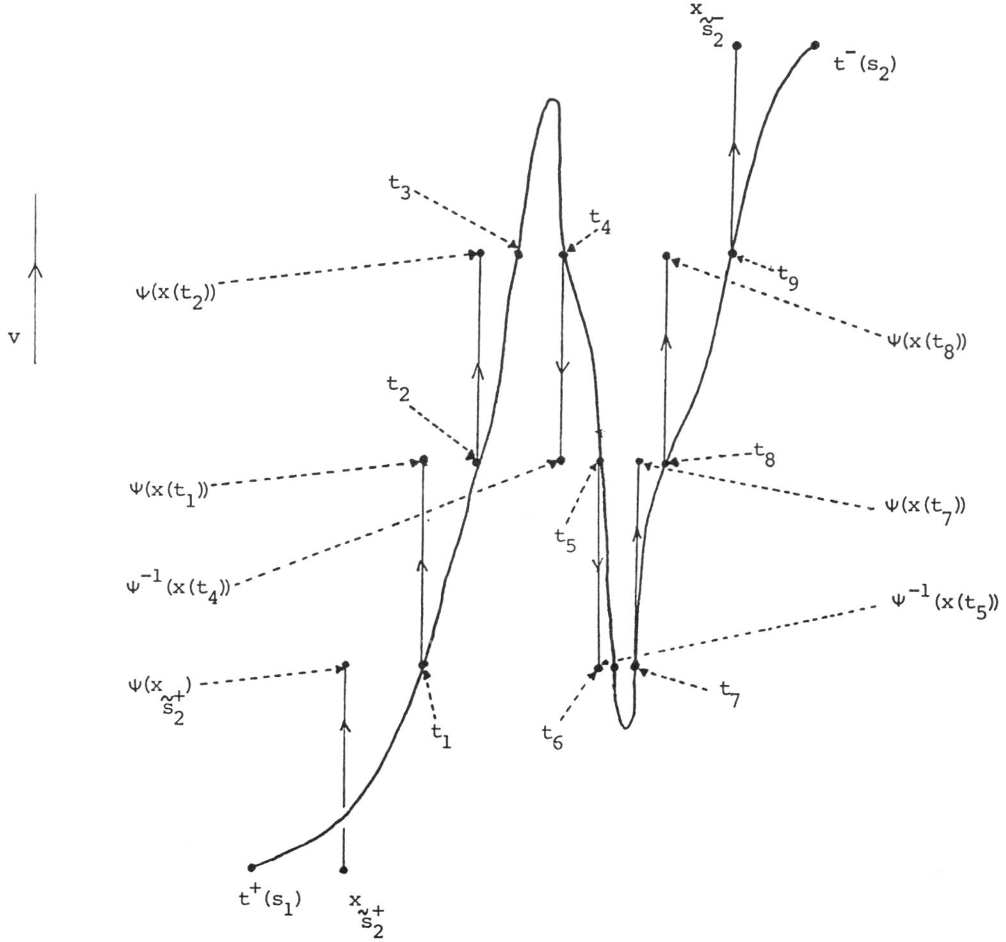

In this drawing, we forget the curve $x_\varepsilon(\,.\,)$ and draw only the wings tangent to v:

$$\psi \quad \overline{x}_{\underset{\sim}{s}_2^-} \quad Z*(\widetilde{s}_2^-) = \begin{bmatrix} 1+\varepsilon' \\ 0 \\ 0 \end{bmatrix}$$

$$x_{\underset{\sim}{s}_2^+} \quad Z*(\widetilde{s}_2^+) = \begin{bmatrix} 1+\varepsilon \\ 0 \\ 0 \end{bmatrix}$$

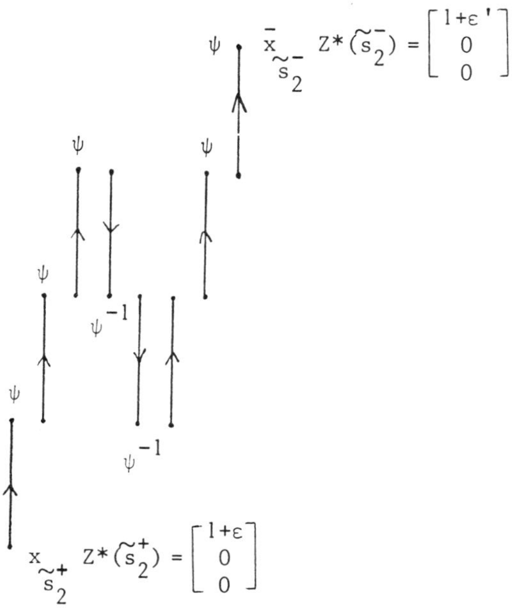

To understand this drawing better, we extend all the pieces in the direction of their arrows and iterate ψ along them:

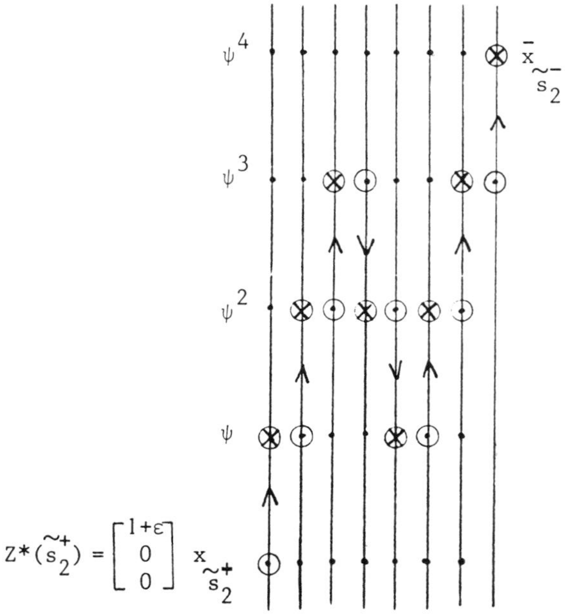

$$\overline{x}_{\underset{\sim}{s}_2^-} \quad Z*(\widetilde{s}_2^-) = \begin{bmatrix} 1+\varepsilon' \\ 0 \\ 0 \end{bmatrix}$$

$\left(\cdot\right)$ are the starting points of the pieces

$\bigotimes$ are the end points of the pieces; hence, a jump is made there (to another piece)

$\bullet$ are the ψ^i of the starting points

$$Z*(\widetilde{s}_2^+) = \begin{bmatrix} 1+\varepsilon \\ 0 \\ 0 \end{bmatrix} \quad x_{\underset{\sim}{s}_2^+}$$

Roughly speaking, each jump occurring at distinguished points $\otimes$ amounts to a loss of order $K(\frac{1}{\sqrt{\omega}} + \Delta t)$, where Δt is the time spent to run along the previous piece of curve.

We want to bring back all these movements along v on a unique v-orbit starting at $x_{\tilde{s}_2}^+$.

For this purpose, we need to carry all our estimates back to this v-orbit. This operation requires hypotheses whose essential content is to ensure that estimates holding between two consecutive distinguished points $\otimes$ and $\odot$ do propagate along the pieces through ψ^i; $i \in \mathbb{Z}$.

These hypotheses are:

(A_2) $\exists\, \underline{k}$ and $\bar{k} > 0$, $\exists\, \varepsilon_5 > 0$, $\forall\, i \in \mathbb{Z}$, $\underline{k}d(x,y) \leq d(\psi^i(x), \psi^i(y)) \leq \bar{k}d(x,y)$ $\forall\, x, y \in M$ such that $d(x,y) < \varepsilon_5$.

(A_3) Let $\lambda_i : M \to \mathbb{R}$ defined by $\lambda_i(x)\,[(\psi^i)^{\star}\alpha]_x = \alpha_x$, $i \in \mathbb{Z}$.
We ask that there exist uniform constants ε_6, ε_3, $\varepsilon_5 > 0$ such that:

$$\begin{cases} |\lambda_i(x) - \lambda_i(y)| \leq \alpha_5 d(x,y) & \text{as soon as } d(x,y) < \varepsilon_6 \\ 0 < \alpha_3 \leq |\lambda_i(x)| \leq \alpha_4, & \forall\, i \in \mathbb{Z},\ \forall\, x \in M. \end{cases}$$

Remark: We point out here that in view of the analysis described in Section 7.1, these hypotheses would be unnecessary if we had a control on the number of pieces $[t_i, t_{i+1}]$. There is a hope of obtaining such a control through a further study of the differential equation (4.9) and the properties of this flow-line or the flow-line for another J-decreasing deformation. In fact, a deformation bringing the number of zeros of b or $\dot{b}$ under control would allow us to get rid of these hypotheses and shorten greatly the computations we are facing.

Remark: Hypotheses (A_2)-(A_3) are constraining. They are satisfied in the case of the S^1-fiber bundles. Indeed, in such a case ψ to a certain power is the identity.

We have for instance $\psi^4 = \mathrm{Id}$ in the case of $S^3 \to S^2$.

Another interesting situation is the case of the recurrent vector fields v.

We then have the following semi-local result between two concentration intervals, under (A_2) and (A_3), which bring back all the estimates on a unique v-orbit.

Lemma 19: Under (A_2) and (A_3), hypothesis (H) is satisfied with $\delta'_0 = \frac{2\alpha_3}{3}$ and $\beta'_0 = \frac{4\alpha_4^2}{3}$ uniformly at the t_i's for ω large enough and ε_1 small enough.

Furthermore, let $I_1(s_1)$ and $I_1(s_2)$ be two consecutive concentration intervals. Let j be the total number of intervals between $t^+(s_1)$ and $t^-(s_2)$. Let k_1 be the number of wings oriented by ψ in the representation (7.26) and k_2 the number of wings oriented by ψ^{-1}.

Then $k_2 - k_1$ is even.

Moreover we have:

$$d(\psi^{k_1-k_2}(x_{\tilde{s}_2^+}), \bar{x}_{\tilde{s}_2^-}) < K[\frac{j}{\sqrt{\omega}} + t^-(s_2) - t^+(s_1))]$$

$$|\lambda_{k_1-k_2}(x_{\tilde{s}_2^+}) - 1| < \varepsilon(\varepsilon_1, \omega) \quad \text{where } \varepsilon \xrightarrow[\substack{\varepsilon_1 \to 0 \\ \omega \to +\infty}]{} 0$$

$$\left|\lambda_{k_1-k_2}(x_{\tilde{s}_2^+}) - \frac{\bar{a}^\star(\tilde{s}_2^-)}{a^\star(\tilde{s}_2^+)}\right| < K\left|\frac{j}{\sqrt{\omega}} + t^-(s_2) - t^+(s_1))\right|.$$

Lemma 19 is summed up in the following drawing:

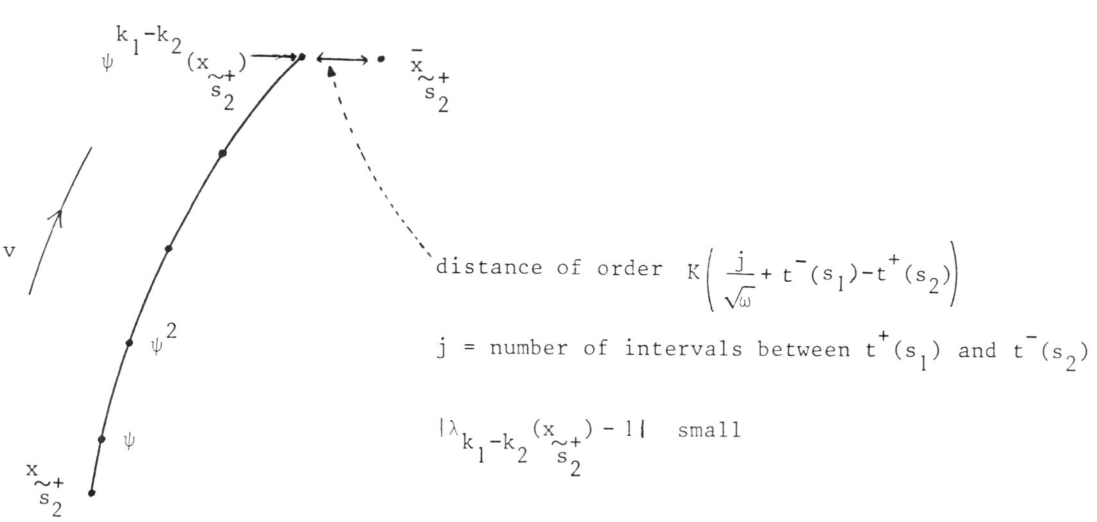

Points $x_{\tilde{s}_2^+}$ and $\bar{x}_{\tilde{s}_2^-}$ are, because of Lemma 19, very constrained: between $x_{\tilde{s}_2^+}$ and $\psi^{k_1-k_2}(x_{\tilde{s}_2^+})$, the second point being very close to $\bar{x}_{\tilde{s}_2^-}$, the form α has made an even number of revolutions and has almost come back on itself <u>as a form</u>. This means that $x_{\tilde{s}_2^+}$ must lie in a neighbourhood of the conjugate points of the form α.

Given x in M, under (A_1), there are a discrete number of coincidence points of the form α: these are the $\psi^i(x)$; $i \in \mathbb{Z}$.

Among these points, the existence of one such that $\lambda_i(x) = 1$ raises a constraint. For i given in $\mathbb{Z}$, such points are expected to belong to a hypersurface Γ_i of M. Such a phenomenon is generical:

Let us examine the case of S^3 with $\alpha = \lambda\alpha_0$ and $\alpha_0 = \Sigma(x_i dy_i - y_i dx_i)$. In this case, with v defining a Hopf fibration, α makes one revolution along a v-orbit after having described 1/4 of a v-orbit. The first oriented coincidence point is obtained after 1/2 of a v-orbit. On the remaining v-orbit, there is a coincidence point at 3/4 of the orbit (reverse orientation) and a conjugate point when the orbit is completely described. Thus:

$$\psi^2(x_0) = -x_0; \quad \psi^4(x_0) = x_0. \tag{7.28}$$

Here, coincidence points which are conjugate points are either x_0 itself or $-x_0$ under the condition:

$$\lambda(-x_0) = \lambda(x_0) \tag{7.29}$$

where λ defines α with respect to α_0.

Consequently, the curves x_ε are of the following type:

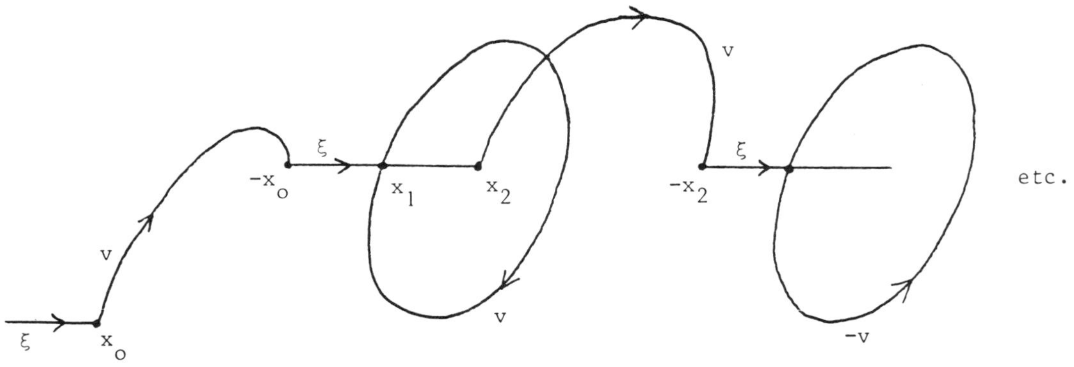

The points x_i satisfy $\lambda(x_i) = \lambda(-x_i)$ for $i = 0$; $i = 2$.

Thus the points x_i belong to:

$$\Gamma = \{x \in S^3 \,|\, \lambda(x) = \lambda(-x)\} \tag{7.30}$$

which is generically a hypersurface of S^3.

Summing up our analysis, we have:

[1] Inside the concentration intervals, we defined two points (x_i^+ and x_i^-) such that:

$$x_i^- = \psi^{k_i'}(x_{i-1}^+); \quad x_{i-1}^+ = (x_{\tilde{s}_2^+})_{i-1}; \quad k_i' = (k_1 - k_2)_{i-1} \tag{7.31}$$

(x^+_{i-1}, x^-_i) corresponds to the pair $(x_{\tilde{s}^+_2}, \psi^{k_1-k_2}(x_{\tilde{s}^+_2}))$ of Lemma 19. k'_i is even and $\lambda_{k'_i}(x^+_i)$ is close to 1.

[2] On a concentration interval, we approximated the curve, at order $\sqrt{\alpha_1}\,\delta t_i$ by a curve z_i, tangent to $a\xi$ during the time δt_i, and starting at $y(t^-_i(s_i))$. This was derived in Lemma 9, with $\delta t_i = t^+_i(s_i) - t^-_i(s_i)$, i being the index of the concentration interval.

We proved in Lemma 15' that $y(t^-_i(s_i))$ is close to $(x_{\tilde{s}^-_2})_i$ at the order $\frac{1}{\sqrt{\omega}} + \sqrt{\alpha_1}\,\delta t_i$; while in Lemma 19, we proved that $(x_{\tilde{s}^-_2})_i$ is close to $x^-_i = \psi^{k'_i}(x^+_{i-1})$ at the order $\frac{\bar{j}_i}{\sqrt{\omega}} + \delta t^-_i$ where $\bar{j}_i$ is the total number of distinguished intervals between $t^+_i(s_{i-1})$ and $t^-_i(s_i)$; and $\delta t^-_i = t^-_i(s_i) - t^+_i(s_{i-1})$. Finally, we know, through Lemma 15, that $(x_{\tilde{s}^+_2})_i$ is close to $y(t^+_i(s_i))$ at the order $\frac{1}{\sqrt{\omega}} + \sqrt{\alpha_1}\,\delta t_1$.

We thus have:

Lemma 20: Consider the curve $\tilde{z}_i$ tangent to $a\xi$ during the time δt_i and starting at $x^-_i = \psi^{k'_i}(x^+_{i-1}) = \psi^{k'_i}(x_{(\tilde{s}^+_2)_{i-1}})$. Then the extremity $\bar{x}^+_i$ of this curve is close to $x^+_i = (x_{\tilde{s}^+_2})_i$ at order $\frac{\bar{j}_i+1}{\sqrt{\omega}} + \delta t^-_i + \sqrt{\alpha_1}\,\delta t_i$.

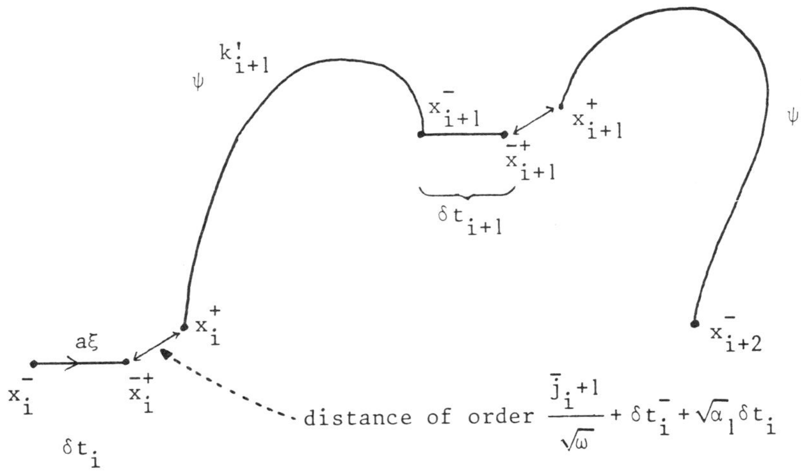

(7.32)

k'_i even; $|\lambda_{k'_{i+1}}(x^+_i) - 1|$ small.

We now have the following theorem which shows that α is more and more (as $\varepsilon \to 0$ or $\omega \to +\infty$) globally preserved along the curve x_ε :

Theorem 6: Let n_0 (depending on ε_1 and ω) be the total number of concentration intervals. Let $\lambda_{k'_i}(x^+_i)$ be the characteristic numbers of $\psi^{k'_i}$, at points x^+_i. We have:

$$\left|\prod_{i=1}^{m} \lambda_{k'_i}(x^+_{i-1}) - 1\right| \to 0 \quad \text{uniformly for } m \leq n_0.$$

The significance of Theorem 6 is that the curve drawn in (7.32) preserves the form α more and more. Indeed, it first implies that, for each $\psi^{k'_i}$, $\lambda_{k'_i}(x^+_{i-1})$ goes to 1. Hence α is sent back on itself from x^+_{i-1} to $\psi^{k'_i}(x^+_{i-1})$. On the other hand, the pieces tangent to ξ preserve α, as $L_\xi \alpha = 0$. Finally, the global influence of the $\psi^{k'_i}$ on α is small by Theorem 6. In this sense, we may say that (7.32) is an "α-geodesic relative to v".

We are ready now for the convergence theorems:

They are obtained through a "rearrangement" process of the curve (7.32) and a control of the oscillation phenomenon between two concentration intervals.

This again will involve new hypotheses $(\mathfrak{A}_1)$ and $(\mathfrak{A}_2)(x)$, which are likely to disappear if we had a better control on the phenomena along a flow-line (4.9).

The "rearrangement" process of (7.32) consists of defining, close to the curve of (7.32), a new almost closed, continuous, curve, made up with pieces, some of them tangent to ξ, the others tangent to v between end points x and $\psi^{k'_i}(x)$ which are almost conjugate (i.e. $(\lambda_{k'_i}(x) - 1) \to 0)$.

These curves will converge to closed curves as ω goes to $+\infty$.

The device consists in cumulating all losses between $\bar{x}_i^+$ and x_i^+ due to concentration intervals on a unique concentration interval. Indeed, if we add all these losses, we are left with something of the order: $\Sigma(\frac{\bar{j}_{i+1}}{\sqrt{\omega}} + \delta t_i^- + \sqrt{\overline{\alpha_1}}\ \delta t_i)$; which is less than $\frac{j_1+j_2}{\sqrt{\omega}} + \Sigma\,\delta t_i^- + \sqrt{\overline{\alpha_1}}$.

This last quantity goes to zero as $\omega \to +\infty$ and $\alpha_1 \to 0$, as by Lemmas 5 and 18, $\frac{j_1+j_2}{\sqrt{\omega}}$ and $\Sigma\,\delta t_i^-$ go then to zero.

Consequently, if we are able to simply cumulate all these losses without multiplying them by factors going to $+\infty$, we will obtain an almost closed curve. Through this operation, we lose Theorem 6, which is not proven on the new curve. This is why, in the following theorems, we will retain the points x_i^+ satisfying Theorem 6. It is probable that, with a better rearrangement process, Theorem 6 will still hold on the almost closed curve.

For our purpose here, we leave aside this problem. However, we show in Theorem 7 that Theorem 6 is not lost in a generical situation.

There are two ways to obtain an almost closed curve.

The first way is direct and involves a strong hypothesis on the one-parameter groups of the vector fields $D\psi^{2k}(\xi)$; $k \in \mathbb{Z}$. However, there are two arguments in favour of this way: the phenomenon is clearly seen and this allows us to overcome the case M fibers in S^1 over a surface.

The second way takes into account more the behaviour of α along the involved v-orbits and its dynamics. It also relies on a generical local situation.

We will at the end give a convergence theorem which mixes both ways and we will show that the limit curves are to be equipped with a certain parametrization and live in a certain part of M (Theorem 5; Theorem 8).

The first approach is the following:

We assume:

$$(\mathcal{A}_1)\quad \begin{cases} (A_2) \text{ and } (A_3) \text{ are satisfied} \\ \text{Let } \theta_s^{2k} \text{ be the one-parameter groups of } \xi_k = D\psi^{2k}(\xi) \quad (k \in \mathbb{Z}). \\ \text{We assume the following hypothesis of local equicontinuity on the } \theta_s^{2k}; \\ \exists \alpha_6, s_0 > 0 \text{ such that } \forall k \in \mathbb{Z} \text{ and } \forall s, |s| \leq s_0, \forall x,y \in M, \\ \qquad d(\theta_s^{2k}(x), \theta_s^{2k}(y)) \leq (1 + \alpha_6 |s|) d(x,y)). \end{cases}$$

We then have:

Theorem $\mathcal{A}_1$: Under $(\mathcal{A}_1)$, the curve (7.32) may be rearranged in a curve $\hat{x}_\varepsilon$, $\varepsilon > 0$ given arbitrary, made up with pieces $[\hat{x}_0^-, \hat{x}_0^+]$, $[\hat{x}_0^+, \hat{x}_1^-]$, $[\hat{x}_1^-, \hat{x}_1^+], \ldots,$ $[\hat{x}_{n_0-1}^-, \hat{x}_{n_0-1}^+]$, $[\hat{x}_{n_0-1}^+, \hat{x}_{n_0}^-]$. On $[\hat{x}_i^-, \hat{x}_i^+]$, the rearranged curve is tangent to ξ during the time $a\delta t_i$. On $[\hat{x}_{i-1}^+, \hat{x}_i^-]$, the curve is tangent to v and we have:

$$\hat{x}_i^- = \psi^{k_i'}(\hat{x}_{i-1}^+); \quad k_i' \text{ even}; \quad |\lambda_{k_i'}(\hat{x}_{i-1}^+) - 1| < \varepsilon$$

$$d(\hat{x}_{n_0}^-, \hat{x}_0^-) < \varepsilon.$$

Moreover, we have:

$$d(\hat{x}_i^\pm, x_i^\pm) \leq K_2 \left(\sum_{k=0}^{i} \frac{\bar{j}_k + 1}{\sqrt{\omega}} + \delta t_k^- + \sqrt{\alpha_1}\, \delta t_k \right) \to 0$$

uniformly on i, where $x_i^\pm$ are the points of (7.32).

The second approach of the "rearrangement" is the following:

We assume:

$(\mathcal{A}_2)(x)$ $\begin{cases} (A_2) \text{ and } (A_3) \text{ are satisfied.} \\ \text{Moreover, we make a local hypothesis: for any sequence of points} \\ \text{of type } x_i^+ \text{ converging to } x \in M,\ x \text{ satisfies:} \\ \exists\, \varepsilon_7 > 0 \text{ so that the equation} \\ \qquad |\lambda_{2k}(x) - 1| < \varepsilon_7;\quad k \in \mathbb{Z} \qquad (7.33) \\ \text{has only a finite number of solutions } q_0, \ldots, q_{r_0}. \end{cases}$

This set of hypotheses will be denoted $(\mathcal{A}_2)(x)$. These results we prove under $(\mathcal{A}_2)(x)$ are somewhat better than the ones proved under $(\mathcal{A}_1)$. In particular, we clear up some properties of the characteristic exponents k_i'. So, it seems interesting to extend slightly hypothesis $(\mathcal{A}_2)(x)$ as follows, so that it covers the fiber cases:

$(\mathcal{A}_2)(x)$ $\begin{cases} \text{when } v \text{ is the vector field tangent to an } S^1\text{-fiber bundle } M, \text{ with} \\ \psi^{2m} = Id,\ m \in N, \text{ we modify } (7.33): \\ \qquad |\lambda_{2k}(x) - 1| < \varepsilon_7;\quad k \in \mathbb{Z}_m = \mathbb{Z}/m\mathbb{Z} \qquad (7.33)' \\ \text{has only a finite number of solutions } q_0, \ldots, q_{r_0}. \end{cases}$

Given a vector field in $\ker \alpha$, the previous hypothesis is natural: the equation $\lambda_{2k}(x) = 1$ defines, generically on the contact form α (i.e. on the contact forms of type $\lambda\alpha_0$; $\lambda \in C^\infty(M, \mathbb{R}^+)$) a hypersurface Γ_k, if k is not zero. The intersection of three such hypersurfaces is generically empty. Hence, the equation $\lambda_{2k}(x) = 1$, for a given x, will be satisfied for at most three distinct non zero k's.

In case v is not the vector field tangent to the fibers of an S^1-bundle, condition (7.33) is somewhat more constraining. It implies that 1 is not an accumulation point for the λ_{2k} when k runs in $\mathbb{Z}$.

Assume that $(\mathcal{A}_2)(x)$ is satisfied.

Then we have:

Theorem $(\mathcal{A}_2)(x)$: Under $(\mathcal{A}_2)(x)$, there exists δ positive so that for j satisfying:

$$\sum_{r=1}^{j-1} \delta t_{i+r} < \delta, \tag{7.34}$$

the characteristic summands:

$$k'_{i+1};\ k'_{i+1} + k'_{i+2};\ \ldots\ ;\ k'_{i+1} + \ldots + k'_{i+k};\ \ldots\ ;\ k'_{i+1} + \ldots + k'_{i+j}$$

are in the set $\{2q_0, \ldots, 2q_{r_0}\}$ (modulo m in the case of fiber bundles in S^1).

The curve (7.32) can be rearranged from x_i^+ to $\bar{x}_{i+j}^+$ according to the process of theorem $(\mathcal{A}_1)$ with similar estimates where the constants depend only on $q_0, \ldots, q_{r_0}$.

In particular, if j_0 is the Sup of the j's satisfying (7.34), the curve can be rearranged from x_i^+ to $\bar{x}_{i+j_0}^+$, i.e. it can be rearranged on a time Δt such that:

$$\Delta t \geq \delta t_{i+1} + \ldots + \delta t_{i+j_0} \geq \inf(\delta, \tfrac{1}{2});$$

δ being a constant not depending on x, $q_0, \ldots, q_{r_0}$, but only on ε_7.

Once the curve has been straightened up, we seek now to understand, under hypothesis $(\mathcal{A}_2)(x)$, what happens locally. For this purpose, we have the following definition:

Definition 8: The situation is said to be generic at x if:

1) the equations $\lambda_{2k}(x) = 1$ have at most one non zero solution k_0;
2) on a neighbourhood of x, the equation $\lambda_{2k_0}(y) = 1$ defines a hypersurface Σ, to which x belongs, if a solution to this equation exists;
3) $\xi(x)$ is non tangent to Σ at x and $\xi(\psi^{2k_0}(x))$ is non tangent to

$\psi^{2k_0}(\Sigma)$ at $\psi^{2k_0}(x)$.

It can be proved that such a situation is indeed generical in the following sense; with perturbing α in a neighbourhood of x, in the contact forms $\lambda\alpha$, we may realize the situation of Definition 8, with a λ is close to 1, in the C^∞ sense, as wanted.

We then have the following theorem:

Theorem 7: Let x_i^+, corresponding to a non zero k'_{i+1} and converging to x which satisfies $(\mathcal{A}_2)(x)$, be such that the situation is generic at x.

Then there exists a non zero k_0 such that $\lambda_{2k_0}(x) = 1$ and

1 $\exists\, \delta_1 > 0$ such that for any j satisfying

$$\sum_{r=1}^{j-1} \delta t_{i+r} < \delta_1, \tag{7.35}$$

we have:

$$k'_{i+1} = 2k_0;\ k'_{i+2} = -2k_0, \dots, k'_{i+k} = (-1)^{k+1}.2k_0, \dots, k'_{i+j} = (-1)^{j+1}.2k_0$$

(modulo m in the case where we are dealing with S^1-fiber bundles).

2 Let j_0 be the supremum of the j's satisfying (7.35). We have:

$$\sum_{r=1}^{j_0-1} \delta t_{i+r} \to 0 \quad \text{when } \varepsilon_1 \to 0 \text{ and } \omega \to +\infty.$$

3 The straightened curve $\hat{x}$, after having possibly oscillated from r = 1 to $j_0 - 1$ around a piece tangent to v of type $[x, \psi^{2k_0}(x)]$ (forwards and backwards), takes the ξ-direction. Thus, in the geometric sense of convergence, the curve $\hat{x}$ converges to one of the following curves, in a neighbourhood of x:

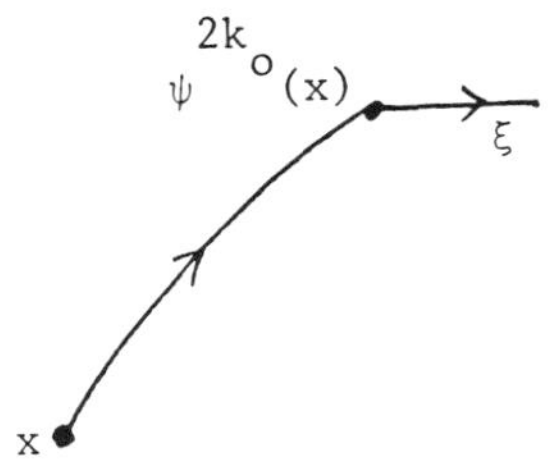

I

limit movement:

the piece $[x, \psi^{2k_0}(x)]$ can be described infinitely many times backwards and forwards "before a final movement" where it is described from x to $\psi^{2k_o}(x)$. Thereafter, the curve is tangent to ξ during a strictly positive time.

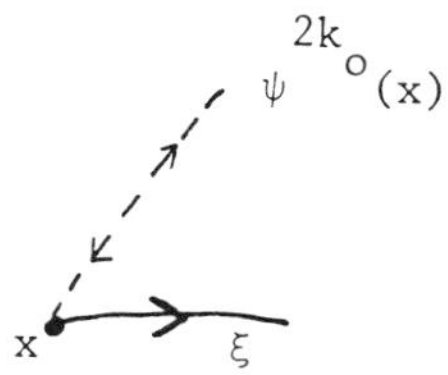

II

limit movement:

the piece $[x, \psi^{2k_0}(x)]$ can be described infinitely many times. The resulting movement is trivial: we come back from x to x. Thereafter, the curve is tangent to ξ during a strictly positive time.

(See Corollary 2 of Theorem 8 to discriminate between I and II.)

<u>Remark</u>: The hypothesis $k'_{i+1} \neq 0$ is not constraining. Indeed, when $k'_{i+1} = 0$, there are no pointwise oscillations along v, starting at x_i^+; hence the straightened up curve $\hat{x}$ is tangent to ξ in a neighbourhood of x_i^+ and x_i^+ is not a remarkable point.

For the oscillation phenomenon, we have the following drawing which illustrates what happens in case I and II of Theorem 7.

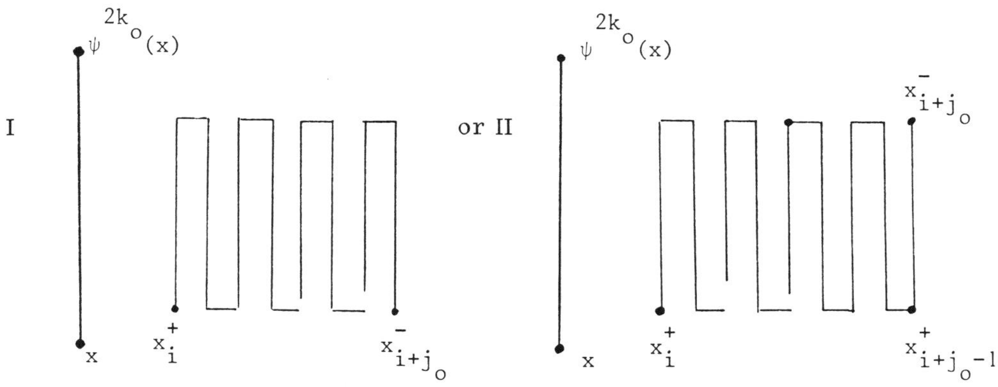
I
$\psi^{2k_o}(x)$
x
x_i^+
$x_{i+j_o}^-$
or II
$\psi^{2k_o}(x)$
x
x_i^+
$x_{i+j_o}^-$
$x_{i+j_o-1}^+$

8 Estimates

Lemma 5: Let $\varepsilon_0 > 0$ be given. Let

$$A = \{t \in S^1 \mid \frac{b^2}{\omega}(t) \geq \varepsilon_0 \text{ or } \frac{|\dot{b}|}{\omega}(t) \geq \varepsilon_0\}.$$

Then

$$\mu(A) \xrightarrow[\omega \to +\infty]{} 0 \quad (\mu \text{ is the Lebesgue measure of } A).$$

Lemma 5 is a "concentration" lemma, which tells us that b^2/ω and $|\dot{b}|/\omega$ are going to zero almost everywhere.

Lemma 6: Let $\alpha_1 > 0$ and $s \in [0,1]$ be given such that:

$$\frac{|\dot{b}|}{\omega}(s) \geq \sqrt{\alpha_1} > 0, \quad \frac{b^2}{\omega}(s) \leq \frac{\varphi(\alpha_1)}{2M}, \tag{8.1}$$

where φ is a continuous and increasing function from $\mathbb{R}^+$ to $\mathbb{R}^{+\star}$ such that $\varphi(0) = 0$ and M is a positive constant larger than 1. Let $I(s)$ be the maximal time interval containing s such that:

$$\forall\, t \in I(s) \qquad \frac{b^2}{\omega}(t) \leq \frac{\varphi(\alpha_1)}{M}. \tag{8.2}$$

We then have:

$$\forall\, t \in I(s) \quad (I - \varepsilon(\omega, M))\,|\dot{b}(s)| \leq |\dot{b}(t)| \leq (1 + \varepsilon(\omega, M))\,|\dot{b}(s)| \tag{8.3}$$

where $\varepsilon(\omega, M) \to 0$ when M and $\omega \to +\infty$ (α_1 fixed).

Proof of Lemma 6: Suppose that $\frac{\dot{b}}{\omega}(s) \geq \alpha_1 > 0$ (the other case is similar). Let $J(s)$ be the maximal time interval contained in $I(s)$ and containing s such that:

$$\forall\ t \in J(s) \qquad \dot{b}(t) > 0 \ . \tag{8.4}$$

Let also

$$\beta_s = \operatorname*{Sup}_{t\in J(s)} \frac{|\dot{b}|}{\omega}\ , \quad \gamma_s = \operatorname*{Inf}_{t\in J(s)} \frac{|\dot{b}|}{\omega}\ , \tag{8.5}$$

b is periodic and satisfies (5.38).

$$\ddot{b} + b\left(-\omega a + \frac{b^2}{2} - \frac{\int_0^1 b^2}{2}\right) + a^2 b\tau - ab^2\bar{\mu}_\xi + b\dot{b}\bar{\mu} = 0.$$

Hence

$$\tfrac{1}{2}\frac{d}{dt}\frac{\dot{b}^2}{\omega^2} + \tfrac{1}{2}\frac{d}{dt}\left(-a + \frac{b^2}{2\omega} - \frac{\int_0^1 b^2}{2\omega}\right)^2 + \left(\left[\frac{a^2\tau}{\omega} - \frac{ab\bar{\mu}_\xi}{\omega}\right]b + \bar{\mu}\frac{\dot{b}}{\omega}b\right)\frac{\dot{b}}{\omega} = 0. \tag{8.6}$$

Let $t \in J(s)$ be given. We integrate (8.6) between t and s:

$$\tfrac{1}{2}\left\{\left(\frac{\dot{b}^2}{\omega^2}(t) - \frac{\dot{b}^2}{\omega^2}(s)\right) + \left(-a + \frac{b^2}{2\omega} - \frac{\int_0^1 b^2}{2\omega}\right)^2(t)\right. \tag{8.7}$$

$$\left. - \left(-a + \frac{b^2}{2\omega} - \frac{\int_0^1 b^2}{2\omega}\right)^2(s)\right\} + \int_s^t \left(\frac{a^2\tau}{\omega} - \frac{ab\bar{\mu}_\xi}{\omega}\right)\frac{\dot{b}}{\omega}\,dx$$

$$+ \int_s^t \bar{\mu}\frac{\dot{b}}{\omega}\frac{b}{\omega}\dot{b}\,dx = 0.$$

We estimate:

$$\left|\int_s^t \left[\frac{a^2\tau}{\omega} - \frac{ab\bar{\mu}_\xi}{\omega}\right]\dot{b}\,dx\right| = (1). \tag{8.8}$$

As $\dot{b} > 0$ on $J(s)$ and $\frac{|b|}{\sqrt{\omega}} \leq \sqrt{\frac{\varphi(\alpha_1)}{M}}$ ((8.2)), we have:

$$(1) \leq C\left[\frac{a^2}{\omega}\left|\int_s^t \dot{b}\,dx\right| + a\sqrt{\frac{\varphi(\alpha_1)}{M}}\frac{1}{\sqrt{\omega}}\left|\int_s^t \dot{b}\,dx\right|\right], \tag{8.9}$$

hence

$$(1) \leq C'\left[\frac{1}{\sqrt{\omega}} + \sqrt{\frac{\varphi(\alpha_1)}{M}}\right]\frac{|b(t) - b(s)|}{\sqrt{\omega}}\ ,$$

hence

$$(1) \leq C_1 \left(\frac{1}{\sqrt{\omega}} + \sqrt{\frac{\varphi(\alpha_1)}{M}} \right) \sqrt{\frac{\varphi(\alpha_1)}{M}}, \text{ using (8.2) again.} \tag{8.10}$$

In the same way, we derive:

$$\left(a - \frac{b^2}{2\omega} + \frac{\int_0^1 b^2}{2\omega} \right)^2 (t) - \left(a - \frac{b^2}{2\omega} + \frac{\int_0^1 b^2}{2\omega} \right)^2 (s) = (2). \tag{8.11}$$

$$(2) \leq \operatorname*{Sup}_{J(s)} \left(\left| 2a - \frac{b^2}{\omega} + \frac{\int_0^1 b^2}{\omega} \right| \right) \left| \frac{b^2}{2\omega}(t) - \frac{b^2}{2\omega}(s) \right|,$$

hence

$$(2) \leq C_2 \frac{\varphi(\alpha_1)}{M} \left(1 + \frac{\varphi(\alpha_1)}{M} \right), \tag{8.12}$$

where we used (8.2), the fact that a is bounded and that $\int_0^1 b^2/\omega$ goes to zero. At last, we have:

$$\left| \int_0^t \bar{\mu} \frac{\dot{b}}{\omega} \frac{b}{\omega} \dot{b} \, dx \right| \leq C \operatorname*{Sup}_{J(s)} \frac{|\dot{b}|}{\omega} \sqrt{\frac{\varphi(\alpha_1)}{M}} \left| \int_s^t \dot{b} \, dx \right| \times \frac{1}{\sqrt{\omega}} \tag{8.13}$$

hence

$$\left| \int_0^t \bar{\mu} \frac{\dot{b}}{\omega} \frac{b}{\omega} \dot{b} \, dx \right| \leq C_3 \beta_s \frac{\varphi(\alpha_1)}{M}. \tag{8.14}$$

Inequalities (8.10), (8.12) and (8.14) imply, via (8.7):

$$\frac{1}{2} \left| \frac{\dot{b}^2}{\omega^2}(t) - \frac{\dot{b}^2}{\omega^2}(s) \right| \leq C_1 \sqrt{\frac{\varphi(\alpha_1)}{M}} \left(\frac{1}{\sqrt{\omega}} + \sqrt{\frac{\varphi(\alpha_1)}{M}} \right) \tag{8.15}$$

$$+ C_2 \frac{\varphi(\alpha_1)}{M} \left(1 + \frac{\varphi(\alpha_1)}{M} \right) + C_3 \beta_s \frac{\varphi(\alpha_1)}{M},$$

$\forall\, t \in J(s)$.

We deduce from (8.15):

$$\forall\, t \text{ and } t_1 \in J(s): \tag{8.16}$$

$$\left|\frac{\dot{b}^2}{\omega^2}(t) - \frac{\dot{b}^2}{\omega^2}(t_1)\right| \leq C_1\sqrt{\frac{\varphi(\alpha_1)}{M}}\left(\frac{1}{\sqrt{\omega}} + \sqrt{\frac{\varphi(\alpha_1)}{M}}\right) + C_2\frac{\varphi(\alpha_1)}{M}\left(1 + \frac{\varphi(\alpha_1)}{M}\right)$$

$$+ C_3\beta_s\frac{\varphi(\alpha_1)}{M} .$$

Hence

$$|\beta_s^2 - \gamma_s^2| \leq C_1\sqrt{\frac{\varphi(\alpha_1)}{M}}\left(\frac{1}{\sqrt{\omega}} + \sqrt{\frac{\varphi(\alpha_1)}{M}}\right) + C_2\frac{\varphi(\alpha_1)}{M}\left(1 + \frac{\varphi(\alpha_1)}{M}\right)$$

$$+ C_3\beta_s\frac{\varphi(\alpha_1)}{M} . \tag{8.17}$$

On the other hand:

$$\beta_s^2 \geq \frac{\dot{b}^2}{\omega^2}(s) \geq \alpha_1 . \tag{8.18}$$

From (8.17) we deduce:

$$\gamma_s^2 \geq \frac{\beta_s^2}{2} - \frac{C_3^2}{4}\frac{\varphi(\alpha_1)^2}{M^2} - C_1\sqrt{\frac{\varphi(\alpha_1)}{M}}\left(\frac{1}{\sqrt{\omega}} + \sqrt{\frac{\varphi(\alpha_1)}{M}}\right) \tag{8.19}$$

$$- C_2\frac{\varphi(\alpha_1)}{M}\left(1 + \frac{\varphi(\alpha_1)}{M}\right) ,$$

hence:

$$\gamma_s^2 \geq \frac{\alpha_1^2}{2} - \frac{C_3^2}{4}\frac{\varphi(\alpha_1)^2}{M^2} - C_1\sqrt{\frac{\varphi(\alpha_1)}{M}}\left(\frac{1}{\sqrt{\omega}} + \sqrt{\frac{\varphi(\alpha_1)}{M}}\right) \tag{8.20}$$

$$- C_2\frac{\varphi(\alpha_1)}{M}\left(1 + \frac{\varphi(\alpha_1)}{M}\right) \geq \frac{\alpha_1^2}{2} - \frac{C_3^2}{4}\varphi(\alpha_1)^2 - C_1\sqrt{\varphi(\alpha_1)}\left(\frac{1}{\sqrt{\omega}} + \sqrt{\varphi(\alpha_1)}\right)$$

$$- C_2\varphi(\alpha_1)(1 + \varphi(\alpha_1)) ,$$

using the fact that $M \geq 1$.

As ω goes to $+\infty$, one can easily choose a function $\varphi(\alpha_1)$ such that the second member of (8.20) is a function larger than $\alpha_1^2/4$. Then:

$$\gamma_s^2 \geq \frac{\alpha_1^2}{4} > 0. \tag{8.21}$$

Inequality (8.21) implies that $J(s) = I(s)$ as $\frac{\dot{b}^2}{\omega^2} \geq \frac{\alpha_1^2}{4}$ on all of $J(s)$. Hence γ_s and β_s are respectively the minimum and the maximum of $|b|/\omega$ over $I(s)$. Turning back to (8.17), we have:

$$\left|1-\frac{\gamma_s^2}{\beta_s^2}\right| \leq C_3 \frac{\varphi(\alpha_1)}{M\beta_s} + \frac{1}{\beta_s^2}\left[C_1\sqrt{\frac{\varphi(\alpha_1)}{M}}\left(\frac{1}{\sqrt{\omega}}+\sqrt{\frac{\varphi(\alpha_1)}{M}}\right)+C_2\frac{\varphi(\alpha_1)}{M}\left(1+\frac{\varphi(\alpha_1)}{M}\right)\right]$$

$$\leq C_3 \frac{\varphi(\alpha_1)}{M\alpha_1} + \frac{1}{\alpha_1^2}\left[C_1\sqrt{\frac{\varphi(\alpha_1)}{M}}\left(\frac{1}{\sqrt{\omega}}+\sqrt{\frac{\varphi(\alpha_1)}{M}}\right)+C_2\frac{\varphi(\alpha_1)}{M}\left(1+\frac{\varphi(\alpha_1)}{M}\right)\right] \quad (8.22)$$

and (8.22) clearly implies (8.3).

The choice of φ continuous and increasing is possible by (8.20). ■

Proof of Lemma 5: As $\int_0^1 b^2/\omega \underset{\omega\to+\infty}{\to} 0$, the proof reduces to the fact that for a given $\varepsilon_1 > 0$, the Lebesgue measure of $A_1 = \{t\in S^1 \,|\, \frac{b^2}{\omega}(t) \leq \varepsilon_1$ and $\frac{|\dot{b}|}{\omega}(t) \geq \varepsilon_0\}$ goes to zero. This ε_1 we can choose depending on ε_0.

ε_0 being given, we choose α_1:

$$\sqrt{\alpha_1} = \varepsilon_0 \quad (8.23)$$

and choose $\varepsilon_1 = \frac{\varphi(\alpha_1)}{2M}$ with M large enough (as well as ω) to ensure:

$$\varepsilon(\omega, M) \leq \tfrac{1}{2} \quad (\varepsilon(\omega, M) \text{ is defined in } (8.3)). \quad (8.24)$$

Let $s \in A_1$.

We have:

$$\frac{|\dot{b}|}{\omega}(s) \geq \varepsilon_0 = \sqrt{\alpha_1}, \quad \frac{b^2}{\omega}(s) \leq \varepsilon_1 = \frac{\varphi(\alpha_1)}{2M}. \quad (8.25)$$

Lemma 6 provides us with an interval $I(s)$ on which (8.3) holds, i.e.

$$\forall\, t \in I(s) \qquad \frac{b^2(t)}{\omega} \leq 2\varepsilon_1 = \frac{\varphi(\alpha_1)}{M}. \quad (8.26)$$

Using (8.26) and the compactness of A_1, we can find a covering of A_1 by a finite number (depending on ω) of intervals of type $I(s)$, $I(s_1), \dots, I(s_n)$. These intervals are disjoint, by the maximality property of each of them (see Lemma 6). Then:

$$\mu(A_1) \leq \sum_{i=1}^{n} \int_{I(s_i)} dt \tag{8.27}$$

and all $I(s_i)$ are disjoint.

We estimate $\int_{I(s_i)} dt$.

From (8.24), we have on $I(s_i)$:

$$\tfrac{1}{2}|\dot{b}(s_i)| \leq |\dot{b}(t)| \leq \frac{3}{2}|\dot{b}(s_i)| . \tag{8.28}$$

In particular, $\dot{b}(t)$ is never zero and:

$$\int_{I(s_i)} dt \leq \left| \int_{I(s_i)} \frac{dt}{db}\, db \right| \leq \frac{2}{|\dot{b}(s_i)|} \left| \int_{I(s_i)} db \right| . \tag{8.29}$$

On the other hand, if $I(s_i) = [s_i^-, s_i^+]$,

$$\int_{I(s_i)} \frac{b^2}{\omega}\, dt \geq \left| \int_{I(s_i)} \frac{b^2}{\omega} \frac{dt}{db}\, db \right| \geq \frac{2}{3|\dot{b}(s_i)|} \left| \int_{I(s_i)} b^2\, db \right| . \frac{1}{\omega} . \tag{8.30}$$

Hence

$$\int_{I(s_i)} \frac{b^2}{\omega}\, dt \geq \frac{2}{3|\dot{b}(s_i)|} \left| \frac{b^3}{3\omega}(s_i^+) - \frac{b^3}{3\omega}(s_i^-) \right| \tag{8.31}$$

$$\geq \frac{2}{3|\dot{b}(s_i)|} \left| \frac{b^2}{3\omega}(s_i^+) + \frac{b^2}{3\omega}(s_i^-) \right| \left| \int_{I(s_i)} db \right| ,$$

hence, using (8.29), we derive

$$\int_{I(s_i)} \frac{b^2}{\omega}\, dt \geq \frac{1}{9}\left(\frac{b^2}{\omega}(s_i^+) + \frac{b^2}{\omega}(s_i^-) \right) \int_{I(s_i)} dt . \tag{8.32}$$

We then distinguish two cases:

1st case: $I(s_i) \neq [0,1]\ \forall\ i$.

In this case, $\frac{b^2}{\omega}(s_i^+) = \frac{b^2}{\omega}(s_i^-) = 2\varepsilon_1$. Hence:

$$\int_{I(s_i)} dt \leq \frac{9}{\varepsilon_1} \int_{I(s_i)} \frac{b^2}{\omega} dt, \tag{8.33}$$

From the fact the $I(s_i)$ are disjoint, we deduce:

$$\mu(A_1) \leq \sum_{i=1}^{n} \int_{I(s_i)} dt \leq \frac{9}{\varepsilon_1} \int_0^1 \frac{b^2}{\omega} dt \xrightarrow[\omega \to +\infty]{} 0 \quad (\varepsilon_1 \text{ given}). \tag{8.34}$$

2nd case: $n = 1$ and $I(s) = [0,1]$ for a given s.

In this case, $\forall\ t \in [0,1]$, we have, using (8.28):

$$|\dot{b}(t)| \geq \tfrac{1}{2}|\dot{b}(s)| \geq \tfrac{1}{2}\varepsilon_0 \omega, \tag{8.35}$$

but (8.35) is impossible, as $\dot{b}$ has to be zero somewhere (indeed b is periodic); which proves Lemma 5.

Points in A^c are concentration points. Indeed, when ε_0 goes to zero, $\int_0^1 a\, dt = a$ concentrates exactly on A^c. In what follows, we will analyse the situation near such points.

Lemma 7 (convexity lemma): There exists a constant $\alpha_0 > 0$ such that the function b^2 is convex in every point s such that:

$$\frac{|\dot{b}|}{\omega}(s) < \alpha_0, \quad \frac{b^2}{2\omega}(s) < \alpha_0 \qquad (\omega \geq \omega_0). \tag{8.36}$$

Proof of Lemma 7: We have:

$$\ddot{b} + b\left(-\omega a + \frac{b^2}{2} - \int_0^1 \frac{b^2}{2} + a\tau^2 - ab\bar{\mu}_\xi + \bar{\mu}\dot{b}\right) = 0,$$

hence:

$$\ddot{b}\, b = \omega b^2 \left(a - \frac{b^2}{2\omega} + \frac{\int_0^1 b^2}{2\omega} - \frac{a^2\tau}{\omega} + \frac{a\bar{\mu}_\xi b}{\omega} - \bar{\mu}\frac{\dot{b}}{\omega}\right) \tag{8.37}$$

hence

$$\ddot{b}b(s) \geq \omega b^2 \left(a - \alpha_0 - \frac{C_1 a^2}{\omega} - \frac{C_2 a}{\sqrt{\omega}} \sqrt{\alpha_0} - C_3 \alpha_0 \right), \tag{8.38}$$

where C_1, C_2, C_3 are positive constants with upper bound $|\tau|$, $|\bar{\mu}_\xi|$, $|\bar{\mu}|$. Our hypothesis is that a is larger than a strictly positive a_0 and ω goes to $+\infty$. We easily derive from this, when ω is large, the existence of α_0 such that $\ddot{b}b(s)$ is positive as soon as (8.36) is satisfied.

But then

$$(\ddot{b^2})(s) = 2\dot{b}^2 + 2b\ddot{b}(s) > 0, \tag{8.39}$$

which proves Lemma 7.

<u>Lemma 8</u>: Let $\overline{M} > 1$ and ω_1 be chosen such that $\varepsilon(\omega, \overline{M}) \leq \frac{1}{2}$ for $\omega \geq \omega_1$. Let $\varepsilon_1 > 0$ be small enough such that:

$$\varepsilon_1 < \frac{\alpha_0}{2}; \quad \sqrt{\varphi^{-1}(4\overline{M}\varepsilon_1)} < \alpha_0. \tag{8.40}$$

Let $\alpha_1 = \varphi^{-1}(4\overline{M}\varepsilon_1)$. Let $0 < \varepsilon_2 < \inf(\alpha_0, \frac{\sqrt{\alpha_1}}{3})$ be given. Let $s \in [0,1]$ be such that:

$$\frac{b^2(s)}{\omega} \leq \varepsilon_1 = \frac{\varphi(\alpha_1)}{4\overline{M}} \quad \text{and} \quad \frac{|\dot{b}(s)|}{\omega} < \varepsilon_2. \tag{8.41}$$

Let $I_1(s)$ be the maximal time interval containing s such that $\forall\, t \in I_1(s)$,

$$\frac{b^2(t)}{\omega} \leq 2\varepsilon_1.$$

Under these hypotheses, the function b^2 is convex over all $I_1(s)$. Moreover if $I_1(s) = [t^-(s), t^+(s)]$, we have:

$$\frac{b^2}{\omega}(t^+) = 2\varepsilon_1; \quad \frac{|\dot{b}|}{\omega}(t^+) < \sqrt{\alpha_1}; \quad \frac{b^2}{\omega}(t^-) = 2\varepsilon_1; \quad \frac{|\dot{b}|}{\omega}(t^-) < \sqrt{\alpha_1}; \tag{8.42}$$

$$\frac{b^2}{\omega}(t) \leq 2\varepsilon_1; \quad \frac{|\dot{b}|}{\omega}(t) < \sqrt{\alpha_1}, \quad \forall\, t \in I_1(s).$$

Finally, there exists a unique point $t_1(s)$ in $I_1(s)$ such that:

$$\text{either } b(t_1(s)) = 0 \text{ or } \dot{b}(t_1(s)) = 0. \tag{8.43}$$

Proof of Lemma 8: We first prove that $\frac{b^2}{\omega}(t) < \alpha_0$ and $\frac{|\dot{b}|}{\omega}(t) < \sqrt{\alpha_1}$, $\forall\, t \in I_1(s)$.

Indeed, over $I_1(s)$, $\frac{b^2}{\omega}(t) \leq 2\varepsilon_1 < \alpha_0$ by (8.40); which proves the first inequality.

For the second one, we argue by contradiction.

Let $t^1 \in I_1(s)$ be such that:

$$\frac{|\dot{b}|}{\omega}(t^1) \geq \sqrt{\alpha_1}. \tag{8.44}$$

As, on the other hand, we have

$$\frac{b^2}{\omega}(t^1) \leq 2\varepsilon_1 = \frac{\varphi(\alpha_1)}{2\overline{M}}, \tag{8.45}$$

we may apply Lemma 6 and, as long as $\frac{b^2(t)}{\omega} \leq \frac{\varphi(\alpha_1)}{\overline{M}}$, (8.3) holds.

As $\frac{b^2(t)}{\omega} \leq 2\varepsilon_1 = \frac{\varphi(\alpha_1)}{2\overline{M}} < \frac{\varphi(\alpha_1)}{\overline{M}}$ over $I_1(s)$, (8.3) holds on this interval. In particular, using the fact that $\varepsilon(\omega, \overline{M}) \leq \frac{1}{2}$ we have:

$$\tfrac{1}{2}|\dot{b}(t^1)| \leq |\dot{b}(s)| \leq \frac{3}{2}|\dot{b}(t^1)|, \tag{8.46}$$

hence

$$\frac{|\dot{b}(t^1)|}{\omega} \leq 2\frac{|\dot{b}(s)|}{\omega} \leq 2\varepsilon_2 < \frac{2}{3}\sqrt{\alpha_1} < \sqrt{\alpha_1}, \tag{8.47}$$

which contradicts (8.44).

Accordingly, on all of $I_1(s)$, we have:

$$\frac{b^2}{\omega}(t) < 2\varepsilon_1 < \alpha_0 \text{ and } \frac{|\dot{b}|}{\omega}(t) < \sqrt{\alpha_1} = \sqrt{\varphi^{-1}(4\overline{M}\varepsilon_1)} < \alpha_0 \tag{8.48}$$

by (8.40).

We then use Lemma 7 to deduce the convexity of b^2 over $I_1(s)$. Inequalities (8.42) are also deduced from (8.48). The equalities follow from the maximality property of $I_1(s)$.

The convexity of b^2 and the fact that $b^2(t^+) = b^2(t^-)$ imply the existence of a point $t_1(s)$ in $I_1(s)$ such that $b\dot{b}(t_1(s)) = 0$. The unicity of $t_1(s)$ is due to the strict convexity of b^2. ∎

Lemmas 5, 6, 7, 8 were technical lemmas.

We now start a more delicate process, which consists of straightening up our curve along the $I_1(s)$ intervals, given by Lemma 8, where there is concentration. The idea is as follows:

The curve $x(.)$ on these intervals looks like the following:

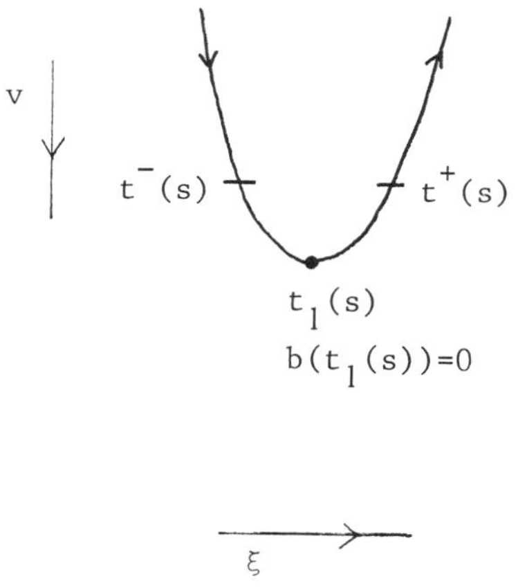

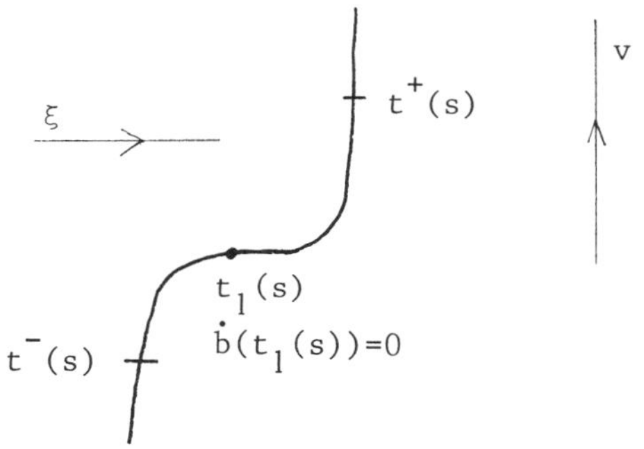

Indeed, on such an interval $[t^+(s), t^-(s)]$, b^2/ω is small. Hence, the curve has the "tendency to be tangent to ξ", in a heuristic sense which will be made more precise later on.

On the other hand, when t is larger than $t^+(s)$ or t is less than $t^-(s)$, but t is close enough to these values, $b^2/\omega(t)$ is larger than $2\varepsilon_1$. As ω goes to $+\infty$ and $\dot{x} = a\xi + bv$ with a bounded a, the curve is rather tangent to v.

Along these intervals $[t^+(s), t^-(s)]$, we will push the curve along v to straighten it up. This means we are going to do the following transformations:

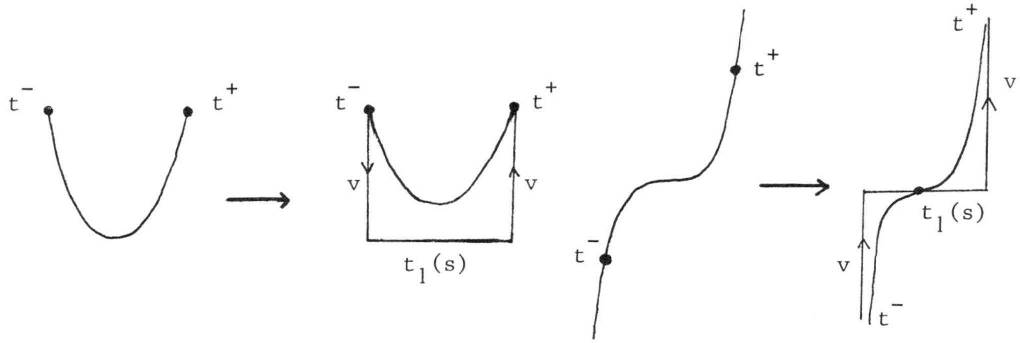

With this transformation, we introduce two pieces (the prior one and the posterior one) tangent to v and a piece between them $\bar{y}(.)$ which we will prove to be very close to a piece of ξ-orbit.

This will be the first step of our straightening-up process. We will then control the transport equation introduced in Section 5.3, along the pieces tangent to v.

We remember that $\phi(s,.)$ is the one-parameter group generated by v. Let

$$u(t) = -\left[\int_{t_1(s)}^{t} b(x)\,dx + \bar{c}(t_1(s))\right], \tag{3.49}$$

where $\bar{c}(t_1(s))$ is an appropriate constant we will choose later on.

We impose on $\bar{c}(t_1(s))$, for the time being, the following condition:

$$|\bar{c}(t_1(s))| \leq C\sqrt{\alpha_1}\ ; \tag{8.50}$$

C a given constant and α_1 defined as in Lemma 8.

We introduce the following definition:

<u>Definition 7</u>: We define the "straightened-up and incomplete" curve (on $I_1(s)$) to be the curve $y(t) = \phi(u(t), x(t))$ for t in $(t^-(s), t^+(s))$, where $x(.)$ is the critical curve to Z_ε.

We define the "complete straightened up" curve (on $I_1(s)$) to be the curve obtained from $y(.)$ when adding to this curve two v-pieces; the first one starting at $x(t^-(s))$ and tangent to v during the time $\int_{t_1}^{t^-(s)} b(x)\,dx + \bar{c}(t_1(s))$;

the second one starting at $x(t^+(s))$ and tangent to v during the time $\int_{t_1}^{t^+(s)} b(x)\,dx + \bar{c}(t_1(s))$.

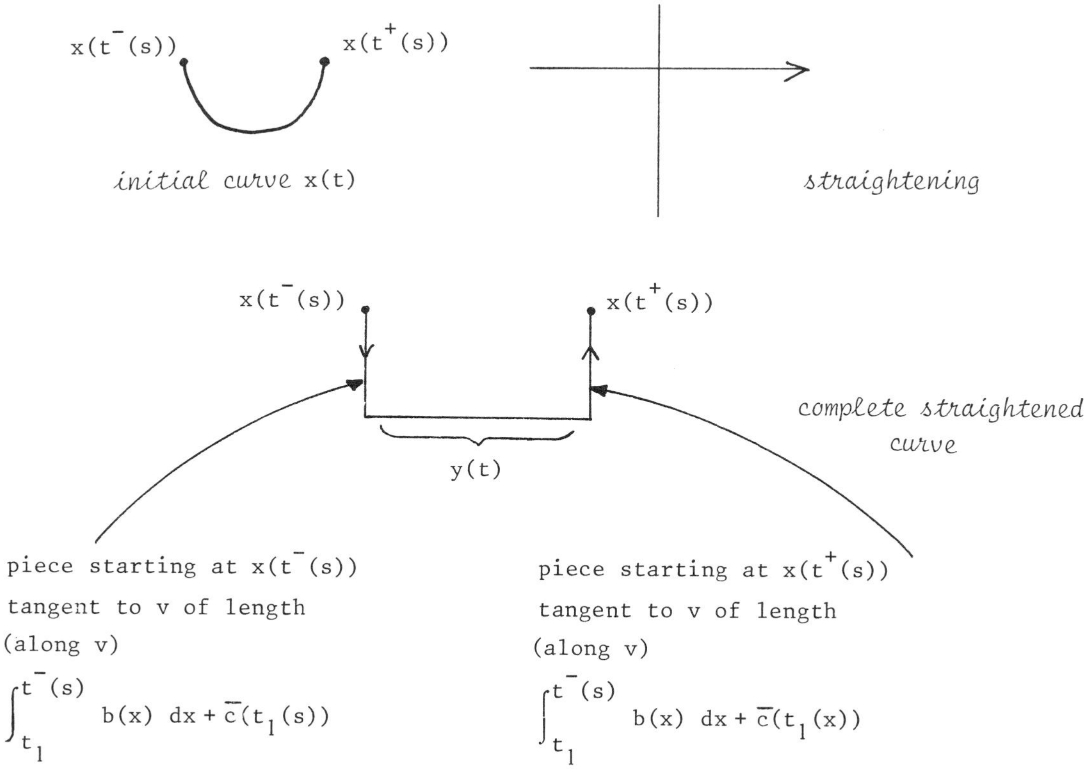

We will prove in a first step that the incomplete straightened curve $y(t)$ is almost tangent to ξ; and also that, if we replace it by the curve $z(t)$ starting at $y(t^-(s))$ and tangent to $a\xi$ on $[t^-(s), t^+(s)]$, the L^∞-distance of y to z remains small.

We remember that we have already chosen a distance d on M as well as a norm $\| \;\|$ on the tangent space.

Lemma 9: Let ε_1 be defined as in Lemma 8. There exist γ, K_1, K_2, $K_3 > 0$ depending only on a_0 $(0 < a_0 \leq a)$ and on the geometry such that if $0 < \varepsilon_1 < \gamma$, and $\omega \geq \omega_2$, we have:

$$u_s = \operatorname*{Sup}_{t \in I_1(s)} |u(t)| \leq K_1 \sqrt{\alpha_1} \tag{8.51}$$

(α_1 as in Lemma 8);

$$\Delta_s = \operatorname*{Sup}_{t \in I_1(s)} \|\dot{y}(t) - a\xi(y(t))\| \leq K_2 \sqrt{\alpha_1}; \tag{8.52}$$

$$\delta_s = \operatorname*{Sup}_{t \in I_1(s)} d(y(t), z(t)) \leq K_3 \sqrt{\alpha_1}\, \mu(I_1(s)) \tag{8.53}$$

$$= K_3 \sqrt{\alpha_1}(t^+(s) - t^-(s))$$

where $z(t)$ is defined by: $z(t^-(s)) = y(t^-(s))$; $\dot{z}(t) = a\xi(z(t))$, $t \in [t^-(s), t^+(s)]$;

ω_2 is uniform on all intervals of type $I_1(s)$.

Proof of Lemma 9: b satisfies:

$$\ddot{b} + b\left(-\omega a + \frac{b^2}{2} - \frac{\int_0^1 b^2}{2} + a^2\tau - ab\bar{\mu}_\xi + \bar{\mu}\dot{b}\right) = 0,$$

thus

$$\frac{\ddot{b}}{\omega a} = b\left(1 - \frac{b^2}{2\omega a} + \frac{\int_0^1 b^2}{2\omega a} + \frac{a\tau}{\omega} - \frac{ab\bar{\mu}_\xi}{\omega} + \frac{\bar{\mu}\dot{b}}{\omega a}\right). \tag{8.54}$$

On $I_1(s)$, one knows that $b^2/\omega > 2\varepsilon_1$, and $|\dot{b}|/\omega a < \sqrt{\alpha_1}/a$.

$\alpha_1 = \varphi^{-1}(4\bar{M}\varepsilon_1)$ goes to zero as ε_1 goes to zero.

On the other hand, a is larger than $a_0 > 0$ and $\int_0^1 b^2/\omega$ goes to zero.

This implies the existence of two constants γ, ω_2, such that, if $0 < \varepsilon_1 < \gamma$ and $\omega > \omega_2$, we have:

$$\tfrac{1}{2} \leq 1 - \frac{b^2}{2\omega a} + \frac{\int_0^1 b^2}{2\omega a} + \frac{a\tau}{\omega} - \frac{ab\bar{\mu}_\xi}{\omega} + \frac{\bar{\mu}\dot{b}}{\omega a} \leq \frac{3}{2} \tag{8.55}$$

$\forall\, t \in I_1(s)$.

The choice of ω_2 is uniform, as it does not depend on the convergence of $\int_0^1 b^2/\omega$ to zero.

On the other hand, integrating equation (8.54) between $t_1(s)$ and t, we obtain:

$$\frac{\dot{b}}{\omega a}(t) - \frac{\dot{b}}{\omega a}(t_1(s)) = \int_{t_1(s)}^{t} b\left(1 - \frac{b^2}{2\omega a} + \frac{\int_0^1 b^2}{2\omega a} + \frac{a\tau}{\omega} - \frac{ab\bar{\mu}}{\omega}\xi + \frac{\bar{\mu}\dot{b}}{\omega a}\right) dx. \quad (8.56)$$

We also know, by the choice of $t_1(s)$ as the unique point in $I_1(s)$ where $b\dot{b}$ is zero, that b has a given sign on $[t_1(s), t]$, for any t in $I_1(s)$. We then deduce, using (8.55)

$$\left|\int_{t_1(s)}^{t} b(x)\,dx\right| \leq 2\left|\frac{\dot{b}}{\omega a}(t) - \frac{\dot{b}}{\omega a}(t_1(s))\right| \quad (8.57)$$

hence, via (8.50)

$$|u(t)| \leq \left|\int_{t_1}^{t} b(x)\,dx\right| + |\bar{c}(t_1(s))| \leq \left(C + \frac{2}{a}\right)\sqrt{\alpha_1} \leq \left(C + \frac{2}{a_0}\right)\sqrt{\alpha_1}, \quad (8.58)$$

as $\frac{|\dot{b}|}{\omega a} < \frac{\sqrt{\alpha_1}}{a}$ on $I_1(s)$, which proves (8.51).

(8.52) and (8.53) are tied estimates we will prove together.

The length of the curve $z(.)$ is bounded by $Ca(t^+(s) - t^-(s))$, as $\dot{z} = a\xi(z(t))$. We can thus cover this curve by a finite number of open sets $U_0, \ldots, U_{n_0}$ where ξ reads as a constant vector through charts $\varphi_0, \ldots, \varphi_{n_0}$.

With these charts, the one-parameter group ϕ_s transforms on $\varphi_i(U_i)$ in $\psi_s^i = \varphi_i \circ \phi_s \circ \varphi_i^{-1}$. Of course, we do have $\psi_0^i = D\psi_0^i = \mathrm{Id}$ as $\phi_0 = D\phi_0 = \mathrm{Id}$. We may also suppose that $D\psi_s^i$ satisfies:

$$|D\psi_s^i - \mathrm{Id}| \leq k_i|s|. \quad (8.59)$$

Indeed, ψ_s^i is C^∞ with respect to s and $\psi_0 = D\psi_0 = \mathrm{Id}$. Similarly, we may assume there exist positive constants $\alpha_i, \beta_i, \alpha_i^1, \beta_i^1$, such that, if $|\ |_e$ is the euclidean norm on $\varphi_i(U_i)$:

$$\beta_i d(x,y) \leq |\varphi_i(x) - \varphi_i(y)|_e \leq \alpha_i d(x,y) \quad \text{if } x, y \in U_i; \quad (8.60)$$

$$\beta_i^1 \|u\| \leq |D\varphi_{i_x}(u)|_e \leq \alpha_i^1 \|u\|, \text{ where } x \in U_i \text{ and } u \text{ is} \quad (8.61)$$

tangent to M on x.

As the set of ξ-orbits of length upperbounded by T ($= Ca$, in our case) is compact, we may take n_0 to depend only on the manifold M and on T and not on the precise curve $z(.)$ we are dealing with; we may also take the constants k_i, α_i, β_i, α_i^1, β_i^1 as uniform constants on M. We thus denote them k, α, β, $\bar{\alpha}$, $\bar{\beta}$ and we thus have

$$|D\psi_s^i - Id| \leq k|s| \quad \text{on } \varphi_i(U_i) \quad \forall\, i; \tag{8.59}$$

$$\beta\, d(x,y) \leq |\varphi_i(x) - \varphi_i(y)|_e \leq \alpha\, d(x,y) \quad \forall\, x, y \in U_i \quad \forall\, i; \tag{8.60}$$

$$\bar{\beta}\, \|u\| \leq |D\varphi_x^i(u)|_e \leq \bar{\alpha}\, \|u\| \quad \forall\, x \in U_i,\ u \text{ tangent on} \tag{8.61}$$

x to M and $\forall\, i$.

The norm in (8.59) is the norm in the matrix spaces of $\mathbb{R}^3$ with respect to the euclidian norm.

We end these preliminaries by choosing points $z_0 = z(t^-(s))$, $z(t_1) = z_1, \ldots, z_i = z(t_i), \ldots, z_{n_0} = z(t^+(s))$ such that:

$$z_0 \in U_0;\ z_{n_0} \in U_{n_0};\ z_i \in U_{i-1} \cap U_i. \tag{8.62}$$

The piece of curve between z_i and z_{i+1} is contained in U_i. (8.63)
We will denote this piece of curve $[z_i, z_{i+1}]$.

As this piece of curve is compact, contained in U_i, there exists $\delta_i > 0$ such that a δ_i-neighbourhood V_{δ_i} of $[z_i, z_{i+1}]$ is still in U_i. δ_i also can be taken to be a uniform constant on M and will be simply denoted δ. We thus have:

$$V_\delta([z_i, z_{i+1}]) \in U_i \quad \forall\, i. \tag{8.64}$$

Hence, using (8.60), we see that if $y_i = \varphi_i(y)$ is such that:

$$y_i \in V_{\delta\beta}(\varphi_i([z_i, z_{i+1}])), \text{ then } y \in U_i \tag{8.65}$$

where $V_{\delta\beta}(\varphi_i([z_i, z_{i+1}]))$ is a neighbourhood of order $\delta_1 = \delta\beta$ of $\varphi_i([z_i, z_{i+1}])$ in the euclidian norm $|\ |_e$ on $\varphi_i(U_i)$.

After these preliminaries, we turn now to the proof of (8.52) and (8.53). For this proof, we assume $d(y(t_i), z_i) = d(y(t_i), z(t_i)) = d_i$ is given and is less than δ, so that $|\varphi_i(y(t_i)) - \varphi_i(z(t_i))| < \delta_1$. For a short time after t_i, the curve $y(t)$ will then remain in $V_\delta([z_i, z_{i+1}])$, hence in U_i. During this time, we show that $|\varphi_i(y(t)) - \varphi_i(z(t))|_e$ satisfies a differential equation. This differential equation implies, if d_i is small enough, that $\varphi_i(y(t))$ remains in $V_\delta([z_i, z_{i+1}])$ until the time t_{i+1} and provides then an estimate of d_{i+1} in function of d_i; hence allowing us to repeat the process with U_{i+1} and $V_\delta([z_{i+1}, z_{i+2}])$. Iterating, we arrive at d_{n_0} and derive global estimates.

Let us suppose thus that $d(y(t_i) z_i) = d(y(t_i), z(t_i)) = d_i$ is given such that:

$$d_i < \delta; \quad \alpha d_i < \delta_1, \tag{8.66}$$

and such that, using (8.60), we have:

$$|\varphi_i(y(t_i)) - \varphi_i(z(t_i))|_e < \delta_1. \tag{8.67}$$

Then $\varphi_i(y(t_i)) \in \varphi_i(U_i)$.
We set

$$\begin{aligned} \tilde{y}(t) &= \varphi_i(y(t)) \quad t \text{ in a neighbourhood of } t_i \\ \tilde{z}(t) &= \varphi_i(z(t)). \end{aligned} \tag{8.68}$$

By continuity $\tilde{y}(t) \in \varphi_i(U_i)$ for a small time after t_i.

We compute $\dot{\tilde{y}}(t) - \dot{\tilde{z}}(t)$.

We have:

$$y(t) = \phi(u(t), x(t)) \quad \text{and} \quad \dot{x}(t) = a\xi + bv.$$

Thus:

$$\dot{y}(t) = D\phi_{u(t)}(\dot{x}(t)) + \dot{u}(t)\, v \tag{8.69}$$

(ϕ_s is the one-parameter group generated by v).
Thus:

$$\dot{y}(t) = D\phi_{u(t)}(a\xi + bv) + \dot{u}(t)v = a\,D\phi_{u(t)}(v) + b\,D\phi_{u(t)}(v) - bv. \qquad (8.70)$$

Now, $D\phi_s(v) = v$ as v generates $\phi(s,.)$. Thus:

$$\dot{y}(t) = a\,D\phi_{u(t)}(\xi) + bv - bv = a\,D\phi_{u(t)}(\xi). \qquad (8.71)$$

As $\tilde{y}(t) = \varphi_i(y(t))$, $\dot{\tilde{y}}(t) = D\varphi_i(\dot{y}(t))$,

$$\dot{\tilde{y}}(t) = a\,D\varphi_i \circ D\phi_{u(t)}(\xi) = a\,D\varphi_i \circ D\phi_{u(t)} \circ D\varphi_i^{-1}(D\varphi_i(\xi)) \qquad (8.72)$$
$$= a\,D\psi^i_{u(t)}(D\varphi_i(\xi));$$

$D\varphi_i(\xi)$ is a constant vector in U_i, by the choice of U_i. (Indeed, on $\varphi_i(U_i)$ $D\varphi_i(\xi)$ is constant as ξ is seen as a constant vector modulo φ_i).
We denote this vector:

$$u_i = D\varphi_i(\xi). \qquad (8.73)$$

On the other hand:

$$\dot{\tilde{z}}(t) = D\varphi_i(\dot{z}(t)) = D\varphi_i(a\xi) = a\,D\varphi_i(\xi) = au_i. \qquad (8.74)$$

Thus:

$$\dot{\tilde{y}}(t) - \dot{\tilde{z}}(t) = a[D\psi^i_{u(t)}(u_i) - u_i] = a[D\psi^i_{u(t)} - \mathrm{Id}]\,u_i. \qquad (8.75)$$

We then use (8.59) and deduce:

$$|\dot{\tilde{y}}(t) - \dot{\tilde{z}}(t)|_e \leq a\,|D\psi^i_{u(t)} - \mathrm{Id}|\;|u_i|_e \leq ka\,|u(t)|\;|u_i|_e \qquad (8.76)$$
$$\leq \bar{\alpha}\,ka\,|u(t)|\;\|\xi\| \qquad \text{(by (8.61))}.$$

Thus:

$$|\dot{\tilde{y}}(t) - \dot{\tilde{z}}(t)|_e \leq \bar{\alpha}\,ka\,\|\xi\| \sup_{t\in[t^-(s),\,t^+(s)]} |u(t)| \leq k_1\,a\,u_s \qquad (8.77)$$

$$\leq k_1(aK_1)\sqrt{\alpha_1} \quad \text{by (8.51)}$$

as long as $|\tilde{y}(t) - \tilde{z}(t)|_e < \delta_1$.

Using (8.77), we see that we have:

$$|\tilde{y}(t) - \tilde{z}(t)|_e \leq a K_1' \sqrt{\alpha_1}[t - t_i] + |\tilde{y}(t_i) - \tilde{z}(t_i)|_e \quad (8.78)$$

Thus

$$|\tilde{y}(t) - \tilde{z}(t)|_e \leq a K_1' \sqrt{\alpha_1}[t - t_i] + \alpha d_i, \quad (8.79)$$

as by (8.60) we have:

$$|y(t_i) - z(t_i)|_e \leq \alpha\, d(y(t_i), z(t_i)) \leq \alpha\, d_i.$$

and using again (8.60), (8.79) implies:

$$d(y(t), z(t)) \leq \frac{1}{\beta} |\tilde{y}(t) - \tilde{z}(t)|_e \leq \frac{1}{\beta}(a K_1' \sqrt{\alpha_1}(t-t_i) + \alpha\, d_i). \quad (8.80)$$

We thus have:

$$d(y(t), z(t)) \leq \frac{1}{\beta}(a K_1' \sqrt{\alpha_1}(t - t_i) + \alpha\, d_i) \quad (8.81)$$

as long as

$$\begin{cases} |y(t) - z(t)|_e < \delta_1 \text{ and if } d_i \text{ satisfies (8.66), i.e.} \\ \qquad d_i < \delta, \quad \alpha d_i < \delta_1. \end{cases}$$

Now $|\tilde{y}(t) - \tilde{z}(t)|_e \leq \alpha d(y, z)$ by (8.60).

Thus, we have:

$$\begin{cases} d(y(t), z(t)) \leq \frac{1}{\beta}(a K_1' \sqrt{\alpha_1}(t - t_i) + \alpha\, d_i) & (8.81) \\ \text{as long as} \\ d(y(t), z(t)) < \frac{\delta_1}{\alpha} = \delta_2 \text{ and if } d_i < \delta_3 = \inf(\delta, \delta_1). & (8.82) \end{cases}$$

But we also have: $d_i = d(y(t_i), z(t_i))$.

So we may transform condition (8.82) as follows:

$$d(y(t), z(t)) < \delta_4 = \inf(\delta_2, \delta_3) \tag{8.83}$$

and we finally have:

$$\begin{cases} d(y(t), z(t)) \le \frac{1}{\beta}(a\, K_1' \sqrt{\alpha_1}(t - t_i) + \alpha d_i) \quad \text{for } t \in [t_i, t_{i+1}] & (8.81) \\ \text{as long as} \\ d(y(\tau), z(\tau)) < \delta_4 \quad t^-(s) \le \tau \le t \quad t \in [t_i, t_{i+1}]. & (8.83) \end{cases}$$

We can make a final transformation of (8.83)-(8.82), noticing that if (8.83) is valid for $t \in [t_i, t_{i+1}]$, it is <u>a fortiori</u> valid for $t \in [t_j, t_{j+1}]$ with $j \le i$.

Consequently, applying (8.81) for $j \le i$, we derive:

$$d(y(t), z(t) \le \frac{1}{\beta}(a\, K_1' \sqrt{\alpha_1}(t - t_j) + \alpha\, d_j) \tag{8.84}$$

for $t \in [t_j, t_{j+1}]$, $j \le i$.

In particular, with $t = t_{j+1}$ we have:

$$d_{j+1} = d(y(t_{j+1}), z(t_{j+1})) \le \frac{1}{\beta}(a K_1' \sqrt{\alpha_1}(t_{j+1} - t_j) + \alpha d_j), \tag{8.85}$$

$j \le i$ or with $j - 1$

$$d_j \le \frac{1}{\beta}(a\, K_1' \sqrt{\alpha_1}(t_j - t_{j-1}) + \alpha d_{j-1}). \tag{8.86}$$

Combining (8.84) and (8.86), we derive:

$$d(y(t), z(t)) \le \frac{1}{\beta}(a K_1' \sqrt{\alpha_1}(t - t_j) + \frac{\alpha}{\beta}(a K_1' \sqrt{\alpha_1}(t_j - t_{j-1}) + \alpha\, d_{j-1})) \tag{8.87}$$

for $t \in [t_j, t_{j+1}]$.

We then reapply (8.85) with $j = j-1$; this yields an estimate of d_{j-1} in function of d_{j-2}. Iterating this, we derive, using also the fact that $d_0 = d(y(t^-(s)), z(t^-(s))) = 0$:

$$\begin{cases} d(y(t), z(t)) \le a \frac{K_1'}{\alpha} \sqrt{\alpha_1} \left(\frac{\alpha}{\beta}(t - t_j) + \frac{\alpha^2}{\beta^2}(t_j - t_{j-1}) + \ldots + \left(\frac{\alpha}{\beta}\right)^{j+1}(t_1 - t^-(s))\right) \\ \qquad = \sqrt{\alpha_1}\, H_j(t); \ \forall\, t, \ t \in [t_j, t_{j+1}] \end{cases} \tag{8.88}$$

under the following condition:

$$d(y(\tau), z(\tau)) < \delta_4; \quad t^-(s) \leq \tau \leq t. \tag{8.89}$$

As, by (8.60), $\frac{\alpha}{\beta} \geq 1$,

$$H_j(t) \leq H = \frac{aK_1'}{\alpha} \sum_{j=0}^{n_0 - 1} \left(\frac{\alpha}{\beta}\right)^{n_0 - j} (t_{j+1} - t_j) \quad , \tag{8.90}$$

and as n_0 is bounded as well as α, $\frac{\alpha}{\beta}$, and a $(a \leq a_1)$, we obtain:

$$H_j(t) \leq C(t_{n_0} - t_0) = C(t^+(s) - t^-(s)). \tag{8.91}$$

And then, (8.88) gives:

$$d(y(t), z(t)) \leq C\sqrt{\alpha_1}\,(t^+(s) - t^-(s)) \tag{8.92}$$

if

$$d(y(\tau), z(\tau)) < \delta_4; \; t^-(s) \leq \tau \leq t. \tag{8.89}$$

Now α_1 goes to zero with ε_1; so that, with a proper choice of the constant γ that we introduced, we may assume:

$$C\sqrt{\alpha_1}\,(t^+(s) - t^-(s)) \leq C\sqrt{\alpha_1} < \delta_4 \quad \text{for } 0 < \varepsilon_1 < \gamma. \tag{8.93}$$

With this choice of γ, (8.89) implies (8.92) implies in turn that (8.89) holds for a small time after t. Hence, (8.92) holds for all time $t \in [t^-(s), t^+(s)]$ and we thus have:

$$d(y(t), z(t)) \leq C\sqrt{\alpha_1}\,(t^+(s) - t^-(s)), \tag{8.93}$$

which implies (8.53).

Now, under (8.93), (8.77) holds also.

Thus,

$$|\dot{\tilde{y}}(t) - \dot{\tilde{z}}(t)|_e \leq a\,K_1'\,\sqrt{\alpha_1} \tag{8.77}$$

or

$$\left| D\varphi_i(\dot{y}(t)) - a\,D\varphi_i(\xi(z(t)))\right|_e \leq a\,K_1\sqrt{\alpha_1}\,.$$

Now $D\varphi_i(\xi)$ is constant in $\varphi_i(U_i)$ by choice of (U_i, φ_i); hence

$$D\varphi_i(\xi(z(t))) = D\varphi_i(\xi(y(t)))$$

and we have:

$$\left| D\varphi_i(\dot{y}(t)) - a\,D\varphi_i(\xi(y(t)))\right|_e \leq a\,K_1'\sqrt{\alpha_1}\,. \tag{8.94}$$

Then we use (8.61) to deduce:

$$\|\dot{y}(t) - a\,\xi(y(t)\| \leq \frac{aK_1'}{\bar{\beta}}\sqrt{\alpha_1}\,, \tag{8.95}$$

which implies (8.52). ∎

Lemma 9 gives an insight into what happens in the $I_1(s)$ intervals, although we still need to understand what happens with the prior and posterior pieces tangent to v. However, the main question which arises now is to understand the situation between two intervals of type $I_1(s)$, i.e. between two concentration points. The important facts which take place between two such points are as follows:

1). the curve on such pieces is "almost tangent to v";

2). between two limit points in such pieces, the form α has completed k ($k \in \mathbb{Z}$) revolutions and has come back on itself as a form:

i.e.

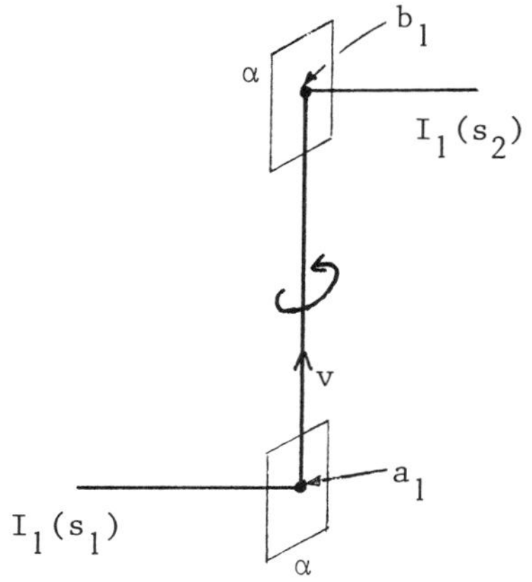

In the transport equation defined in Chapter 5, α has turned along the v-piece and came back on itself as a form.

For ε_1 given, now let $I(s_1)$ and $I(s_2)$ be two consecutive concentration intervals (given through Lemma 8).

Let

$$\begin{cases} t^+(s_1) < t_1 < t_2 < \ldots < t_k < t^-(s_2) \\ \dot{b}(t_i) = 0 . \end{cases} \qquad (8.96)$$

$t_1, \ldots, t_k$ are the zeros of $\dot{b}$ on the interval $]t^+(s_1), t^-(s_2)[$.

Between t_i and t_{i+1}, b takes at most once the value zero as the t_i's cover all zeros of b.

We distinguish thus two types of intervals:

The type (I) intervals where b never takes the value zero (8.97) in $[t_i, t_{i+1}]$. We draw the curve between these two times as follows:

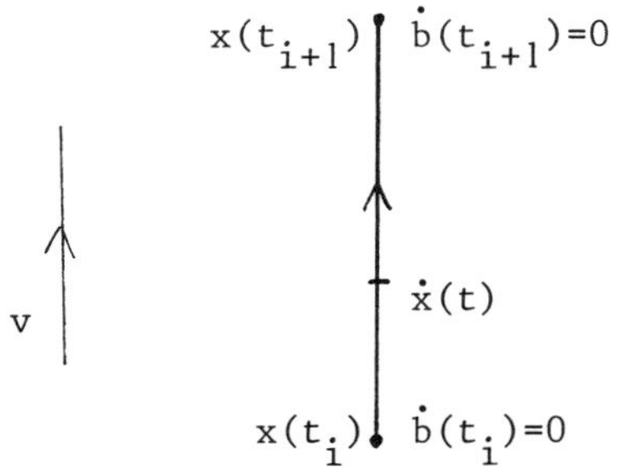

The type (II) intervals, where b is zero somewhere between t_i and t_{i+1}. We draw the curve then as follows: (8.98)

(if $b(t_i) > 0$)

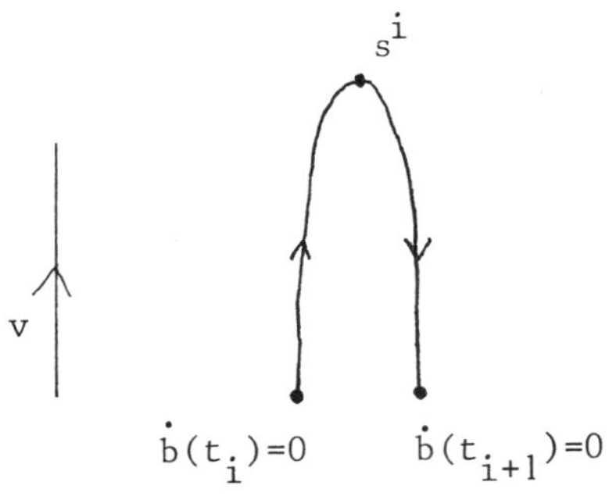

s^i is such that $b(s^i) = 0$

Finally there exist two intervals, the initial one and the final one, we will analyse separately.

Dealing with intervals of type (I)

If $[t_i, t_{i+1}]$ is a type (I) interval, (8.41) is not verified for any t in $[t_i, t_{i+1}]$, (otherwise, we would be in a concentration interval).

$$\forall\, t \in [t_i, t_{i+1}] \quad \frac{b^2(t)}{\omega} \geq \varepsilon_1 \quad \text{or} \quad \frac{|\dot{b}(t)|}{\omega} \geq \varepsilon_2 . \tag{8.99}$$

In particular, at time t_i and t_{i+1}, as $\dot{b}(t_i) = \dot{b}(t_{i+1}) = 0$, we necessarily have:

$$\frac{b^2(t_i)}{\omega} \geq \varepsilon_1 \qquad \frac{b^2(t_{i+1})}{\omega} \geq \varepsilon_1 . \tag{8.100}$$

As $\dot{x}(t_i) = a\,\xi(x(t_i)) + b(t_i)\,v(x(t_i))$ and as $|b(t_i)| \geq \varepsilon_1 \sqrt{\omega}$, where $\omega \to +\infty$, $\dot{x}(t_i)$ is rather collinear to $v(x(t_i))$. The same remark is valid with t_{i+1} instead of t_i. We consider then the curve $\bar{z}(t)$ defined as follows:

$$\begin{cases} \bar{z}(t_i) = x(t_i) \qquad \dot{\bar{z}}(t) = b(t)\,v(\bar{z}(t)) \\ t \in [t_i, t_{i+1}] ; \end{cases} \tag{8.101}$$

Let

θ_s be the one-parameter group generated by ξ. (8.102)

Let

$$\bar{u}(t) = -\int_{t_i}^{t} a(\tau)\,d\tau = -a(t - t_i) \qquad (a(\tau) = \text{Cte} = a). \tag{8.103}$$

Finally, let

$\bar{y}(t)$ be the curve defined by (8.102)'

$$\bar{y}(t) = \theta_{\bar{u}(t)}(x(t)).$$

We have then the following:

<u>Lemma 10:</u> Assume $\left|\int_{t_i}^{t_{i+1}} b(t)\,dt\right| \leq M_3$; where M_3 is a given constant. There exist $K_4 > 0$ and ω_4 such that for $\omega \geq \omega_4$, we have:

$$\sup_{t\in[t_i,t_{i+1}]} d(x(t),\bar{y}(t)) \leq K_4(t_{i+1} - t_i) \tag{8.104}$$

$$\sup_{t\in[t_i,t_{i+1}]} d(\bar{y}(t),\bar{z}(t)) \leq K_4(t_{i+1} - t_i)^2 \tag{8.105}$$

$$\sup_{t\in[t_i,t_{i+1}]} d(x(t),\bar{z}(t)) \leq K_4[(t_{i+1} - t_i) + (t_{i+1} - t_i)^2] \tag{8.106}$$

ω_4 is uniform on all intervals of type (I).

<u>Proof of Lemma 10</u>:

As $\bar{y}(t) = \theta_{\bar{u}(t)}(x(t))$, where θ_s is the group generated by ξ, we have:

$$d(x(t),\bar{y}(t)) \leq \|\xi\|\,|\bar{u}(t)|. \tag{8.107}$$

From (8.103), $|\bar{u}(t)| = a|t - t_i| \leq a(t_{i+1} - t_i)$, whence (8.104) with $K_4' \geq a\|\xi\|$. Inequality (8.106) is deduced from (8.104) and (8.105).

To prove (8.105), we make the following change of variables on $[t_i, t_{i+1}]$

$$s(t) = \int_{t_i}^{t} b(x)\,dx, \quad t(s) \text{ is the inverse function,} \tag{8.108}$$

which is well-defined as b is never zero on $[t_i, t_{i+1}]$.
The curves $x(t)$, $\bar{y}(t)$ and $\bar{z}(t)$ then yield:

$x_1(s)$, $y_1(s)$, x_s starting at the same point $x(t_i)$ as: (8.109)

$$x_1(0) = x(t_i) \tag{8.110}$$

$$y_1(0) = \bar{y}(t_i) = \theta_{\bar{u}(t_i)}(x(t_i)) = \theta_0(x(t_i)) = x(t_i)$$

and $x_0 = \bar{z}(t_i) = \bar{y}(t_i) = x(t_i)$ (by (8.101).

On the other hand:

$$\frac{\partial x_1}{\partial s} = \dot{x}(t(s)) \frac{\partial t}{\partial s} = \frac{1}{b(t(s))}(a\xi + bv) = v(x(s)) + \frac{a}{b(t(s))}\xi(x_1(s)) \tag{8.111}$$

$$\frac{\partial x_s}{\partial s} = \frac{\partial \bar{z}}{\partial t}\frac{\partial t}{\partial s} = b(t(s))\, v(x_s) \frac{1}{b(t(s))} = v(x_s) \tag{8.112}$$

and

$$\frac{\partial y_1}{\partial s} = D\theta_{\bar{u}(t(s))}(\dot{x}(t(s)) \frac{\partial t}{\partial s} + \frac{\partial \bar{u}}{\partial s} \cdot \frac{\partial t}{\partial s}\xi(\theta_{\bar{u}(t(s))}(x(t(s)))) \tag{8.113}$$

$$= D\theta_{\bar{u}(t(s))}(v + \frac{a}{b(t(s))}\xi(x_1(s))) - \frac{a}{b(t(s))}\xi(\theta_{\bar{u}(t(s))}(x_1(s)))$$

$$= D\theta_{\bar{u}(t(s))}(v) + \frac{a}{b(t(s))} D\theta_{\bar{u}(t(s))}(\xi(x_1(s)) - \frac{a}{b(t(s))}\xi(\theta_{\bar{u}}(x_1(s)))$$

$$= D\theta_{\bar{u}(t(s))}(v) + \frac{a}{b(t(s))}\xi(\theta_{\bar{u}}(x_1(s))) - \frac{a}{b(t(s))}\xi(\theta_{\bar{u}}(x_1(s)))$$

(as $D\theta_s(\xi_x) = \xi(\theta_s(x)))$

$$= D\theta_{\bar{u}(t(s))}(v) = D\theta_{u_1(s)}(v(x_1(s))),$$

where we have set:

$$\bar{u}(t(s)) = u_1(s). \tag{8.114}$$

Finally, the interval $[t_i, t_{i+1}]$ transforms into:

$$\left[0, \int_{t_i}^{t_{i+1}} b(t)\, dt\right] \quad \text{through } (8.108). \tag{8.115}$$

This interval is thus contained in $[-M_3, M_3]$ and therefore is uniformly bounded.

The proof of Lemma 10 reduces then to the proof of (8.52) and (8.53) of Lemma 9, with ξ replaced by v, x replaced by x_1, y by y_1 and z by x_s, $D\phi$ is replaced by $D\theta$ and $[t^-, t^+]$ by $[0, \int_{t_i}^{t_{i+1}} b(t)\, dt]$.

We then derive estimates on $d(x_1(s), y_1(s))$, $d(y_1(s), x_2)$ and $d(x_1(s), x_s)$ which, when we reparametrize, yield (8.104), (8.105), (8.106).

We point out here that the role played by $\sqrt{\alpha_1}$ in the proof of Lemma 9 is played here by $(t_{i+1} - t_i)$ as now $|u_1(s)| = |\bar{u}(t(s))|$ is upper bounded by $a(t_{i+1} - t_i)$. In the proof of Lemma 9, α_1 had to be chosen small enough (or equivalently $\varepsilon_1 < \gamma$ small enough). This possibility is replaced here by taking $(t_{i+1} - t_i)$ small enough; this is allowed by the fact that $[t_i, t_{i+1}]$ is not a concentration interval and hence, $t_{i+1} - t_i$ and even $t^-(s_2) - t^+(s_2)$ which upperbounds this quantity goes to zero by Lemma 5 when ω goes to $+\infty$. Hence the existence of ω_4 such that Lemma 10 holds as soon as ω is larger than ω_4.

This ω_4 can be chosen uniformly on all type (I) intervals as the total measure of the non-concentration set they are contained in goes to zero with ε.

We point out here that, in the preceding proof, the fact that $\dot{b}(t_i) = b(t_{i+1}) = 0$ is not important. The only conditions which matter are related to the possibility of changing variables with (8.108) and the fact that the interval $[t_i, t_{i+1}]$ has small enough length.

This allows the statement of the following lemma:

<u>Lemma 10'</u>: Let $[t^1, t^2]$ be a time interval such that b is never 0 in $]t^1, t^2[$. Assume $\left|\int_{t^1}^{t^2} b(x)\, dx\right| \leq C$, where C is a given constant.

There exist then constants $\varepsilon_0(C) > 0$ and $\bar{K}$ such that if $|t^2 - t^1| < \varepsilon_0(C)$, one has:

$$\sup_{t\in[t^1,t^2]} d(x(t),\bar{y}(t)) \leq \bar{K}\,|t^2 - t^1| \tag{8.104'}$$

$$\sup_{t\in[t^1,t^2]} d(\bar{y}(t),\bar{z}(t)) \leq \bar{K}\,|t^2 - t^1|^2 \tag{8.105'}$$

$$\sup_{t\in[t^1,t^2]} d(x(t),\bar{z}(t)) \leq \bar{K}[\,|t^2 - t^1| + |t^2 - t^1|^2] \tag{8.106'}$$

Here $\bar{y}$ and $\bar{z}$ are defined by (8.101) and (8.104) where we replace the interval $[t_i, t_{i+1}]$ by $[t^1, t^2]$.

Finally, the condition $\omega \geq \omega_4$ of Lemma 10 is here to be replaced by $|t^2 - t^1| < \varepsilon_0(C)$. Lemma 10' is useful for the remainder of this book.

Now to each type (I) interval $[t_1, t_{i+1}]$ we associate a curve $\bar{z}(t)$ defined in (8.101), tangent to v and close to $x(t)$ provided $|\int_{t_i}^{t_{i+1}} b(x)\,dx| \leq M_3$.

We give in what follows a condition which ensures that this last inequality holds and we compare, under this condition, the equations (5.12) and (5.13) which govern the transport of forms along the pieces $\bar{z}(t)$ and $x(t)$.

Thus, we consider a type (I) interval $[t_i, t_{i+1}]$. We assume the following inequalities hold:

<u>Hypothesis</u> (H) $\quad \exists\, \delta_0 > 0,\ \exists\, \beta_0 > 0,\ \beta_0 \geq \left| 1 - \dfrac{b^2(t_i)}{2\omega a} + \dfrac{\int_0^1 b^2}{2\omega a} \right| \geq \delta_0.$

We assume that

$$b(t_i) > 0. \tag{8.116}$$

The case where $b(t_i) < 0$ is identical to this one.

Let x_s be the curve starting from $x(t_0)$ tangent to v $(s \in [0,+\infty[)$. Let

$$x_{\bar{s}(x(t_i))} = \psi(x(t_i)) \tag{8.117}$$

$\bar{s} : M \to R$ has been defined in Definition 3.

Let us consider now the solution of the transport equation governed by v, along the line x_s, given by (5.12), with initial data

$$Z_0^\star = \begin{bmatrix} 1 - \frac{b^2(t_i)}{2\omega a} + \frac{\int_0^1 b^2}{2\omega a} \\ 0 \\ 0 \end{bmatrix}$$

i.e.

$$\begin{cases} \dot{Z}^\star = -{}^t\Gamma(x_s) Z^\star \\ Z^\star(0) = Z_0^\star \end{cases} \tag{8.118}$$

where

$${}^t\Gamma(x) = \begin{bmatrix} 0 & 0 & 1 \\ 0 & 0 & 0 \\ -1 & \bar{\mu}_\xi(x) & \bar{\mu}(x) \end{bmatrix} \qquad x \in M.$$

In (8.118), $Z^\star = Z^\star(s)$ and $\dot{Z}^\star = \frac{\partial Z^\star}{\partial s}$.

As we have seen in Section 5.3, the differential equation (5.38) satisfied by b can be written under the matrix form:

$$\dot{Z}_1^\star = -b(t)\, {}^t\Gamma(x(t))\, Z_1^\star + \begin{bmatrix} 0 \\ -\frac{\dot{b}(t)}{\omega} \\ -\frac{ab\tau(x(t))}{\omega} \end{bmatrix} \tag{8.119}$$

(see (5.58)) with

$$Z_1^\star(t) = \begin{bmatrix} 1 - \frac{b^2}{2\omega a}(t) + \frac{\int_0^1 b^2}{2\omega a} \\ -\frac{b(t)}{\omega} \\ \frac{\dot{b}(t)}{\omega a} \end{bmatrix} .$$

Here, $\dot{Z}_1^\star = \frac{\partial Z_1}{\partial t}$.

As b is never zero on $[t_i, t_{i+1}]$, we can use the change of variables (8.108) in this interval:

$$s(t) = \int_{t_i}^{t} b(x)\,dx . \tag{8.108}$$

We set:

$$Z_2^\star(s) = Z_1^\star(t(s)) \qquad s \in \left[0, \int_{t_i}^{t_{i+1}} b(x)\,dx\right], \tag{8.120}$$

where $t(s)$ is the inverse function of $s(t)$.

$Z_2^\star$ satisfies:

$$Z_2^{\star}(s) = \frac{\partial Z_2^\star}{\partial s} = {}^t\Gamma(x_1(s))\,Z_2^\star + \begin{bmatrix} 0 \\ -\dfrac{\dot b(t(s))}{\omega b(t(s))} \\ -\dfrac{a\tau(x(t(s)))}{\omega} \end{bmatrix} \tag{8.121}$$

$$Z_2^\star(s) = \begin{bmatrix} 1 - \dfrac{b^2}{2\omega a}(t(s)) + \dfrac{\int_0^1 b^2}{2\omega a} \\ -\dfrac{b(t(s))}{\omega} \\ \dfrac{\dot b(t(s))}{\omega a} \end{bmatrix}$$

$x_1(s)$ has been defined in (8.109) as the curve obtained from $x(t)$ when we reparametrize $x(t)$ along (8.108).

We have then:

Lemma 11: Assume hypothesis (H) is satisfied, there exist $\delta_1 > 0$, $\overline{M}_3 > 0$, $K > 0$ and $\overline{\omega}_5 > 0$ such that, if $\omega \geq \overline{\omega}_5$, we have:

$$0 < \delta_1 \leq \int_{t_i}^{t_{i+1}} b(x)\,dx \leq \overline{M}_3 \tag{8.122}$$

$$d(x(t_{i+1}), x_{\bar{s}(x(t_i))}) = d(x(t_{i+1}), \psi(x(t_i))) \leq K(t_{i+1} - t_i + \frac{1}{\sqrt{\omega}}) \tag{8.123}$$

$$\begin{aligned} \|Z_1^\star(t_{i+1}) - Z^\star(x_{\bar{s}(x(t_i))})\| &= \|Z_1^\star(t_{i+1}) - Z^\star(\psi(x(t_i)))\| \\ &\leq K[t_{i+1} - t_i + \frac{1}{\sqrt{\omega}}]. \end{aligned} \tag{8.124}$$

Here $\left\| \begin{bmatrix} a \\ b \\ c \end{bmatrix} \right\| = |a| + |b| + |c|$. By $Z^\star(x_{\bar{s}(x(t_i))})$ we mean $Z^\star(\bar{s}(x(t_i))$ although this is a misuse of notation.

Remark: If $b(t_i)$ is negative, we replace $x_{\bar{s}(x(t_i))} = \psi(x(t_i))$ by $x_{\underline{s}(x(t_i))} = \psi^{-1}(x(t_i))$ in Lemma 11.

Remark: $\overline{\omega_5}$ is uniform on all type (I) intervals satisfying (H) with δ_0 and β_0.

Lemma 11 is a consequence of the following more general lemmas:

Consider

$$\text{an interval }]t_1, t_2[\text{ where } b \text{ and } \dot{b} \text{ do not vanish.} \tag{8.127}$$

(We point out here that t_2 can be smaller than t_1 in the notation $]t_1, t_2[$.)

We consider on $[t_1, t_2]$ the analogous curve to (8.101)

$$\bar{z} : \begin{cases} \bar{z}(t_1) = x(t_1) \\ \dot{\bar{z}}(t) = b(t)\ v(\bar{z}(t)) \end{cases} \qquad t \in [t_1, t_2]. \tag{8.128}$$

Consider the following change of variable, which is analogous to (8.108)

$$s(t) = \int_{t_1}^{t} b(x)\ dx \qquad t \in [t_1, t_2], \tag{8.129}$$

$t(s)$ being the inverse function.

We will assume that b is positive on $[t_1, t_2]$; the other case being symmetric.

We set, as in (8.109):

$$x_1(s) \text{ the reparametrized curve } x(t) \text{ according to (8.129);} \tag{8.130}$$

$$x_s \text{ the reparametrized curve } \bar{z}(t) \text{ according to (8.129).} \tag{8.131}$$

We then have $\dot{x}_s = v(x_s)$ as can be easily checked.

We consider the solution $Z^\star$ of the transport equation by v, along the line

x_s, given by (5.12) with initial data:

$$Z_0^\star = \begin{bmatrix} 1 - \dfrac{b^2(t_1)}{2\omega a} + \dfrac{\int_0^1 b^2}{2\omega a} \\ 0 \\ \dfrac{\dot b(t_1)}{\omega a} \end{bmatrix} .$$

$Z^\star$ satisfies:

$$\begin{cases} \dot Z^\star = -{}^t\Gamma(x_s)\, Z^\star \\ Z^\star(0) = Z_0^\star = \begin{bmatrix} 1 - \dfrac{b^2(t_1)}{2\omega a} + \dfrac{\int_0^1 b^2}{2\omega a} \\ 0 \\ \dfrac{\dot b(t_1)}{\omega a} \end{bmatrix} \end{cases} \tag{8.132}$$

Here $\dot Z^\star = \dfrac{\partial Z^\star}{\partial s}$; $\dfrac{\partial}{\partial s}$ is the time-derivative along x_s which is tangent to v.
Using again (5.58), we have, if we set:

$$Z_1^\star(t) = \begin{bmatrix} 1 - \dfrac{b^2(t)}{2\omega a} + \dfrac{\int_0^1 b^2}{2\omega a} \\ -\dfrac{b(t)}{\omega} \\ \dfrac{\dot b(t)}{\omega a} \end{bmatrix} \tag{8.133}$$

$$\dot Z_1^\star = -b(t)\, {}^t\Gamma(x(t))\, Z_1^\star + \begin{bmatrix} 0 \\ -\dfrac{\dot b}{\omega} \\ -\dfrac{ab\tau}{\omega} \end{bmatrix} . \tag{8.134}$$

By making the change of variables (8.129), (8.133)-(8.134) give:

$$\begin{cases} Z_2^\star(s) = Z_1^\star(t(s))\,; \\ \dot{Z}_2^\star = \dfrac{\partial Z_2^\star}{\partial s} = -{}^t\Gamma(x_1(s))\,Z_2^\star + \begin{bmatrix} 0 \\ -\dfrac{\dot{b}(t(s))}{\omega b(t(s))} \\ -\dfrac{a\tau(x_1(s))}{\omega} \end{bmatrix} \end{cases} \qquad (8.135)$$

We remember that we assumed b to be positive on $[t_1, t_2]$. We denote, following Definition 3,

$\bar{s}(x(t))$ the first time $s > 0$ such that x_s is a coincidence point of $x_0 = x(t_1)$. (8.136)

Furthermore, let

$$\bar{\alpha} = \inf_{x \in M} \left(\frac{\bar{s}(x)}{3}\right), \qquad (8.137)$$

$\bar{s}$ defined as in Definition 3.

We then have:

Lemma 12: Let α_0 and β_0 be given positive. Assume

$$\delta_0 \leq \frac{|\dot{b}(t_1)|}{\omega a} + \left|1 - \frac{b^2(t_1)}{2\omega a} + \frac{\int_0^1 b^2}{2\omega a}\right| \leq \beta_0 \qquad (H_{\delta_0}^{\beta_0}).$$

There exist then constants ω_5, $\bar{\varepsilon}_0$, $\bar{K}_1$, δ, K_5, $\bar{K}_5$ and $\bar{\delta} > 0$ which depend on the considered interval $[t_1, t_2]$ such that

$$\int_{t_1}^{t_2} b(x)\,dx \leq \bar{s}(x(t_1)) + \bar{\alpha} \qquad (8.138)$$

$$\delta \leq \|Z^\star(s)\| \leq \bar{K}_1, \quad \forall s \in [0, s(t_2)] \qquad (8.139)$$

$$\sup_{t \in [t_1, t_2]} \|Z^\star(s(t)) - Z_1^\star(t)\| \leq K_5\left[\frac{1}{\sqrt{\omega}} + |t_2 - t_1|\left|\int_{t_1}^{t_2} b(x)\,dx\right|\right] \qquad (8.140)$$

$$\bar{\delta}\,\|Z_1^\star(t_2) - Z_1^\star(t_1)\| \leq \left|\int_{t_1}^{t_2} b(x)\,dx\right| + \bar{K}_5\left[\frac{1}{\sqrt{\omega}} + |t_2 - t_1|\left|\int_{t_1}^{t_2} b(x)\,dx\right|\right]. \qquad (8.141)$$

On the other hand, let:

$$x_s \text{ a v-orbit, } s \in \left]-\infty, +\infty\right[, \text{ passing through } x_0 \text{ in } M. \tag{8.142}$$

Let

$$Z^\star = \begin{bmatrix} a^\star \\ b^\star \\ c^\star \end{bmatrix}(s) \tag{8.143}$$

be the solution of the transport equation (5.12)

$$\dot{Z}^\star(s) = -{}^t\Gamma(x_s)\, Z^\star,$$

with initial data $\begin{bmatrix} a_0^\star \\ 0 \\ c_0^\star \end{bmatrix}$.

Then we have the following:

<u>Lemma 13</u>: Let $s_1 \in \mathbb{R}$; M and m given positive such that:

$$|a^\star| + |c^\star|(s_1) \leq M, \qquad |a^\star|(s_1) \geq m > 0. \tag{8.144}$$

There exist then $\chi(M, m) > 0$, $\theta(M, m) > 0$, $\mu(M, m) > 0$, $K_6(M, m) > 0$, $K_7(M, m) > 0$ such that if $|c^\star(s_1)| < \chi(M, m)$, the function $c^\star(s)$ vanishes once and only once in the time interval $]s_1 - \theta(M, m), s_1 + \theta(M, m)[$.
If s_2 is this point, $s_2 \in]s_1 - K_6|c^\star|(s_1), s_1 + K_6|c^\star|(s_1)[$.
Furthermore, we have

$$\begin{cases} c^\star(s_1 - \theta(M, m))\, c^\star(s_1 + \theta(M, m)) < 0 \\ |c^\star(s_1 \pm \theta(M, m))| \geq \mu(M, m) \\ |a^\star|(s) + |c^\star(s)| \leq M + 1 \text{ over }]s_1 - \theta(M, m), s_1 + \theta(M, m)[\end{cases} \tag{8.145}$$

$$\begin{cases} \|Z^\star(s_2) - Z^\star(s_1)\| \leq K_6 |c^\star(s_1)| \\ |c^\star(s)| = |c^\star(s) - c^\star(s_2)| \geq K_7 |s - s_2| \end{cases} \tag{8.145'}$$

$\forall\ s \in]s_1 - \theta(M, m), s_1 + \theta(M, m)[$.

We defer the proof of Lemmas 12 and 13.

Proof of Lemma 11: We set $t_1 = t_i$, $t_2 = t_{i+1}$.
We know that $\dot{b}(t_i) = 0$ and that t_i satisfies hypothesis (H):

$$\text{(H)} \qquad \delta_0 \le \left| 1 - \frac{b^2(t_i)}{2\omega a} + \frac{\int_0^1 b^2}{2\omega a} \right| \le \beta_0 .$$

Thus:

$$\delta_0 \le \frac{|\dot{b}(t_i)|}{\omega} + \left| 1 - \frac{b^2(t_i)}{2\omega a} + \frac{\int_0^1 b^2}{2\omega a} \right| \le \beta_0 .$$

$\overline{\omega}_5 \ge \omega_5$ is chosen large enough so that $\omega \ge \overline{\omega}_5$ implies $|t_{i+1} - t_i| < \overline{\varepsilon}_0$. We are then allowed to apply Lemma 12.

By (8.138), we then have:

$$\left| \int_{t_i}^{t_{i+1}} b(x)\,dx \right| \le \bar{s}(x(t_i)) + \bar{\alpha} \le \overline{M}_3 = \sup_{x \in M} (\bar{s}(x) + \bar{\alpha}), \tag{8.146}$$

which establishes one of the inequalities of (8.122).

By (8.140), we have

$$\begin{aligned} \| z^{\star}(s(t_{i+1})) - z_1^{\star}(t_{i+1}) \| &\le K_5 \left[\frac{1}{\sqrt{\omega}} + [t_{i+1} - t_i] \left| \int_{t_i}^{t_{i+1}} b(x)\,dx \right| \right] \\ &\le K_5 \left[\frac{1}{\sqrt{\omega}} + [t_{i+1} - t_i] \overline{M}_3 \right] . \end{aligned} \tag{8.147}$$

We explicitly know $$z_1^{\star}(t_{i+1}) = \begin{bmatrix} 1 - \dfrac{b^2(t_{i+1})}{2\omega a} + \dfrac{\int_0^1 b^2}{2\omega a} \\ -\dfrac{b(t_{i+1})}{\omega} \\ \dfrac{\dot{b}(t_{i+1})}{\omega a} \end{bmatrix} = \begin{bmatrix} 1 - \dfrac{b^2(t_{i+1})}{2\omega a} + \dfrac{\int_0^1 b^2}{2\omega a} \\ -\dfrac{b(t_{i+1})}{\omega} \\ 0 \end{bmatrix}$$

as $\dot{b}(t_{i+1}) = 0$.

We set:

$$Z^\star(s) = \begin{bmatrix} a^\star \\ b^\star \\ c^\star \end{bmatrix}(s) . \tag{8.148}$$

By definition, $Z^\star(0) = Z_0^\star = \begin{bmatrix} a_0^\star \\ b_0^\star \\ c_0^\star \end{bmatrix} = \begin{bmatrix} 1 - \dfrac{b^2(t_i)}{2\omega a} + \dfrac{\int_0^1 b^2}{2\omega a} \\ 0 \\ 0 \end{bmatrix}$, by (8.118).

With the notations of (8.148), (8.147) then becomes:

$$\left| a^\star(s(t_{i+1}) - 1 + \frac{b^2(t_{i+1})}{2\omega a} - \frac{\int_0^1 b^2}{2\omega a} \right| + \left| b^\star(s(t_{i+1})) + \frac{b(t_{i+1})}{\omega} \right|$$
$$+ \left| c^\star(s(t_{i+1})) \right| \le K_5\left[\frac{1}{\sqrt{\omega}} + [t_{i+1} - t_i]\overline{M}_3\right], \tag{8.149}$$

which implies:

$$\boxed{\left| c^\star(s(t_{i+1})) \right| \le K_5\left[\frac{1}{\sqrt{\omega}} + [t_{i+1} - t_i]\,\overline{M}_3\right] .} \tag{8.150}$$

On the other hand, we have (8.139):

$$\forall s \in [0, s(t_{i+1})], \quad \delta \le \|Z^\star(s)\| \le \overline{K}. \tag{8.139}$$

Thus:

$$\delta \le \|Z^\star(s(t_{i+1}))\| \le \overline{K} \tag{8.151}$$

i.e. $\quad \delta \le |a^\star(s(t_{i+1}))| + |b^\star(s(t_{i+1}))| + |c^\star(s(t_{i+1}))| \le \overline{K}.$

As $b_0^\star = 0$ and as $Z^\star = \begin{bmatrix} a^\star \\ b^\star \\ c^\star \end{bmatrix}$ satisfies (8.118), we have:

$$b^\star \equiv 0. \tag{8.152}$$

Indeed, $\dot{Z}^\star = -{}^t\Gamma Z^\star \iff \begin{bmatrix} \dot{a}^\star \\ \dot{b}^\star \\ \dot{c}^\star \end{bmatrix} = - \begin{bmatrix} 0 & 0 & 1 \\ 0 & 0 & 0 \\ -1 & \bar{\mu}_\xi & \bar{\mu} \end{bmatrix} \begin{bmatrix} a^\star \\ b^\star \\ c^\star \end{bmatrix}$, hence $\dot{b}^\star = 0$;

hence $b^\star = b_0^\star = 0$.

Inequality (8.151) then becomes:

$$\delta \leq |a^\star(s(t_{i+1}))| + |c^\star(s(t_{i+1}))| \leq \overline{K}$$

which implies:

$$|a^\star(s(t_{i+1}))| + |c^\star(s(t_{i+1}))| \leq \overline{K}; \tag{8.153}$$

we have also taken into account (8.150)

$$|a^\star(s(t_{i+1})| \geq \delta - K_5\left[\frac{1}{\sqrt{\omega}} + [t_{i+1} - t_i]\,\overline{M}_3\right]. \tag{8.154}$$

Let us choose $\overline{\omega}_5$ large enough so that, as soon as $\omega \geq \overline{\omega}_5$, we have:

$$K_5\left[\frac{1}{\sqrt{\omega}} + [t_{i+1} - t_i]\,\overline{M}_3\right] < \inf\left(\frac{\delta}{2}, \chi(\overline{K}, \frac{\delta}{2})\right), \tag{8.155}$$

where $\chi(M, m)$ has been defined in Lemma 13.

Inequality (8.155) holds as $t_{i+1} - t_i \to 0$ when ω goes to $+\infty$. (We indeed are outside the concentration intervals.) Here again, the choice of $\overline{\omega}_5$ can be made uniform on all type (I) intervals, as the total measure of these intervals goes to zero when ω goes to $+\infty$. Inequality (8.150) then becomes:

$$|c^\star(s(t_{i+1}))| < \chi(\overline{K}, \frac{\delta}{2}). \tag{8.156}$$

Inequality (8.153) then becomes:

$$|a^\star(s(t_{i+1}))| \geq \frac{\delta}{2}, \tag{8.157}$$

and we have:

$$|a^\star(s(t_{i+1}))| + |c^\star(s(t_{i+1}))| \leq \overline{K}. \tag{8.153}$$

Consequently, setting $s_1 = (s(t_{i+1}))$, $M = \overline{K}$, $m = \delta/2$, $Z = \begin{bmatrix} a^\star \\ 0 \\ c^\star \end{bmatrix}$ satisfies the transport equation and (8.145)-(8.146) of Lemma 13. This implies that $c^\star$ vanishes once and only once in the interval $]s_1 - K_6|c^\star|(s(t_{i+1}), s_1 + K_6|c^\star|(s(t_{i+1}))[$ at a point s_2 $(s_1 = (s(t_{i+1}))$.

Furthermore we have:

$$\|Z^{\star}(s_2) - Z^{\star}(s(t_{i+1}))\| \leq K_6 |c^{\star}|(s(t_{i+1})). \tag{8.i}$$

Consequently, by (8.150), we have:

$$|s_2 - s(t_{i+1})| \leq K_6 K_5 \left[\frac{1}{\sqrt{\omega}} + [t_{i+1} - t_i]\overline{M}_3\right] \tag{8.ii}$$

and

$$\begin{aligned}\|Z^{\star}(s_2) - Z^{\star}(s(t_{i+1}))\| &\leq K_6 |c^{\star}|(s(t_{i+1})) \\ &\leq K_6 K_5 \left[\frac{1}{\sqrt{\omega}} + [t_{i+1} - t_i]\overline{M}_3\right].\end{aligned} \tag{8.iii}$$

To conclude, by Lemma 10 applied with $M_3 = \overline{M}_3$, we derive through (8.106)

$$\sup_{t \in [t_i, t_{i+1}]} d(x(t), \bar{z}(t)) \leq K_4(\overline{M}_3)\,[t_{i+1} - t_i + (t_{i+1} - t_i)^2]. \tag{8.iv}$$

In particular, for $t = t_{i+1}$, we have (notice that $\bar{z}(t_{i+1}) = x_{s(t_{i+1})}$):

$$d(x(t_{i+1}), x_{s(t_{i+1})}) \leq K_4(\overline{M}_3)\,[t_{i+1} - t_i + (t_{i+1} - t_i)^2]. \tag{8.158}$$

We thus have the following inequalities:

$$|s_2 - s(t_{i+1})| \leq K_6 K_5 \left[\frac{1}{\sqrt{\omega}} + [t_{i+1} - t_i]\overline{M}_3\right] \tag{8.ii}$$

$$\|Z^{\star}(s_2) - Z^{\star}(s(t_{i+1}))\| \leq K_6 K_5 \left[\frac{1}{\sqrt{\omega}} + [t_{i+1} - t_i]\overline{M}_3\right] \tag{8.iii}$$

$$d(x(t_{i+1}), x_{s(t_{i+1})}) \leq K_4(\overline{M}_3)\,[t_{i+1} - t_i + (t_{i+1} - t_i)^2]. \tag{8.iv}$$

Let x_s be the v-orbit starting at $x_0 = x(t_i)$; and let $Z^{\star}(s)$ be the solution of the transport equation (8.118) with initial data

$$Z_0^{\star} = \begin{bmatrix} 1 - \dfrac{b^2(t_i)}{2\omega a} + \dfrac{\int_0^1 b^2}{2\omega a} \\ 0 \\ 0 \end{bmatrix}.$$ From the definition of s_2, we then know that the third component of $Z^{\star}(s)$, $c^{\star}(s)$ vanishes at s_2.

From (8. ii), we derive:

$$d(x_{s_2}, x_{s(t_{i+1})}) \leq [\operatorname*{Sup}_{x \in M} \|v\|] K_6 K_5 \left[\frac{1}{\sqrt{\omega}} + [t_{i+1} - t_i]\overline{M}_3\right]. \tag{8.159}$$

Indeed, x_s is tangent to v and goes from x_{s_2} to $x_{s(t_{i+1})}$ in a time $|s_2 - s(t_{i+1})|$, upperbounded, (see (8. ii), by $K_6 K_5 \left[\frac{1}{\sqrt{\omega}} + [t_{i+1} - t_i]\overline{M}_3\right]$.

(8. 159) and (8. 158) then imply:

$$d(x_{s_2}, x(t_{i+1})) \leq [\operatorname*{Sup}_{x \in M} \|v\|] K_6 K_5 \left[\frac{1}{\sqrt{\omega}} + [t_{i+1} - t_i]\overline{M}_3\right] \tag{8.160}$$
$$+ K_4(\overline{M}_3)[t_{i+1} - t_i + (t_{i+1} - t_i)^2],$$

hence

$$d(x_{s_2}, x(t_{i+1})) \leq K\left[t_{i+1} - t_i + \frac{1}{\sqrt{\omega}}\right]; \tag{8.161}$$

where K is an appropriate constant.

Similarly, with an appropriate constant K, (8.iii) and (8. 149) imply:

$$\|Z_1^{\star}(t_{i+1}) - Z^{\star}(s_2)\| \leq K\left[t_{i+1} - t_i + \frac{1}{\sqrt{\omega}}\right]. \tag{8.162}$$

(8. 161) and (8. 162) will imply (8. 123) and (8. 126) if we prove that:

$$s_2 = \bar{s}(x(t_i)). \tag{8.163}$$

We cannot prove equation (8. 163) directly. First assume that the second inequality in (8. 122) has been proved, i. e.

$$0 < \delta_1 \leq \left|\int_{t_i}^{t_{i+1}} b(x)\, dx\right| \underset{b>0}{\Longleftrightarrow} 0 < \delta_1 \leq \int_{t_i}^{t_{i+1}} b(x)\, dx. \tag{8.164}$$

Assume (8. 164) proven and let us prove (8. 163).

For this, we remember that if we consider:

$$\dot{Z}^\star = -{}^t\Gamma(x_s)\,Z^\star, \quad Z_0^\star = \begin{bmatrix} 1 - \dfrac{b^2(t_i)}{2\omega a} + \dfrac{\int_0^1 b^2}{2\omega a} \\ 0 \\ 0 \end{bmatrix}, \tag{8.165}$$

(x_s tangent to v and starting at $x(t_i)$) and if we set $Z^\star(s) = \begin{bmatrix} a^\star(s) \\ 0 \\ c^\star(s) \end{bmatrix}$, (we already noticed that the second component of $Z^\star$ remains zero), then $c^\star(s_2) = 0$:

$$c^\star(s_2) = 0. \tag{8.166}$$

Furthermore, we already proved:

$$|s_2 - s(t_{i+1})| \leq K_6 K_5 \left[\frac{1}{\sqrt{\omega}} + [t_{i+1} - t_i]\overline{M}_3\right]. \tag{8.155}$$

$$s(t_{i+1}) = \int_{t_i}^{t_{i+1}} b(x)\,dx \leq \overline{s}(s(t_i)) + \overline{\alpha}, \tag{8.146}$$

and we suppose proven:

$$0 < \delta_1 \leq \int_{t_i}^{t_{i+1}} b(x)\,dx. \tag{8.164}$$

Hence we deduce:

$$\begin{aligned} \delta_1 - K_6 K_5 \left[\frac{1}{\sqrt{\omega}} + (t_{i+1} - t_i)\overline{\mathbf{M}}_3\right] < s_2 < \overline{s}(x(t_i)) + \overline{\alpha} + \\ + K_6 K_5 \left[\frac{1}{\sqrt{\omega}} + (t_{i+1} - t_i)\overline{M}_3\right]. \end{aligned} \tag{8.167}$$

We choose $\overline{\omega}_5$ large enough so that:

$$\delta_1 - K_6 K_5 \left[\frac{1}{\sqrt{\omega}} + (t_{i+1} - t_i)\overline{M}_3\right] \geq \frac{\delta_1}{2} \tag{8.168}$$

$$\overline{\alpha} + K_6 K_5 \left[\frac{1}{\sqrt{\omega}} + (t_{i+1} - t_i)\overline{\mathbf{M}}_3\right] \leq \frac{2}{3} \inf_{x \in M} (\overline{s}(x)). \tag{8.169}$$

Inequality (8.168) is clearly possible as $t_{i+1} - t_i$ goes to zero and ω goes to

$+\infty$. (The uniformity of $\overline{\omega}_5$ is here again possible as the total measure of the type (I) intervals goes to zero.)

In (8.169), we remember that we have by (8.137):

$$\overline{\alpha} = \inf_{x \in M} \frac{(\overline{s}(x))}{3} > 0. \tag{8.170}$$

Thus (8.169) holds as ω goes to $+\infty$.

With this choice of $\overline{\omega}_5$, we will have for $\omega \geq \overline{\omega}_5$:

$$0 < \frac{\delta_1}{2} < s_2 < \overline{s}(x(t_i)) + \frac{2}{3} \inf_{x \in M} (\overline{s}(x)) \tag{8.171}$$

hence

$$0 < \frac{\delta_1}{2} < s_2 < \overline{s}(x(t_i)) + \overline{s}(x(t_{i+1})). \tag{8.172}$$

Equations (8.172) and (8.166) imply (8.163): indeed, if we consider (8.165), we know it is the equation of a transported form. As

$$Z_0^\star = \begin{bmatrix} 1 - \dfrac{b^2(t_i)}{2\omega a} + \dfrac{\int_0^1 b^2}{2\omega a} \\ 0 \\ 0 \end{bmatrix},$$

this form is nothing else than

$$\left(1 - \frac{b^2(t_i)}{2\omega a} + \frac{\int_0^1 b^2}{2\omega a}\right) \alpha_{x_0} = \left(1 - \frac{b^2(t_i)}{2\omega a} + \frac{\int_0^1 b^2}{2\omega a}\right) \alpha_{x(t_i)}$$

transported along the piece x_s.

We express the fact that the form α_{x_s} has done k complete revolutions between $s = 0$ and $s = s_0$ (as a field of planes) by the fact that the third component $c^\star(s)$ of $Z^\star(s)$, which is the solution of (8.165), vanishes at s_0.

These zeros occur at $x(t_i)$, $\psi(x(t_i))$, $\psi^2(x(t_i)), \ldots$; or equivalently at $s = 0$, $s = \overline{s}(x(t_i))$, $s = \overline{s}(x(t_i)) + \overline{s}(x(t_{i+1})), \ldots$

s_2 is one of these zeros. Being strictly positive and less than $(\bar{s}(x(t_i)) + \bar{s}(x(t_{i+1}))$ by (8.172), s_2 is necessarily equal to $\bar{s}(x(t_i))$; hence the proof of (8.163) and, via (8.161) and (8.162), the proof of (8.123) and (8.124).

It remains to show (8.164).

For this we apply Lemma 12 again with the interval $[t_i, t_{i+1}]$ replaced by the following $[t_i, t^i]$ interval:

As $\dot{b}(t_i) = \dot{b}(t_{i+1}) = 0$, there exists $t^i \in]t_i, t_{i+1}[$ such that

$$\ddot{b}(t^i) = 0; \tag{8.173}$$

hence by (5.38):

$$\ddot{b}(t^i) = -b(t^i)\left(1 - \frac{b^2(t^i)}{2\omega a} + \frac{\int_0^1 b^2}{2\omega a} - \frac{a\tau}{\omega} - \frac{b(t^i)\bar{\mu}_\xi}{\omega} + \bar{\mu}\frac{\dot{b}(t^i)}{\omega a}\right) = 0. \tag{8.174}$$

As $b(t^i)$ is different from zero, as b does not vanish on $[t_i, t_{i+2}]$, we have:

$$\boxed{1 - \frac{b^2(t^i)}{2\omega a} + \frac{\int_0^1 b^2}{2\omega a} - \frac{a\tau}{\omega} - \frac{b(t^i)\bar{\mu}_\xi}{\omega} + \bar{\mu}\frac{\dot{b}(t^i)}{\omega a} = 0.} \tag{8.175}$$

We may apply Lemma 12 to $[t_i, t^i]$. Indeed b and $\dot{b}$ are never zero on $]t_i, t^i[$; t_i satisfies hypothesis (H) and $|t^i - t_i|$ is smaller than $|t_{i+1} - t_i|$, hence than $\bar{\varepsilon}_0$, by the choice of $\bar{\omega}_5$.

In particular, by (8.141), we have

$$\left|\int_{t_i}^{t^i} b(x)\,dx\right| \geq \bar{\delta}\left\|Z_1^\star(t^i) - Z_1^\star(t_i)\right\| - K_5\left[\frac{1}{\sqrt{\omega}} + |t_i - t^i|\bar{M}_3\right], \tag{8.176}$$

(we use here the fact that $\left|\int_{t_i}^{t^i} b(x)\,dx\right| \leq \left|\int_{t_i}^{t_{i+1}} b(x)\,dx\right| \leq \bar{M}_3$ to derive (8.176) from (8.141)).

We have: $Z_1^\star(t^i) = \begin{bmatrix} 1 - \dfrac{b^2(t^i)}{2\omega a} + \dfrac{\int_0^1 b^2}{2\omega a} \\ -\dfrac{b(t^i)}{\omega} \\ \dfrac{\dot{b}(t^i)}{\omega a} \end{bmatrix}$; $Z_1^\star(t_i) = \begin{bmatrix} 1 - \dfrac{b^2(t_i)}{2\omega a} + \dfrac{\int_0^1 b^2}{2\omega a} \\ -\dfrac{b(t_i)}{\omega} \\ \dfrac{\dot{b}(t_i)}{\omega a} = 0 \end{bmatrix}$

by (8.133). Hence:

$$\|Z_1^\star(t^i) - Z_1^\star(t_i)\| \geq \left|\frac{\dot{b}(t_i)}{\omega a}\right| + \left| 1 - \frac{b^2(t^i)}{2\omega a} + \frac{\int_0^1 b^2}{2\omega a} - \right. \tag{8.177}$$

$$\left. - \left(1 - \frac{b^2(t_i)}{2\omega a} + \frac{\int_0^1 b^2}{2\omega a}\right) \right| .$$

We first remark that we may assume $\frac{|b(t^i)|}{\sqrt{\omega}}$ is bounded by a constant depending only on the δ_0 and β_0 of the (H) hypothesis.

Indeed, arguing by contradiction, we would have $\frac{b^2(t^i)}{2\omega a}$ very large. Then by the (H) hypothesis, $\left|1 - \frac{b^2(t_i)}{2\omega a} + \frac{\int_0^1 b^2}{2\omega a}\right| \leq \beta_0$. Hence (8.177) would imply, with an appropriate lower bound of $\frac{b^2(t^i)}{2\omega a}$ the following inequality:

$$\|Z_1^\star(t^i) - Z_1^\star(t_i)\| \geq 1. \tag{8.178}$$

And then, by (8.176):

$$\left|\int_{t_i}^{t^i} b(x)\, dx\right| \geq \bar{\delta} - K_5\left[\frac{1}{\sqrt{\omega}} + |t_i - t^i|\overline{M}_3\right]. \tag{8.179}$$

It would be enough then to choose ω_5 large enough so that $\omega \geq \omega_5$ would imply:

$$K_5\left[\frac{1}{\sqrt{\omega}} + [t_{i+1} - t_i]\overline{M}_3\right] < \frac{\bar{\delta}}{2}, \tag{8.180}$$

and (8.179) would give, using the fact that b is positive on $[t_i, t_{i+1}]$:

$$\left|\int_{t_i}^{t_{i+1}} b(x)\, dx\right| \geq \frac{\bar{\delta}}{2}. \tag{8.181}$$

This implies (8.164) with $\delta_1 = \frac{\bar{\delta}}{2}$.

Thus we are left with the case where:

$$\frac{|b(t^i)|}{\sqrt{\omega}} \leq C(\beta_0). \tag{8.182}$$

Inequality (8.177) holds and implies:

$$\|Z_1^\star(t^i) - Z_1^\star(t_i)\| \geq$$

$$\geq \frac{1}{1+\operatorname*{Sup}_{x\in M}|\bar{\mu}(x)|}\left[\left|\bar{\mu}(x(t^i))\frac{\dot{b}(t^i)}{\omega a}\right| + \left|1 - \frac{b^2(t^i)}{2\omega a} + \frac{\int_0^1 b^2}{2\omega a} - \right.\right.$$

$$\left.\left. - \left(1 - \frac{b^2(t_i)}{2\omega a} + \frac{\int_0^1 b^2}{2\omega a}\right)\right|\right] .$$

Hence:

$$\|Z_1^\star(t^i) - Z_1^\star(t_i)\| \geq \tag{8.184}$$

$$\geq \eta\left[\left|1 - \frac{b^2(t^i)}{2\omega a} + \frac{\int_0^1 b^2}{2\omega a} + \bar{\mu}(x(t^i))\frac{\dot{b}(t^i)}{\omega a} - \left(1 - \frac{b^2(t_i)}{2\omega a} + \frac{\int_0^1 b^2}{2\omega a}\right)\right|\right],$$

where we set:

$$\eta = \frac{1}{1+\operatorname*{Sup}_{x\in M}|\bar{\mu}(x)|} . \tag{8.185}$$

Inequality (8.184) implies via (8.175) and (8.182):

$$\|Z_1^\star(t^i) - Z_1^\star(t_i)\| \geq \eta\left|1 - \frac{b^2(t_i)}{2\omega a} + \frac{\int_0^1 b^2}{2\omega a} - \frac{a\tau}{\omega} - \frac{b(t^i)}{\omega}\bar{\mu}_\xi\right| \tag{8.186}$$

$$\geq \eta\left[\left|1 - \frac{b^2(t_i)}{2\omega a} + \frac{\int_0^1 b^2}{2\omega a}\right| - \frac{a\bar{\tau}}{\omega} - \frac{1}{\sqrt{\omega}}\operatorname*{Sup}_{x\in M}|\bar{\mu}_\xi(x)|\,C(\beta_0)\right],$$

where $\bar{\tau} = \operatorname*{Sup}_{x\in M}|\tau(x)|$.

On the other hand, by hypothesis (H), we have:

$$\left|1 - \frac{b^2(t^i)}{2\omega a} + \frac{\int_0^1 b^2}{2\omega a}\right| \geq \delta_0 .$$

Hence:

$$\|Z_1^\star(t^i) - Z_1^\star(t_i)\| \geq \eta\left(\delta_0 - \frac{a\bar{\tau}}{\omega} - \frac{1}{\sqrt{\omega}}C_1(\beta_0)\right) ; \tag{8.187}$$

where $C_1(\beta_0) = C(\beta_0) \sup_{x \in M} |\bar{\mu}_\xi(x)|$.

Hence, via (8.176):

$$\left|\int_{t_i}^{t^i} b(x)\,dx\right| \geq \bar{\delta}\eta\left(\delta_0 - \frac{a\bar{\tau}}{\omega} - \frac{1}{\sqrt{\omega}}C_1(\beta_0)\right) - K_5\left[\frac{1}{\sqrt{\omega}} + |t_i - t^i|\bar{M}_3\right]. \quad (8.188)$$

We choose then $\bar{\omega}_5$ large enough, uniform on all type (I) intervals such that $\omega \geq \bar{\omega}_5$ implies:

$$\bar{\delta}\eta\left(\frac{a\bar{\tau}}{\omega} + \frac{1}{\sqrt{\omega}}C_1(\beta_0)\right) + K_5\left[\frac{1}{\sqrt{\omega}} + [t_{i+1} - t_i]\bar{M}_3\right] < \frac{\bar{\delta}\eta\delta_0}{2}. \quad (8.189)$$

Then we have:

$$\left|\int_{t_i}^{t_{i+1}} b(x)\,dx\right| \geq \left|\int_{t_i}^{t^i} b(x)\,dx\right| \geq \frac{\bar{\delta}\eta\delta_0}{2} = \delta_1 > 0, \quad (8.190)$$

b being positive on $[t_i, t_{i+1}]$,

which proves (8.164) and completes the proof of Lemma 11. ■

Remark: The choice of ω_5 is uniform on all intervals of type (I), satisfying hypothesis (H) with given δ_0 and β_0. It depends in fact only on the choice of ω_5 in Lemma 12, where $[t_1, t_2]$ may be chosen arbitrarily, provided b and $\dot{b}$ are never zero on $]t_1, t_2[$; and it depends also on the fact that $|t_{i+1}, t_i| < \bar{\varepsilon}_0$, which holds uniformly on all type (II) intervals, as these intervals are contained in a set whose measure goes to zero when ω goes to $+\infty$. (It is indeed the complement of a concentration set.)

Proof of Lemmas 12 and 13: We first prove Lemma 13.

We state the differential equation satisfied by $Z^\star$, (8.143):

$$\begin{bmatrix} \dot{a}^\star \\ \dot{b}^\star \\ \dot{c}^\star \end{bmatrix} = -\begin{bmatrix} 0 & 0 & 1 \\ 0 & 0 & 0 \\ -1 & \bar{\mu}_\xi & \bar{\mu} \end{bmatrix}\begin{bmatrix} a^\star \\ b^\star \\ c^\star \end{bmatrix} \qquad Z_0^\star = \begin{bmatrix} a_0^\star \\ 0 \\ c_0^\star \end{bmatrix}.$$

As we already noticed in (8.152), we have $b^\star = 0$ and (8.143) reduces to:

$$\begin{cases} \dot{a}^\star = -c^\star \\ \dot{c}^\star = a^\star - \bar{\mu} c^\star . \end{cases} \tag{8.191}$$

At s_1, we have $|a^\star| + |c^\star|(s_1) \leq M$, $|a^\star|(s_1) \geq m > 0$ by (8.144).
On the other hand, we can find a positive real ν, uniform on M, such that:

$$\text{if } s \in \left]s_1-\nu,\ s_1+\nu\right[,\quad |a^\star|+|c^\star|(s) \leq M+1;\quad |a^\star|(s) \geq \frac{m}{2} > 0. \tag{8.192}$$

Let

$$\chi_1(M, m) = \frac{m}{8} \frac{1}{1+\sup\limits_{x\in M} |\bar{\mu}(x)|} . \tag{8.193}$$

If $|c^\star(s_1)| < \chi_1(M, m)$, we can find a positive real ν_1, uniform on M, such that:

$$\text{if } s \in \left]s_1-\nu_1,\ s_1+\nu_1\right[,\quad |c^\star|(s) < \frac{m}{4} \cdot \frac{1}{1+\sup\limits_{x\in M} |\bar{\mu}(x)|} . \tag{8.194}$$

We set

$$\theta(M, m) = \inf(\nu, \nu_1). \tag{8.195}$$

If $s \in \left]s_1 - \theta(M, m),\ s_1 + \theta(M, m)\right[$, we have:

$$|\dot{c}^\star| \geq |a^\star| - |\bar{\mu}|\ |c^\star| \geq \frac{m}{2} - |\bar{\mu}| \frac{m}{4} \cdot \frac{1}{1+\sup\limits_{x\in M} |\bar{\mu}(x)|} \geq \frac{m}{4}. \tag{8.196}$$

Then, we take:

$$\chi(M, m) = \inf\left(\chi_1(M, m),\ \frac{m}{8}\, \theta(M, m)\right). \tag{8.197}$$

If $|c^\star(s_1)| < \chi(M, m)$, $|c^\star(s_1)| < \chi_1(M, m)$; then (8.196) holds and $|\dot{c}^\star| \geq \frac{m}{4}$.

Assume for instance that $\dot{c}^\star > 0$ (the other case is identical). On the time interval $\left]s_1 - \theta(M, m),\ s_1 + \theta(M, m)\right[$, $c^\star$ takes at most once the value zero as $\dot{c}^\star \geq \frac{m}{4} > 0$.

$c^\star$ must vanish as:

$$c^\star(s_1+\theta(M,m)) \geq c^\star(s_1)+\frac{m}{4}\theta(M,m) \geq -\frac{m}{8}\theta(M,m)+\frac{m}{4}\theta(M,m) \tag{8.198}$$
$$\geq \frac{m}{8}\theta(M,m)$$
$$c^\star(s_1-\theta(M,m)) \leq c^\star(s_1)-\frac{m}{4}\theta(M,m) \leq \frac{m}{8}\theta(M,m)-\frac{m}{4}\theta(M,m)$$
$$\leq -\frac{m}{8}\theta(M,m);$$

(these inequalities are implied by the choice of $\chi(M,m)$).
This yields the existence of s_2 and the proof of (8.145) with $\mu(M,m)=\frac{m}{8}\theta(M,m)$.
Similarly, we have, as $\dot{c}^\star \geq \frac{m}{4}$:

$$|c^\star(s)| = |c^\star(s) - c^\star(s_2)| \geq \frac{m}{4}|s - s_2|, \tag{8.199}$$

$\forall\, s \in\,]s_1 - \theta(M,m),\, s_1 + \theta(M,m)[$,
which proves the second inequality of (8.145)' with $K_7 = \frac{m}{4}$.
In particular, with $s = s_1$, we have:

$$|s_1 - s_2| \leq \frac{4}{m}|c^\star(s_1)|; \tag{8.200}$$

which shows that $s_2 \in\,]s_1 - \frac{4}{m}|c^\star(s_1)|,\, s_1 + \frac{4}{m}|c^\star(s_1)|[$.
Finally,

$$\|Z^\star(s_2)-Z^\star(s_1)\| = |a^\star(s_2)-a^\star(s_1)| + |c^\star(s_2)-c^\star(s_1)| \tag{8.201}$$
$$= |c^\star(s_1)| + |a^\star(s_2-a^\star(s_1)|.$$

As $\dot{a}^\star = -c^\star$ and as $|c^\star| < \chi_1(M,m)$, (8.201) implies via (8.200):

$$\|Z^\star(s_2)-Z^\star(s_1)\| \leq |c^\star(s_1)| + \chi_1(M,m)\,|s_2 - s_1| \tag{8.202}$$
$$\leq |c^\star(s_1)|\left(\frac{4}{m}\chi_1(M,m) + 1\right).$$

We take then $K_6 = \text{Sup}(\frac{4}{m}, \frac{4}{m}\chi_1 + 1)$; which allows us to complete the proof of Lemma 13.

Proof of Lemma 12: Let

$$\bar{t} = \mathrm{Sup}\{t \in [t_1,t_2] \,|\, |\int_{t_1}^{t} b(x)\,dx| \leq \bar{s}(x(t_1)) + \bar{\alpha}\}. \qquad (8.203)$$

Consider the interval $[t_1,\bar{t}]$. Lemma 10' applies to it as b is never zero on $]t_1,t_2[$, hence on $]t_1,\bar{t}[$; C can be taken equal to: $C = \mathrm{Sup}_{x\in M}(\bar{s}(x) + \bar{\alpha})$. As $|\int_{t_1}^{\bar{t}} b(x)\,dx| \leq C$, we derive the existence of $\varepsilon_0(C)$ such that if $|\bar{t}-t_2|<\varepsilon_0(C)$, (8.106)' holds.

Thus, if we impose:

$$|t_2 - t_1| < \varepsilon_0(C), \qquad (8.204)$$

(8.106)' will hold on $[t_1,\bar{t}]$.

This means that we will have:

$$\mathrm{Sup}_{x\in[t_1,\bar{t}]} d(x(t),\bar{z}(t)) \leq \bar{K}[|\bar{t} - t_1| + (\bar{t} - t_1)^2], \qquad (8.205)$$

where $\bar{z}(t)$ is the curve tangent to $b(t)\,v(x(t))$ starting from $x(t_1)$. Thus, by the change of variables (8.129) and according to the notations of (8.119), we have under (8.204):

$$\mathrm{Sup}_{s\in[0,s(\bar{t})]} d(x_1(s),x_s) \leq \bar{K}[|\bar{t} - t_1| + (\bar{t} - t_1)^2] \leq 2\bar{K}|\bar{t} - t_1|, \qquad (8.206)$$

where $x_1(s)$ is the curve $x(t)$ reparametrized according to (8.129) and x_s is the curve tangent to v, reparametrized of $\bar{z}(t)$ following (8.129).

We first prove that the estimates (8.138), (8.139), (8.140), (8.141) hold with t_2 replaced by $\bar{t}$. We then derive that $\bar{t}$ is equal to t_2.

To prove (8.138), with t_2 replaced by $\bar{t}$, we use the very definition of $\bar{t}$.

As for (8.139), with t_2 replaced by $\bar{t}$, i.e. to prove:

$$\delta \leq \|Z^\star(s)\| \leq \bar{K}_1, \qquad \forall s \in [0,s(\bar{t})], \qquad (8.207)$$

we recall the definition of $Z^\star(s)$, given by (8.132):

$$\begin{cases} \dot{Z}^{\star} = -{}^t\Gamma(x_s)\, Z^{\star} \\ Z^{\star}(0) = Z_0^{\star} = \begin{bmatrix} 1 - \dfrac{b^2(t_1)}{2\omega a} + \dfrac{\int_0^1 b^2}{2\omega a} \\ 0 \\ \dfrac{\dot{b}(t_1)}{\omega a} \end{bmatrix} . \end{cases} \tag{8.132}$$

By hypothesis (H), we have:

$$\delta_0 \leq \|Z^{\star}(0)\| = \left| 1 - \frac{b^2(t_1)}{2\omega a} + \frac{\int_0^1 b^2}{2\omega a} \right| + \left| \frac{\dot{b}(t_1)}{\omega a} \right| \leq \beta_0. \tag{8.208}$$

Now $\dot{Z}^{\star} = -{}^t\Gamma(x_s)\, Z^{\star}$ and $[0, s(\bar{t})]$ is contained in $[0, C]$ $(C = \operatorname{Sup}_{x \in M}(s(x) + \bar{\alpha}))$.
Inequality (8.208) and these two last facts then imply the existence of constants δ and $\bar{K}_1 > 0$, uniform on M (i.e. independent of the choice of $x(t_1)$) such that (8.207) holds.

Indeed, we may always consider on an integral curve of v, $x_s(x_0)$, issued from $x_0 \in M$, the differential equation:

$$\dot{Z}^{\star} = -{}^t\Gamma(x_s)\, Z^{\star}. \tag{8.209}$$

We assume:

$$s \in [0, C] \tag{8.210}$$

$$\delta_0 \leq \|Z^{\star}(0, x_0)\| \leq \beta_0 . \tag{8.211}$$

We can find $\bar{K}_1$ and $\delta > 0$ <u>independent of</u> x_0 such that:

$$\delta \leq \inf_{s \in [0, C]} \|Z^{\star}(s, x_0)\| \leq \sup_{s \in [0, C]} \|Z^{\star}(s, x_0)\| \leq \bar{K}_1, \tag{8.212}$$

which proves (8.207).

We also have:

$$\|Z^{\star}(s_1, x_0) - Z^{\star}(0, x_0)\| \leq \sup_{s \in [0, C]} \left\| \frac{\partial Z^{\star}}{\partial s}(s, x_0) \right\| |s_1|, \tag{8.213}$$

$\forall\, s_1 \in [0, C]$,

hence the existence of a constant $\frac{1}{\bar{\delta}}$ such that:

$$\| Z^\star(s_1, x_0) - Z^\star(0, x_0) \| \le \frac{1}{\bar{\delta}} |s_1|, \quad \forall\, s_1 \in [0, C]. \tag{8.214}$$

By the choice of

$$\frac{1}{\bar{\delta}} = \sup_{x_0 \in M} \ \sup_{\delta_0 \le \| Z^\star(0, x_0) \| \le \beta_0} \ \sup_{s \in [0, C]} \left\| \frac{\partial Z^\star}{\partial s}(s, x_0) \right\|, \tag{8.215}$$

we obtain a (geometrical) uniform constant $\bar{\delta}$ such that:

$$\bar{\delta} \, \| Z^\star(s_1, x_0) - Z^\star(0, x_0) \| \le |s_1|, \quad \forall\, s_1 \in [0, C];\ \forall\, x_0 \in M. \tag{8.216}$$

In particular, applying (8.216) with $x_0 = x(t_1)$ and $s_1 = s(\bar{t})$, we obtain for the solution of (8.132);

$$\bar{\delta} \, \| Z^\star(s(\bar{t})) - Z^\star(0) \| \le |s(\bar{t})| = \left| \int_{t_1}^{\bar{t}} b(x)\, dx \right|. \tag{8.217}$$

Inequality (8.217) will imply (8.141) with t_2 replaced by $\bar{t}$ as soon as we have proved (8.140) with t_2 replaced by $\bar{t}$.

Indeed, if we assume that (8.140) is proved with t_2 replaced by $\bar{t}$, we have:

$$\| Z^\star(s(\bar{t})) - Z_1^\star(\bar{t}) \| \le K_5 \left[\frac{1}{\sqrt{\omega}} + |\bar{t} - t_1| \left| \int_{t_1}^{\bar{t}} b(x)\, dx \right| \right]. \tag{8.218}$$

On the other hand, we have by definition of $Z^\star(0)$ and $Z_1^\star(t_1)$, (see (8.132) - (8.135)):

$$Z^\star(0) = \begin{bmatrix} 1 - \dfrac{b^2(t_1)}{2\omega a} + \dfrac{\int_0^1 b^2}{2\omega a} \\ 0 \\ \dfrac{\dot{b}(t_1)}{\omega a} \end{bmatrix} \qquad Z_1^\star(t_1) = \begin{bmatrix} 1 - \dfrac{b^2(t_1)}{2\omega a} + \dfrac{\int_0^1 b^2}{2\omega a} \\ -\dfrac{b(t_1)}{\omega} \\ \dfrac{\dot{b}(t_1)}{\omega a} \end{bmatrix} \tag{8.219}$$

Thus:

$$\| Z^\star(0) - Z_1^\star(t_1) \| \le \frac{|b(t_1)|}{\omega} . \tag{8.220}$$

As, by hypothesis, $\left| 1 - \frac{b^2(t_1)}{2\omega a} + \frac{\int_0^1 b^2}{2\omega a} \right| \le \beta_0$ and as $\frac{\int_0^1 b^2}{\omega}$ goes to zero and ω to $+\infty$, $\frac{|b(t_1)|}{\sqrt{\omega}}$ is uniformly bounded by a constant $C_1(\beta_0)$. This implies:

$$\| Z^\star(0) - Z_1^\star(t_1) \| \le \frac{1}{\sqrt{\omega}} C_1(\beta_0) . \tag{8.221}$$

From (8.217), (8.218) and (8.221) we deduce:

$$\begin{aligned} &\bar{\delta} \| Z_1^\star(\bar{t}) - Z_1^\star(t_1) \| \le \qquad (8.222) \\ &\le \bar{\delta} \| Z_1^\star(\bar{t}) - Z^\star(s(\bar{t})) \| + \bar{\delta} \| Z^\star(s(\bar{t})) - Z^\star(0) \| + \bar{\delta} \| Z^\star(0) - Z_1^\star(t_1) \| \\ &\le \bar{\delta}\left(K_5 \left[\frac{1}{\sqrt{\omega}} + |\bar{t} - t_1| \left| \int_{t_1}^{\bar{t}} b(x)\, dx \right| \right] + \left| \int_{t_1}^{\bar{t}} b(x)\, dx \right| \right) + \bar{\delta} \frac{C_1(\beta_0)}{\sqrt{\omega}} . \end{aligned}$$

(8.222) implies (8.141) with t_2 replaced by $\bar{t}$ in that inequality.

To complete the proof of Lemma 12, it remains to prove that (8.140) holds with t_2 replaced by $\bar{t}$ and that $\bar{t}$ is in fact t_2.

Let us first prove (8.140) with t_2 replaced by $\bar{t}$, i.e.

$$\sup_{t \in [t_1, \bar{t}]} \| Z^\star(s(t)) - Z_1^\star(t) \| \le K_5 \left[\frac{1}{\sqrt{\omega}} + |\bar{t} - t_1| \left| \int_{t_1}^{\bar{t}} b(x)\, dx \right| \right] . \tag{8.223}$$

For this, we have:

$$\dot{Z}^\star = -{}^t\Gamma(x_s) Z^\star ; \quad Z^\star(0) = \begin{bmatrix} 1 - \frac{b^2(t_1)}{2\omega a} + \frac{\int_0^1 b^2}{2\omega a} \\ 0 \\ \frac{\dot{b}(t_1)}{\omega a} \end{bmatrix} .$$

and $Z_1^\star(t) = Z_2^\star(s(t))$ (see (8.135)) and $Z_2^\star$ satisfies:

$$\dot{Z}_2^{\star} = -{}^t\Gamma(x_1(s))\,Z_2^{\star} + \begin{bmatrix} 0 \\ -\dfrac{\dot{b}(t(s))}{\omega b(t(s))} \\ -\dfrac{a\tau(x_1(s))}{\omega b} \end{bmatrix}$$

with $$Z_2^{\star}(0) = \begin{bmatrix} 1 - \dfrac{b^2(t_1)}{2\omega a} + \dfrac{\int_0^1 b^2}{2\omega a} \\ -\dfrac{b(t_1)}{\omega} \\ \dfrac{\dot{b}(t_1)}{\omega a} \end{bmatrix} ;$$

which implies:

$$\left\{ \begin{aligned} &\dot{\overbrace{Z^{\star} - Z_2^{\star}}} = -{}^t\Gamma(x(s))(Z^{\star} - Z_2^{\star}) + ({}^t\Gamma(x(s)) - {}^t\Gamma(x_s))Z^{\star} \\ &\qquad + \begin{bmatrix} 0 \\ -\dfrac{\dot{b}(t(s))}{\omega b(t(s))} \\ -\dfrac{a\tau(x_1(s))}{\omega b(t(s))} \end{bmatrix} \\ &Z^{\star}(0) - Z_2^{\star}(0) = \begin{bmatrix} 0 \\ \dfrac{b(t_1)}{\omega} \\ 0 \end{bmatrix} . \end{aligned} \right. \tag{8.224}$$

On the other hand we have:

$$d(x_1(s), x_2) \le 2\bar{K}|\bar{t} - t_1| , \quad s \in [0, s(\bar{t})] \tag{8.206}$$

$$\delta \le \|Z^{\star}(s)\| \le \bar{K}_1 . \tag{8.207}$$

Let

$$\begin{cases} Q = \operatorname*{Sup}_{x \in M} \| {}^t\Gamma(x) \| \ ; \qquad F = \operatorname*{Sup}_{x \in M,\ y \in M} \dfrac{\| {}^t\Gamma(x) - {}^t\Gamma(y) \|}{d(x,y)} \\ \bar{\tau} = \operatorname*{Sup}_{x \in M} |\tau(x)| \ . \end{cases} \tag{8.225}$$

(8.224), (8.206) and (8.207) imply:

$$\widehat{\| Z^\star - Z_2^\star \|}^{\bullet}(s) \le Q \| Z^\star - Z_2^\star \| + 2F\bar{K}\bar{K}_1 |\bar{t} - t_1| + \left| \frac{\dot{b}(t(s))}{\omega b(t(s))} \right| \tag{8.226}$$
$$+ \frac{|a\bar{\tau}|}{\omega |b|},$$

$\forall\ s \in [0, s(\bar{t})]$; hence:

$$\| Z^\star - Z_2^\star \|(s) \le \| Z^\star(0) - Z_2^\star(0) \| + e^{Qs} \int_0^s \Big[2F\bar{K}\bar{K}_1 |\bar{t} - t_1| \tag{8.227}$$
$$+ \frac{|\dot{b}(t(\alpha))|}{\omega |b(t(\alpha))|} + a\bar{\tau} \frac{1}{\omega |b(t(\alpha))|} \Big] d\alpha \ .$$

As b is positive and as $\dot{b}$ does not vanish, we have:

$$\frac{1}{\omega} \times \int_0^s \left[\left| \frac{\dot{b}(t(\alpha))}{b(t(\alpha))} \right| + \left| \frac{a\bar{\tau}}{b(t(\alpha))} \right| \right] d\alpha = \frac{1}{\omega} \left| \int_0^s \frac{\dot{b}(t(\alpha))}{b(t(\alpha))} d\alpha \right| \tag{8.228}$$
$$+ \frac{a\bar{\tau}}{\omega} \int_0^s \frac{d\alpha}{b(t(\alpha))},$$

by setting $t(\alpha) = x$, we thus have $\alpha = s(x)$ and $d\alpha = b(x)\,dx$, and we have then:

$$\int_0^s \left[\left| \frac{\dot{b}(t(\alpha))}{\omega b(t(\alpha))} \right| + \left| \frac{a\bar{\tau}}{\omega b(t(\alpha))} \right| \right] d\alpha = \frac{1}{\omega} \Big[\left| \int_{t_1}^{t(s)} \dot{b}(x)\,dx \right| \tag{8.229}$$
$$+ a\bar{\tau} \int_{t_1}^{t(s)} dx \Big] = \frac{1}{\sqrt{\omega}} \left| \frac{b(t_1)}{\sqrt{\omega}} - \frac{b(t(s))}{\sqrt{\omega}} \right| + \frac{a\bar{\tau}}{\omega} |t(s) - t_1|$$
$$\le \frac{1}{\sqrt{\omega}} \left| \frac{b(t_1)}{\sqrt{\omega}} - \frac{b(t(s))}{\sqrt{\omega}} \right| + \frac{a\bar{\tau}}{\omega} |\bar{t} - t_1| \ ;$$

$\forall\ s \in [0, s(\bar{t})]$.

If we also take into account the fact that $\| Z^\star(0) - Z_2^\star(0) \| = \dfrac{|b(t_1)|}{\omega}$, inequality (8.229) then implies:

$$\|Z^\star - Z_2^\star\|(s) \tag{8.230}$$

$$\leq e^{Qs}\left[2F\bar{K}\bar{K}_1\,|\bar{t}-t_1|s+\frac{a\bar{\tau}}{\omega}\,|\bar{t}-t_1|+\frac{1}{\sqrt{\omega}}\left|\frac{b(t_1)}{\sqrt{\omega}}-\frac{b(t(s))}{\sqrt{\omega}}\right|\right]+\frac{1}{\sqrt{\omega}}\left|\frac{b(t_1)}{\sqrt{\omega}}\right|;$$

$\forall\, s \in [0, s(\bar{t})]$.

We know that $\|Z^\star(s)\| \leq \bar{K}_1$, $s \leq C$, $\dfrac{|b(t_1)|}{\sqrt{\omega}} \leq C_1(\beta_0)$. On the other hand (see (8.133)-(8.135))

$$\|Z_2^\star\|(s) = \left|1 - \frac{b^2(t(s))}{2\omega a} + \frac{\int_0^1 b^2}{2\omega a}\right| + \left|\frac{b(t(s))}{\omega}\right| + \left|\frac{\dot{b}(t(s))}{\omega a}\right|.$$

Hence:

$$\left|1 - \frac{b^2(t(s))}{2\omega a} + \frac{\int_0^1 b^2}{2\omega a}\right| \leq \|Z_2^\star\|(s) \leq \|Z^\star - Z_2^\star\|(s) + \|Z^\star(s)\| \tag{8.231}$$

$$\leq \bar{K}_1 + e^{QC}\left[(2F\bar{K}\bar{K}_1+\frac{a\bar{\tau}}{\omega})\,|\bar{t}-t_1|s+\frac{1}{\omega}\,|b(t(s))|+\frac{1}{\sqrt{\omega}}C_1(\beta_0)\right]+\frac{1}{\sqrt{\omega}}C_1(\beta_0).$$

Taking into account the fact that $\dfrac{\int_0^1 b^2}{2\omega a}$ goes to zero when ω goes to $+\infty$ and that $a_0 \leq a \leq a_1$, inequality (8.231) implies:

$$\frac{|b(t(s))|}{\sqrt{\omega}} \leq C' \quad \text{for } \omega \geq \omega_5, \tag{8.232}$$

which yields through (8.231), taking into account the fact that $\dfrac{|b(t_1)|}{\sqrt{\omega}} < C_1(\beta_0)$ and that $s \leq s(\bar{t}) = \left|\int_{t_1}^{\bar{t}} b(x)\,dx\right|$:

$$\|Z^\star - Z_2^\star\|(s) \leq \tag{8.233}$$

$$\leq e^{QC}\left[2F\bar{K}\bar{K}_1\,|\bar{t}-t_1|\left|\int_{t_1}^{\bar{t}} b(x)\,dx\right| + \frac{a\bar{\tau}}{\omega}|\bar{t}-t_1|+\frac{1}{\sqrt{\omega}}C_1(\beta_0)+\frac{C'}{\sqrt{\omega}}\right]+\frac{1}{\sqrt{\omega}}C_1(\beta_0),$$

which gives, with an appropriate constant and for $s = s(t)$:

$$\|Z^\star - Z_1^\star\|(s(t)) \leq K_5\left[\frac{1}{\sqrt{\omega}} + |\bar{t}-t_1|\left|\int_{t_1}^{\bar{t}} b(x)\,dx\right|\right] \text{ for } \omega \geq \omega_5. \tag{8.234}$$

Now $Z_2^\star(s(t)) = Z_1^\star(t)$.

Hence:

$$\sup_{t \in [t_1, \bar{t}]} \| Z^\star(s(t)) - Z_1^\star(t) \| \leq K_5 \left[\frac{1}{\sqrt{\omega}} + |\bar{t} - t_1| \left| \int_{t_1}^{\bar{t}} b(x)\, dx \right| \right], \tag{8.235}$$

for $\omega \geq \omega_5$; which is exactly (8.223) or else (8.140) with t_2 replaced by $\bar{t}$.

Arguing by contradiction, we assume that:

$$t_2 > \bar{t}. \tag{8.236}$$

Then, by the very definition of $\bar{t}$, we have:

$$s(\bar{t}) = \int_{t_1}^{\bar{t}} b(x)\, dx = \bar{s}(x(t_1)) + \bar{\alpha}. \tag{8.237}$$

We prove now that (8.137) implies, when ω is large enough and $t_2 - t_1$ is small enough, that $\dot{b}$ takes the value zero in $[t_1, \bar{t}]$; this yields a contradiction.

The idea is the following: as $s(\bar{t}) = \bar{s}(x(t_1)) + \bar{\alpha}$, $s(\bar{t})$ is larger than $\bar{s}(x(t_1))$. Between 0 and $\bar{s}(x(t_1))$, the form α_x has rotated and came back on itself as a field of planes, along the piece x_s tangent to v; indeed, $x_{\bar{s}(x(t_1))} = \psi(x(t_1))$ by the very definition of $\bar{s}$ (see Definition 3).

Consider now an indirect basis $(v, e_1(0), e_2(0))$ at $x_0 = x(t_1)$ and transport it along x_s, thus obtaining a basis $(v, e_1(s), e_2(s))$ at each time s.

The trace $u(s)$ of α_{x_s} in the plane $(e_1(s), e_2(s))$ rotates from e_1 to e_2 when s increases, as seen in Proposition 9:

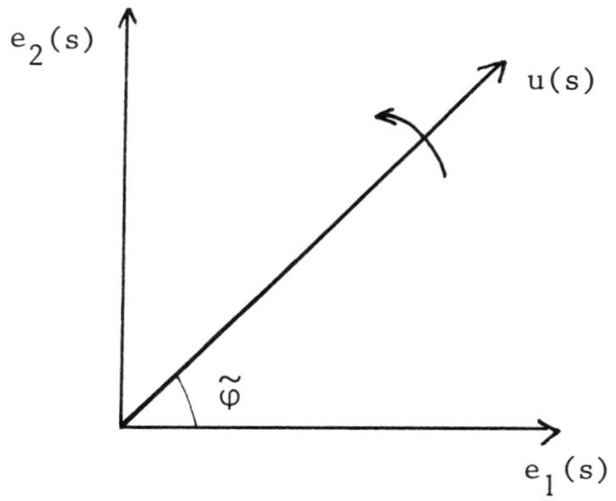

As in Definition 1, we denote by $\tilde{\varphi}$ the angle of $u(s)$ and $e_1(s)$. We know that $\dot{\tilde{\varphi}}$ is positive (see Proposition 9). Between $s = 0$ and $s = \bar{s}(x(t_1)) + \frac{\bar{\alpha}}{2}$, has increased from the initial value $\tilde{\varphi}(0)$ to $\tilde{\varphi}(s) + \pi + \delta\tilde{\varphi}$. We claim that:

$$\delta\tilde{\varphi} \geq \Delta\theta > 0, \tag{8.238}$$

where $\Delta\theta$ is a uniform positive constant on M.

To see (8.238), we choose, when x_0 varies in M, a continuously dependent family of indirect basis $(v(x_0), e_1(0,x_0), e_2(0,x_0))$. We consider $u(s,x_0)$ which is the trace of $\alpha_{x_s(x_0)}$ in $(e_1(s,x_0), e_2(s,x_0))$, where $x_s(x_0)$ is the v-orbit through x_0. We then have a continuous function $\tilde{\varphi}(s,x_0)$ which represents the angle of $u(s,x_0)$ and $e_1(s,x_0)$.

Now, between $s = 0$ and $s = \bar{s}(x_0) + \frac{\bar{\alpha}}{2}$, $\tilde{\varphi}(s,x_0)$ increases from $\tilde{\varphi}(0,x_0)$ to $\tilde{\varphi}(0,x_0) + \pi + \delta\tilde{\varphi}(x_0)$ with $\delta\tilde{\varphi}(x_0) > 0$.

We then set:

$$\Delta\theta = \inf_{x_0 \in M} \delta\tilde{\varphi}(x_0), \tag{8.239}$$

and (8.238) is proved.

Consider now a <u>direction</u> $z(0,x_0)$ at x_0 in the plane $(e_1(0,x_0), e_2(0,x_0))$. We transport it by v along the v-orbit. Let $z(s,x_0)$ be the result of this transport at time s in $(e_1(s,x_0), e_2(s,x_0))$. If $\theta(z(0,x_0),x_0) = \theta_0(z(0,x_0))$ is the initial angle of $z(0,x_0)$ with $e_1(0,x_0)$, $\theta_0(z(0,x_0))$ is also the angle of $z(s,x_0)$ and $e_1(s,x_0)$, as these two vectors are transported vectors.

As $\tilde{\varphi}(s,x_0)$ covers an interval of length larger than $\pi + \Delta\theta$, with $\Delta\theta > 0$, when s runs from 0 to $\bar{s}(x_0) + \frac{\bar{\alpha}}{2}$, we thus derive the existence of an $s_1(z(0,x_0),x_0)$ and a uniform $\delta s > 0$, such that:

$$\bar{s}(x_0) + \frac{\bar{\alpha}}{2} \geq s_1(z(0,x_0),x_0) \geq \delta s > 0$$

and

$$u(s_1(z(0,x_0),x_0) = z(s,x_0) \quad \text{(as directions)},$$

i.e. the trace of α_{x_s} in $(e_1(s), e_2(s))$ will cover $z(s, x_0)$ at a time s_1 larger than $\delta s > 0$ and less than $\bar{s}(x_0) + \frac{\overline{\alpha}}{2}$.

This is merely due to the fact that $\tilde{\varphi}(s, x_0)$ covers an interval of length larger than $\pi + \Delta\theta$, with $\Delta\theta > 0$ and uniform, when s runs from 0 to $\bar{s}(x_0) + \frac{\overline{\alpha}}{2}$. Hence, by continuity, $\tilde{\varphi}(s, x_0)$ will do the same thing on $[\delta s, \bar{s}(x_0) + \frac{\overline{\alpha}}{2}]$, $\delta s > 0$ and uniform; hence covering any transported direction during this time.

Let $\delta\bar{s} = \inf(\delta s, \overline{\alpha}) > 0$.

We thus proved given a direction in $(e_1(0, x_0), e_2(0, x_0))$ the existence of s_1 satisfying

$$\delta\bar{s} \leq s_1 \leq \bar{s}(x_0) + \frac{\overline{\alpha}}{2}, \tag{8.240}$$

such at time s_1, this direction is covered by the trace of α_{x_s} in the plane $(e_1(s, x_0), e_2(s, x_0))$.

We apply this result to our situation.

Let $Z^\star$ be solution of equation (8.132)

$$\begin{cases} \dot{Z}^\star = -{}^t\Gamma(x_s)\, Z^\star \\ Z^\star(0) = Z_0^\star = \begin{bmatrix} 1 - \dfrac{b^2(t_1)}{2\omega a} + \dfrac{\int_0^1 b^2}{2\omega a} \\ 0 \\ \dfrac{\dot{b}(t_1)}{\omega a} \end{bmatrix} \end{cases} \tag{8.132}$$

$Z^\star$ provides us with the components of a transported form along v in the frame $(\alpha_{x_s}, \gamma_{x_s}, -\beta_{x_s})$.

The second component of this form is zero at time 0; by (8.152), it will remain zero. Hence, this transported form vanishes on v, as $\alpha_{x_s}(v) = \beta_{x_s}(v) = 0$. As a field of planes, we can then follow it with its trace in the moving frame $(e_1(s, x(t_1)), e_2(s, x(t_1)))$. In this frame, this trace defines an initial direction $z(0, x(t_1))$ which we transport along x_s. This direction is covered, as we have just proved, at a certain time s_1 satisfying (8.138) by the trace of α_{x_s} in

$(e_1(s, x(t_1)), e_2(s, x(t_1)))$. At this point s_1, $Z^\star(s_1)$ and α_{x_s} are thus dependent as their traces coincide while they are both tangent to v. Hence

$$Z^\star(s_1) = \lambda \begin{bmatrix} 1 \\ 0 \\ 0 \end{bmatrix} \quad \lambda \neq 0, \quad s_1 \in [\delta \bar{s}, \bar{s}(x(t_1)) + \frac{\bar{\alpha}}{2}]. \tag{8.241}$$

We denote $(a^\star, 0, c^\star)$ the components of $Z^\star$ solution of (8.132)

$$Z^\star = \begin{bmatrix} a^\star \\ 0 \\ c^\star \end{bmatrix}. \tag{8.241'}$$

We have:

$$c^\star(s_1) = 0.$$

On the other hand, by (8.207), holding on $[0, s(\bar{t})]$, thus on s_1, as $s(\bar{t}) = \bar{s}(x(t_1)) + \bar{\alpha}$, we have

$$\delta \leq \|Z^\star(s_1)\| = |a^\star(s_1)| + |c^\star(s_1)| \leq \bar{K}_1 . \tag{8.242}$$

Thus

$$|a^\star(s_1)| \geq \delta \qquad (c^\star(s_1) = 0) \tag{8.243}$$

and

$$|c^\star(s_1)| + |a^\star(s_1)| \leq \bar{K}_1 . \tag{8.244}$$

Finally, $\begin{bmatrix} a^\star(0) \\ 0 \\ c^\star(0) \end{bmatrix} = \begin{bmatrix} a_0^\star \\ 0 \\ c_0^\star \end{bmatrix}$ (i.e. the second component is zero).

Consequently, setting $M = K_1$ and $m = \delta$, we apply Lemma 13. As $|c^\star(s_1)|$ is zero here, it is certainly smaller than $\chi(M, m) = \chi(\bar{K}_1, \delta)$. Similarly the point s_2 given by Lemma 13 is in this situation the point s_1 itself. Hence, this lemma implies here that $c^\star(s)$ vanishes once and only once in $]s_1 - \theta(\bar{K}_1, \delta), s_1 + \theta(\bar{K}_1, \delta)[$ and that, by (8.145)',

$$|c^\star(s)| \geq K_7(\bar{K}_1, \delta)\ |s - s_2| = K_7(\bar{K}_1, \delta)\ |s - s_1| \tag{8.245}$$

on this interval.

Let then

$$\underline{\alpha} = \inf \left(\frac{\theta(\overline{K}_1, \delta)}{2}, \frac{\delta \bar{s}}{2}, \frac{\overline{\alpha}}{2} \right) . \qquad (8.246)$$

The interval $[s_1 - \underline{\alpha}, s_1 + \underline{\alpha}]$ is contained in $]s_1 - \theta(\overline{K}_1, \delta), s_1 + \theta(\overline{K}_1, \delta)[$ and also, as $s_1 - \underline{\alpha} \geq \delta\bar{s} - \underline{\alpha} \geq \frac{\delta\bar{s}}{2}$ and $s_1 + \underline{\alpha} \leq \bar{s}(x(t_1)) + \frac{\overline{\alpha}}{2} + \frac{\overline{\alpha}}{4} < \bar{s}(x(t_1)) + \overline{\alpha}$, in $[0, \bar{s}(x(t_1)) + \overline{\alpha}]$, i.e.

$$[s_1 - \underline{\alpha}, s_1 + \underline{\alpha}] \subset]s_1 - \theta(\overline{K}_1 - \delta), s_1 + \theta(\overline{K}_1, \delta)[\cap [0, \bar{s}(x(t_1)) + \overline{\alpha}]. \qquad (8.247)$$

In particular, (8.235) holds on this interval, and by applying it to $s_1 - \underline{\alpha}$ and $s_1 + \underline{\alpha}$, we derive:

$$|c^\star(s_1 - \underline{\alpha})| \geq K_7 \underline{\alpha} > 0 \qquad |c^\star(s_1 + \underline{\alpha})| \geq K_7 \underline{\alpha} > 0. \qquad (8.248)$$

We also know that:

$$c^\star(s_1 - \underline{\alpha})\, c^\star(s_1 + \underline{\alpha}) < 0 \qquad (8.249)$$

as $c^\star$ changes sign once, at s_1, in this interval.

We conclude as follows:

Inequality (8.235) holds for all $t \in [t_1, \bar{t}]$, if $\omega \geq \omega_5$.

Let

$$\underline{t} \in [t_1, \bar{t}] \text{ such that } s(\underline{t}) = s_1 - \underline{\alpha} \qquad (8.250)$$

$$\underline{\underline{t}} \in [t_1, \bar{t}] \text{ such that } s(\underline{\underline{t}}) = s_1 + \underline{\alpha}.$$

The existence of $\underline{t}$ and $\underline{\underline{t}}$ is implied by the fact that $[s_1 - \underline{\alpha}, s_1 + \underline{\alpha}]$ is contained in $[0, \bar{s}(x(t_1)) + \overline{\alpha}] = [0, s(\bar{t})] = [s(t_1), s(\bar{t})]$. From (8.235) and (8.236) we derive:

$$\|Z^\star(s(\underline{t})) - Z_1^\star(\underline{t})\| = \|Z^\star(s_1 - \underline{\alpha}) - Z_1^\star(\underline{t})\| \qquad (8.251)$$
$$\leq K_5 \left[\frac{1}{\sqrt{\omega}} + |\bar{t} - t_1| \left| \int_{t_1}^{\bar{t}} b(x)\, dx \right| \right];$$

$$\|Z^\star(s(\underline{\underline{t}})) - Z_1^\star(\underline{\underline{t}})\| = \|Z^\star(s_1 + \underline{\alpha}) - Z_1^\star(\underline{\underline{t}})\| \tag{8.252}$$

$$\leq K_5\left[\frac{1}{\sqrt{\omega}} + |\bar{t} - t_1| \, \left|\int_{t_1}^{\bar{t}} b(x)\, dx\right|\right].$$

hence, if we come back to (8.133) which defines $Z_1^\star(t)$:

$$\left|c^\star(s_1 - \underline{\alpha}) - \frac{\dot{b}(\underline{t})}{\omega a}\right| \leq \|Z^\star(s_1 - \underline{\alpha}) - Z_1^\star(\underline{t})\| \tag{8.253}$$

$$\leq K_5\left[\frac{1}{\sqrt{\omega}} + |\bar{t} - t_1| \, \left|\int_{t_1}^{\bar{t}} b(x)\, dx\right|\right]$$

$$\leq K_5\left[\frac{1}{\sqrt{\omega}} + |t_2 - t_1| [\bar{s}(x(t_1)) + \bar{\alpha}]\right]$$

$$\leq K_5\left[\frac{1}{\sqrt{\omega}} + |t_2 - t_1| \sup_{x\in M} (\bar{s}(x) + \bar{\alpha})\right];$$

$$\left|c^\star(s_1 + \underline{\alpha}) - \frac{\dot{b}(\underline{\underline{t}})}{\omega a}\right| \leq \|Z^\star(s(\underline{\underline{t}})) - Z_1(\underline{t})\| \tag{8.254}$$

$$\leq K_5\left|\frac{1}{\sqrt{\omega}} + |t_2 - t_1| \sup_{x\in M} (\bar{s}(x) + \bar{\alpha})\right|.$$

Thus:

$$\begin{cases} c^\star(s_1 - \underline{\alpha}) - K_5\left[\frac{1}{\sqrt{\omega}} + |t_2 - t_1| \sup_{x\in M} (\bar{s}(x) + \bar{\alpha})\right] \leq \frac{\dot{b}(\underline{t})}{\omega a} \\ \leq c^\star(s_1 - \underline{\alpha}) + K_5\left[\frac{1}{\sqrt{\omega}} + |t_2 - t_1| \sup_{x\in M} (\bar{s}(x) + \bar{\alpha})\right] \\ c^\star(s_1 + \underline{\alpha}) - K_5\left[\frac{1}{\sqrt{\omega}} + |t_2 - t_1| \sup_{x\in M} (\bar{s}(x) + \bar{\alpha})\right] \leq \frac{\dot{b}(\underline{\underline{t}})}{\omega a} \\ \leq c^\star(s_1 + \underline{\alpha}) + K_5\left[\frac{1}{\sqrt{\omega}} + |t_2 - t_1| \sup_{x\in M} (\bar{s}(x) + \bar{\alpha})\right] \end{cases} \tag{8.255}$$

We now choose ω_5 and $0 < \bar{\varepsilon}_0 < \varepsilon_0(c)$ such that:

$$K_5\left[\frac{1}{\sqrt{\omega}} + \bar{\varepsilon}_0 \sup_{s\in M} (\bar{s}(x) + \bar{\alpha})\right] < \frac{K_7\underline{\alpha}}{2}. \tag{8.256}$$

Then (8.256) and (8.248) imply, if $|t_2 - t_1| < \bar{\varepsilon}_0$ and $\omega \geq \omega_5$:

$$\begin{cases} \dfrac{c^\star(s_1 - \underline{\alpha})}{2} \le \dfrac{\dot b(\underline{t})}{\omega a} \le \dfrac{3c^\star(s_1 - \underline{\alpha})}{2} \\[2ex] \dfrac{3}{2} c^\star(s_1 + \underline{\alpha}) \le \dfrac{\dot b(\underline{\underline{t}})}{\omega a} \le \dfrac{c^\star(s_1 + \underline{\alpha})}{2} \end{cases} \tag{8.257}$$

if $c^\star(s_1 - \underline{\alpha}) > 0$ hence $c^\star(s_1 + \underline{\alpha}) < 0$ by (8.249) or

$$\begin{cases} \dfrac{3}{2} c^\star(s_1 - \underline{\alpha}) \le \dfrac{\dot b(\underline{t})}{\omega a} \le \dfrac{c^\star(s_1 - \underline{\alpha})}{2} \\[2ex] \dfrac{c^\star(s_1 + \underline{\alpha})}{2} \le \dfrac{\dot b(\underline{\underline{t}})}{\omega a} \le \dfrac{3}{2} c^\star(s_1 + \underline{\alpha}) \end{cases} \tag{8.258}$$

if $c^\star(s_1 - \underline{\alpha}) < 0$ hence $c^\star(s_1 + \underline{\alpha}) > 0$ by (8.249) again.

In any case, $\dfrac{\dot b(\underline{t})}{\omega a}$ and $\dfrac{\dot b(\underline{\underline{t}})}{\omega a}$ have opposite signs, as $c^\star(s_1 - \underline{\alpha})$ and $c^\star(s_1 + \underline{\alpha})$ have opposite signs; this forces $\dot b$ to vanish somewhere in $[\underline{t}, \underline{\underline{t}}]$, hence in $]t_1, t_2[$, thus yielding a contradiction.

This implies that $\bar t = t_2$ as soon as $|t_2 - t_1| < \bar\varepsilon_0$ and $\omega \ge \omega_5$. The proof of Lemma 12 is thereby complete.

<u>Remark</u>: One can easily check that ω_5 can be chosen uniform as well as all other constants.

We sum up now what we have proved while analysing the type (I) intervals.

Lemma 11 is expressed by the following drawings:

(8.259)

<u>If t_i satisfies (H)</u>

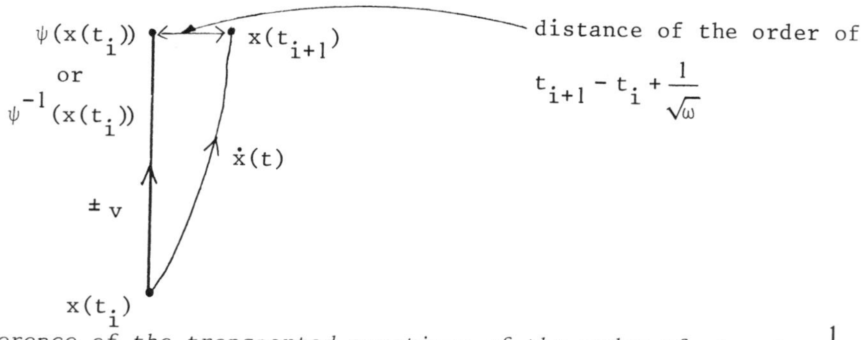

difference of the transported equations of the order of $t_{i+1} - t_i + \frac{1}{\sqrt{\omega}}$

Lemma 12 is expressed in:

(8.260)

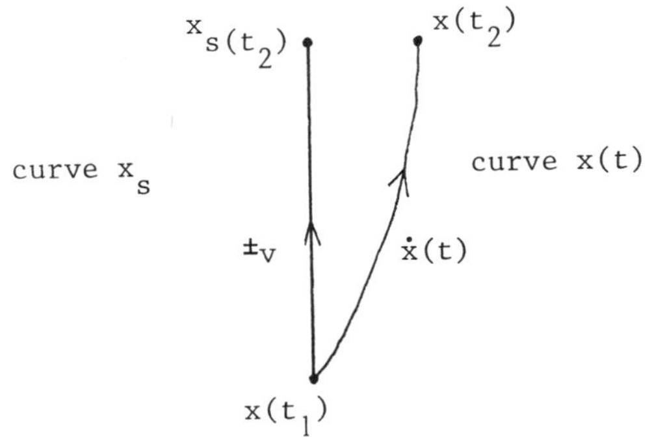

if $|t_2 - t_1|$ is small enough and if b and $\dot{b}$ do not vanish in $]t_1, t_2[$, then:

1) the length along v of $x(t)$ on $[t_1, t_2]$ $|\int_{t_1}^{t_2} b(x)\, dx|$ is bounded by an explicit constant;

2) the difference of the transport equations is of the order $\frac{1}{\sqrt{\omega}} + |t_2 - t_1| \, |\int_{t_1}^{t_2} b(x)\, dx|$;

3) if $Z_1^{\star}(t)$ has considerably changed from t_1 to t_2, $|\int_{t_1}^{t_2} b(x)\, dx|$ too has considerably changed.

Remark: We may have $t_2 < t_1$.

Lemma 12 means that each time α_{x_s} coincides with a transported form (as fields of planes), then, in a small neighbourhood, there is no other coincidence point and α_{x_s} diverges from this transported form backwards and forwards before and after the coincidence time. This is merely Proposition 9 expressed on transported planes with respect to α_{x_s}.

Lemmas 12 and 13 allow us to shorten the analysis of type (II) intervals; we start now.

Analysis of type (II) intervals

Let $[t_i, t_{i+1}]$ be a type (II) interval. Such an interval is characterized by:

$$\dot{b}(t_i) = \dot{b}(t_{i+1}) = 0, \quad b(s^i) = 0 \tag{8.261}$$

$s^i \in]t_i, t_{i+1}[$; $\dot{b} \neq 0$ on $]t_i, t_{i+1}[$.

$[t_i, t_{i+1}]$ does not meet a concentration point.

We already pointed out that s^i is the unique point in $]t_i, t_{i+1}[$ where b vanishes. As $[t_i, t_{i+1}]$ does not meet a concentration interval, we have:

$$\forall\, t \in [t_i, t_{i+1}], \quad \frac{b^2(t)}{\omega} \geq \varepsilon_1 \quad \text{or} \quad \frac{|\dot{b}(t)|}{\omega} \geq \varepsilon_2 ; \tag{8.262}$$

Indeed, by (8.41) of Lemma 8, a concentration interval $I_1(s)$ is defined by the fact it contains s such that

$$\frac{b^2(s)}{\omega} \leq \varepsilon_1 \quad \text{and} \quad \frac{\dot{b}(s)}{\omega} < \varepsilon_2 = \inf\left(\alpha_0, \frac{\sqrt{\alpha_1}}{3}\right).$$

In particular, as $\dot{b}(t_i) = \dot{b}(t_{i+1}) = 0$, we have:

$$\frac{b^2(t_i)}{\omega} \geq \varepsilon_1, \quad \frac{b^2(t_{i+1})}{\omega} \geq \varepsilon_1, \tag{8.263}$$

whereas, in s^i, as $b(s^i) = 0$, we have:

$$\frac{|\dot{b}(s^i)|}{\omega} \geq \varepsilon_2 . \tag{8.264}$$

We will assume that at t_i, hypothesis (H) is satisfied:

$$\boxed{\exists\, \beta_0, \quad \exists\, \delta_0 > 0 \;|\; \beta_0 \geq 1 - \frac{b^2(t_i)}{2\omega a} + \frac{\int_0^1 b^2}{2\omega a} \geq \delta_0 > 0.} \tag{8.265}$$

We have then:

Lemma 14: Assume that in t_i hypothesis (H) is satisfied. There exist δ_3, K_8, M_4, ω_6 positive, M_4, δ_3, K_8 being geometric constants such that, if $\omega \geq \omega_6$, we have:

$$\delta_3 \varepsilon_2 \le \left| \int_{t_i}^{s^i} b(t)\,dt \right| \le M_4 \; ; \; \delta_3 \varepsilon_2 \le \left| \int_{s^i}^{t_{i+1}} b(t)\,dt \right| \le M_4 . \tag{8.266}$$

$$d(x(t_i), x(t_{i+1})) \le \bar{K}_8 \left(t_{i+1} - t_i + \frac{1}{\sqrt{\omega}} \right) \tag{8.267}$$

$$\left\| Z_1^\star(t_{i+1}) - Z_1^\star(t_i) \right\| \le \bar{K}_8 \left(t_{i+1} - t_i + \frac{1}{\sqrt{\omega}} \right) , \tag{8.268}$$

ω_6 is uniform on all type (II) intervals.

As usual $Z_1^\star$ is defined by (8.133).

The proof of Lemma 14 is obtained by applying several times Lemmas 12 and 13.

Proof of Lemma 14: First we take ω_6 large enough such that $\omega \ge \omega_6$ implies $t_{i+1} - t_i < \bar{\varepsilon}_0(c)$.

Such a choice of ω_6 can be uniformly done on all type (II) intervals as their total measure goes to zero.

We consider then the interval $]t_i, s^i[$.

On $]t_i, s^i[$, b and $\dot{b}$ do not vanish. Furthermore, at t_i, we have, by (H) and the fact that $\dot{b}(t_i) = 0$:

$$\alpha_0 \le \frac{|\dot{b}|}{\omega a}(t_i) + \left| 1 - \frac{b^2(t_i)}{2\omega a} + \frac{\int_0^1 b^2}{2\omega a} \right| = \left| 1 - \frac{b^2(t_i)}{2\omega a} + \frac{\int_0^1 b^2}{2\omega a} \right| \le \beta_0 . \tag{8.269}$$

We also have:

$$|s^i - t_i| \le |t_{i+1} - t_i| < \bar{\varepsilon}_0 , \tag{8.270}$$

by the choice of ω_6.

We can then apply Lemma 12 to $[t_i, s^i]$ and we have by (8.138), (8.139), (8.140), if we choose $\omega_6 \ge \omega_5$:

$$\left| \int_{t_i}^{s^i} b(x)\,dx \right| \le \bar{s}(x(t_i)) + \bar{\alpha} \le M_4 = \sup_{x \in M} (\bar{s}(x) + \bar{\alpha}) \; ; \tag{8.271}$$

$$\delta \le \| Z^\star(s) \| \le \bar{K}_1 ; \quad s \in [0, s(s^i)] \; ; \tag{8.272}$$

$$\sup_{t\in[t_i, s^i]} \|Z^\star(s(t)) - Z_1^\star(t)\| \leq K_5 \left[\frac{1}{\sqrt{\omega}} + |s^i - t_i| \left| \int_{t_i}^{s^i} b(x)\,dx \right| \right] \tag{8.273}$$

$$\leq K_5 \left[\frac{1}{\sqrt{\omega}} + |s^i - t_i| M_4\right].$$

In particular, we have:

$$\delta \leq \|Z^\star(s^i))\| \leq \overline{K}_1 ; \tag{8.272)'}$$

$$\|Z^\star(s(s^i)) - Z_1^\star(s^i)\| \leq K_5 \left[\frac{1}{\sqrt{\omega}} + |s^i - t_i| M_4\right], \tag{8.273)'}$$

from which we derive:

$$\left\| Z^\star(s(s^i)) - \begin{bmatrix} 1 - \frac{b^2(s^i)}{2\omega a} + \frac{\int_0^1 b^2}{2\omega a} \\ 0 \\ \frac{\dot{b}(s^i)}{\omega a} \end{bmatrix} \right\| + \left| \frac{b(s^i)}{\omega a} \right| \tag{8.274}$$

$$\leq K_5 \left[\frac{1}{\sqrt{\omega}} + |s^i - t_i| M_4\right],$$

as the second component of $Z^\star$ is zero.

Inequalities (8.274) and (8.272)' imply:

$$\|Z^\star(s(s^i))\| - K_5 \left[\frac{1}{\sqrt{\omega}} + M_4 |s^i - t_i|\right] \leq \tag{8.275}$$

$$\leq \left| 1 - \frac{b^2(s^i)}{2\omega a} + \frac{\int_0^1 b^2}{2\omega a} \right| + \left| \frac{\dot{b}(s^i)}{\omega a} \right| \leq \|Z^\star(s(s^i))\| + K_5 \left[\frac{1}{\sqrt{\omega}} + M_4 |s^i - t_i|\right]$$

$$\delta - K_5 \left[\frac{1}{\sqrt{\omega}} + M_4 |s^i - t_i|\right] \leq \left| 1 - \frac{b^2(s^i)}{2\omega a} + \frac{\int_0^1 b^2}{2\omega a} \right| + \left| \frac{\dot{b}(s^i)}{\omega a} \right| \tag{8.276}$$

$$\leq \overline{K}_1 + K_5 \left[\frac{1}{\sqrt{\omega}} + M_4 |s^i - t_i|\right].$$

We choose a uniform ω_6 such that:

$$\begin{cases} \delta - K_5 \left[\frac{1}{\sqrt{\omega}} + M_4 |s^i - t_i|\right] \geq \frac{\delta}{2} \\ \overline{K}_1 + K_5 \left[\frac{1}{\sqrt{\omega}} + M_4 |s^i - t_i|\right] \leq 2\overline{K}_1 \end{cases} \qquad \text{for } \omega \geq \omega_6.$$

Then we have:

$$\delta'_0 = \frac{\delta}{2} \leq \left| 1 - \frac{b^2(s^i)}{2\omega a} + \frac{\int_0^1 b^2}{2\omega a} \right| + \left| \frac{\dot{b}(s^i)}{\omega a} \right| \leq \beta'_0 = 2\overline{K}_1 . \qquad (8.277)$$

Hence s^i satisfies the hypothesis (H) with $\delta_0 = \delta/2$ and $\beta_0 = 2\overline{K}_1$.

Furthermore, on $]t_i, s^i[$ and on $]s^i, t_{i+1}[$, b and $\dot{b}$ are never zero.

Let $\bar{\varepsilon}'_0$ be the value of $\bar{\varepsilon}_0$ which Lemma 12 provides when (H) is satisfied with δ'_0 and β'_0.

We may assume that we chose $\omega_6 \geq \omega_5$ such that $t_{i+1} - t_i < \bar{\varepsilon}'_0$. This choice of ω_6 is uniform on all type (II) intervals where the (H) hypothesis, with δ_0 β_0, is satisfied.

We are then allowed to apply Lemma 12 to both intervals $[t_i, s^i]$ and $[s^i, t_{i+1}]$, with constants ω'_5, $\overline{K}'_1$, δ', K'_5, $\overline{K}'_5$ and $\overline{\delta}'$ which are geometrical constants. We will take $\omega_6 \geq \omega'_5$, where ω'_5 is also given by Lemma 12.

We assume $b(t_i)$ is negative, while $b(t_{i+1})$ is positive; and also

$$-\int_{t_i}^{s^i} b(t)\, dt \leq \int_{s^i}^{t_{i+1}} b(t)\, dt . \qquad (8.278)$$

This is to make our situation precise; the other cases are equivalent to this precise one.

We draw the curve:

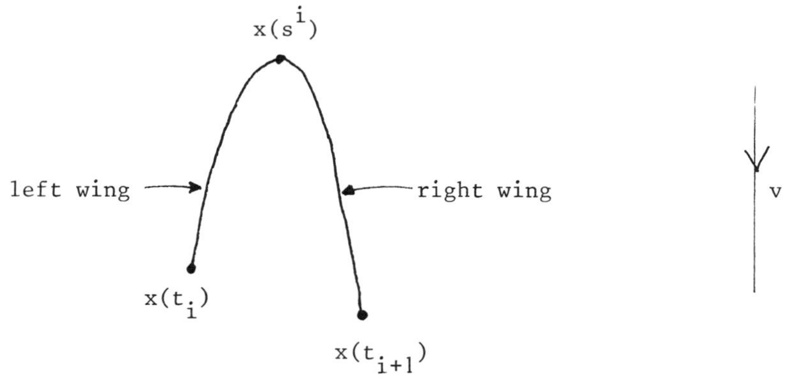

We consider a v-orbit of v, x_s, starting at $x(s^i)$.

We make the changes of variables, on $[s^i, t_i]$ and $[s^i, t_{i+1}]$ respectively, given by (8.120):

$$s_1(t) = \int_{s^i}^{t} b(x)\, dx \; ; \quad t \in [t_i, s^i] \tag{8.279}$$

$$s_2(t) = \int_{s^i}^{t} b(x)\, dx \; ; \quad t \in [s^i, t_{i+1}]. \tag{8.280}$$

We set:

$$x_i^- = x_{s_1(t_i)} \; ; \quad x_i^+ = x_{s_2(t_{i+1})} \, . \tag{8.281}$$

Split the two wings of the curve:

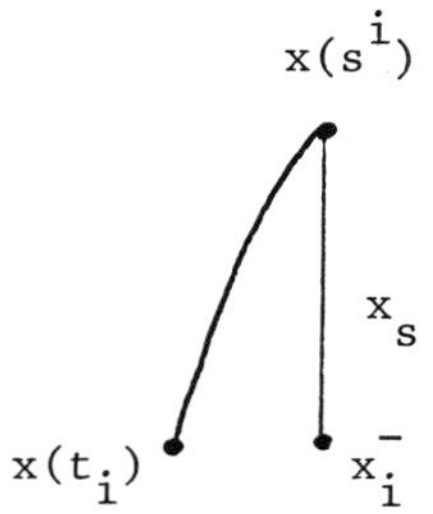

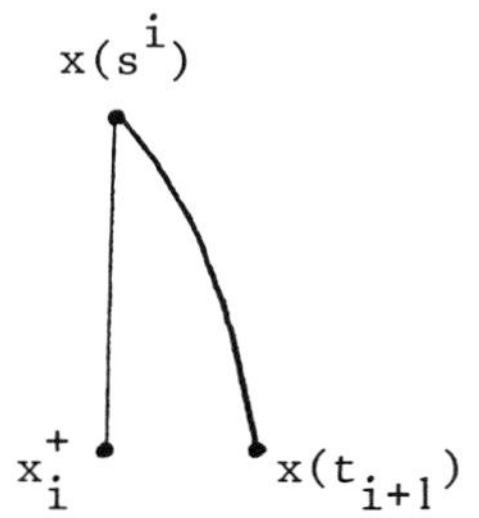

At $x(t_i) = x_1(s_1(t_i))$, where x_1 is the curve x reparametrized along (8.279), $\frac{\dot{b}(t)}{\omega a}$ vanishes.

Hence the third component of $Z_1^\star(t_i) = Z_2^\star(s_1(t_i))$ vanishes; $Z_1^\star$ and $Z_2^\star$ are here defined as usual; which implies via (8.140) of Lemma 12, applied with $t_1 = s^i$ and $t_2 = t_i$:

$$\|Z^\star(s_1(t_i)) - Z_1^\star(t_i)\| \leq K_5'\left[\frac{1}{\sqrt{\omega}} + |t_i - s^i|\,\left|\int_{t_i}^{s^i} b(x)\,dx\right|\right] \tag{8.282}$$

$$\leq K_5'\left[\frac{1}{\sqrt{\omega}} + |t_i - s_i|M_4\right] \quad \text{(via 8.271)},$$

where $Z^\star$ is the solution, as usual, of

$$\dot{Z}^\star = -{}^t\Gamma(x_s)Z^\star;$$

$$Z^\star(0) = \begin{bmatrix} 1 - \frac{b^2(s^i)}{2\omega a} + \frac{\int_0^1 b^2}{2\omega a} \\ 0 \\ \frac{\dot{b}(s^i)}{\omega a} \end{bmatrix} = \begin{bmatrix} 1 + \frac{\int_0^1 b^2}{2\omega a} \\ 0 \\ \frac{\dot{b}(s^i)}{\omega a} \end{bmatrix}.$$

We denote $\begin{bmatrix} a^\star \\ 0 \\ c^\star \end{bmatrix}$ the components of $Z^\star$.

From (8.282), we deduce:

$$|c^\star(s_1(t_i))| \leq K_5'\left[\frac{1}{\sqrt{\omega}} + |t_i - s^i|M_4\right]. \tag{8.283}$$

On the other hand, by (8.139) of Lemma 12, we have:

$$\delta' \leq \|Z^\star\|(s) \leq \overline{K}_4'; \quad \forall s \in [0, s_1(t_i)], \tag{8.284}$$

thus

$$\delta' \leq (|a^\star| + |c^\star|)(s_1(t_i)) \leq \overline{K}_5, \tag{8.285}$$

thus, using (8.283), we deduce:

$$|a^\star| + |c^\star|(s_1(t_i)) \le \overline{K}'_1 \tag{8.286}$$

$$|a^\star|(s_1(t_i)) \ge \overline{K}'_1 - K_5\left[\frac{1}{\sqrt{\omega}} + |t_i - s^i| M_4\right]. \tag{8.287}$$

We choose ω_6 large enough and uniform on all type (II) intervals so that

$$\overline{K}_1 - K'_5\left[\frac{1}{\sqrt{\omega}} + [t_{i+1} - t_i] M_4\right] \ge \frac{\overline{K}'_1}{2} \text{ for } \omega \ge \omega_6. \tag{8.288}$$

We then have:

$$|c^\star(s_1(t_i))| \le K'_5\left[\frac{1}{\sqrt{\omega}} + [t_{i+1} - t_i] M_4\right] \tag{8.289}$$

$$|a^\star(s_1(t_i))| + |c^\star(s_1(t_i))| \le \overline{K}'_1 \tag{8.290}$$

$$|a^\star(s_1(t_i))| \ge \frac{\overline{K}'_1}{2}. \tag{8.291}$$

We apply now Lemma 13. For this, we check that $|c^\star(s_1(t_i))| < \chi\left(\frac{\overline{K}'_1}{2}, \overline{K}'_1\right)$. Indeed, (8.290) allows us to set $M = \overline{K}'_1$ and (8.291) allows us to set $m = \overline{K}'_1/2$ (m and M of Lemma 13).

The inequality holds as soon as

$$K'_5\left[\frac{1}{\sqrt{\omega}} + [t_{i+1} - t_i] M_4\right] < \chi\left(\frac{\overline{K}'_1}{2}, \overline{K}'_1\right) \quad ;$$

which will hold uniformly on all type (II) intervals, if we take $\omega \ge \omega_6$ with a uniform and large enough ω_6.

From Lemma 13, we derive then that $c^\star(s)$ takes the value zero in a neighbourhood of $s_1(t_i)$ of order $K_6 |c^\star(s_1(t_i))|$, that is of order

$$K_6 K'_5\left[\frac{1}{\sqrt{\omega}} + [t_{i+1} - t_i] M_4\right].$$

We then have:

$$|s_1(t_i) - s_i| < K_6 K'_5\left[\frac{1}{\sqrt{\omega}} + [t_{i+1} - t_i] M_4\right]. \tag{8.291'}$$

By (8.145)' we have on $\left]s_i - \theta\left(\frac{\overline{K}'_1}{2}, \overline{K}'_1\right),\ s_i + \theta\left(\frac{\overline{K}'_1}{2}, \overline{K}'_1\right)\right[$:

$$|c\star(s)| \geq K_7 |s - s_i| , \tag{8.292}$$

for any s in this neighbourhood.

Let K_9 be a positive constant such that:

$$K_7 K_9 > K'_5 . \tag{8.293}$$

We take ω_6 large enough and uniform, such that:

$$K_9 \left[\frac{1}{\sqrt{\omega}} + |t_{i+1} - t_i| M_4\right] < \theta\left(\frac{\bar{K}'_1}{2}, \bar{K}'_1\right) . \tag{8.294}$$

Let

$$\bar{s}_i = s_i + K_9 \left[\frac{1}{\sqrt{\omega}} + |t_{i+1} - t_i| M_4\right]; \quad \underline{s}_i = s_i - K_9 \left[\frac{1}{\sqrt{\omega}} + |t_{i+1} - t_i| M_4\right]. \tag{8.295}$$

By (8.294) and (8.292), we have:

$$|c\star(\bar{s}_i)| \geq K_7 K_9 \left[\frac{1}{\sqrt{\omega}} + |t_{i+1} - t_i| M_4\right] \tag{8.296}$$

$$|c\star(\underline{s}_i)| \geq K_7 K_9 \left[\frac{1}{\sqrt{\omega}} + |t_{i+1} - t_i| M_4\right] .$$

On the other hand, as s_i is between $\bar{s}_i$ and $\underline{s}_i$, we know that:

$$c\star(\bar{s}_i) \ c\star(\underline{s}_i) < 0. \tag{8.297}$$

In fact, $\underline{s}_i$ is strictly positive. Indeed, otherwise, $0 \in]\underline{s}_i, \bar{s}_i[$. Hence

$$|c\star(0) - c\star(s_i)| = |c\star(0)| \tag{8.298}$$

$$\leq \sup_{s \in [s_i - \theta, s_i + \theta]} |\dot{c}\star(s)| K_9 \left[\frac{1}{\sqrt{\omega}} + |t_{i+1} - t_i| M_4\right] ,$$

hence

$$|c\star(0)| \leq \sup_{s \in [s_i - \theta, s_i + \theta]} [\,|a\star| + \sup_{x \in M} |\bar{\mu}| \ |c\star|\,] K_9 \left[\frac{1}{\sqrt{\omega}} + |t_{i+1} - t_i| M_4\right],$$

as $\dot{c}\star = a\star - \bar{\mu} c\star$.

By (8.145), we then know that $|a\star| + |c\star| \leq M + 1$ on $[s_i - \theta, s_i + \theta]$. Then

$$\sup_{s\in[s_0-\theta,\, s_i+\theta]} |a\star| + \sup_{x\in M} |\bar{\mu}|\,|c\star| \leq M + 1 + (\sup|\bar{\mu}|)(M+1)$$
$$= (\overline{K}_1' + 1)(1 + \sup|\bar{\mu}|),$$

as here $M = \overline{K}_1'$.

Thus, (8.298) implies:

$$|c\star(0)| \leq \overline{K}_2' K_9 \left[\frac{1}{\sqrt{\omega}} + |t_{i+1} - t_i| M_4\right] = K_{10}\left[\frac{1}{\sqrt{\omega}} + |t_{i+1} - t_i| M_4\right]. \qquad (8.299)$$

Now $c\star(0) = \dfrac{\dot{b}(s^i)}{\omega a}$. Thus, by (8.264) we have:

$$a^{-1} \times \varepsilon_2 \leq |c\star(0)| = \frac{|\dot{b}(s^i)|}{\omega a} \leq K_{10}\left[\frac{1}{\sqrt{\omega}} + |t_{i+1} - t_i| M_4\right]. \qquad (8.300)$$

Choosing ω_6 large enough and uniform, we may ensure that (8.300) is violated, which yields a contradiction and implies $\underline{s}_i > 0$.

We thus have defined $\underline{s}_i$ and $\bar{s}_i$ satisfying:

$$\underline{s}_i > 0;\ \ 0 < \bar{s}_i - \underline{s}_i \leq 2\, K_9\left[\frac{1}{\sqrt{\omega}} + |t_{i+1} - t_i| M_4\right]; \qquad (8.301)$$

$$s_i, \underline{s}_i \in \left]s_i - K_9\left[\frac{1}{\sqrt{\omega}} + |t_{i+1} - t_i| M_4\right],\ s_i + K_9\left[\frac{1}{\sqrt{\omega}} + |t_{i+1} - t_i| M_4\right]\right[\qquad (8.302)$$

$$c\star(\bar{s}_i)\, c\star(\underline{s}_i) < 0;\ \ |c\star(\bar{s}_i)| \geq K_7 K_9\left[\frac{1}{\sqrt{\omega}} + |t_{i+1} - t_i| M_4\right] \qquad (8.303)$$

$$|c\star(\underline{s}_i)| \geq K_7 K_9\left[\frac{1}{\sqrt{\omega}} + |t_{i+1} - t_i| M_4\right].$$

Moreover, we have, by (8.291)':

$$|s_1(t_i) - s_i| < K_6 K_5'\left[\frac{1}{\sqrt{\omega}} + |t_{i+1} - t_i| M_4\right]. \qquad (8.304)$$

We now study, having established the inequalities (8.301)-(8.304), the right wing of the curve $[s^i, t_{i+1}]$. We apply to this piece of curve Lemma 12; indeed we have already noticed that this lemma holds on such a piece.

We then have:

$$\int_{s^i}^{t_{i+1}} b(x)\,dx = \left|\int_{s^i}^{t_{i+1}} b(x)\,dx\right| \leq \bar{s}(x(t_i)) + \bar{\alpha} \leq M_4. \tag{8.305}$$

Furthermore we know that:

$$s_1(t_i) = \int_{s^i}^{t_i} b(x)\,dx \leq \int_{s^i}^{t_{i+1}} b(t)\,dt = s_2(t_{i+1}) \tag{8.306}$$

(see (8.278)-(8.279)-(8.280)).

We want to prove:

$$\int_{s^i}^{t_{i+1}} b(x)\,dx = s_2(t_{i+1}) \leq s_1(t_i) + K_9\left[\frac{1}{\sqrt{\omega}} + |t_{i+1} - t_i|M_4\right]. \tag{8.307}$$

We will argue by contradiction. If (8.307) does not hold, then

$$\bar{s}_i = s_i + K_9\left[\frac{1}{\sqrt{\omega}} + [t_{i+1} - t_i]M_4\right] \leq s_1(t_i) + K_9\left[\frac{1}{\sqrt{\omega}} + [t_{i+1} - t_i]M_4\right] \leq s_2(t_{i+1}).$$

Thus $\bar{s}_i \in [0, s_2(t_{i+1})]$, and $\underline{s}_i \in [0, s_2(t_{i+1})]$ as $0 < \underline{s}_i < \bar{s}_i$ ($0 < \underline{s}_i$ by (8.301)). Thus there exist $\underline{t}, \underline{\underline{t}} \in]s_i, t_{i+1}[$ such that:

$$\bar{s}_i = s_2(\underline{t}); \quad \underline{s}_i = s_2(\underline{\underline{t}}).$$

Applying (8.140) of Lemma 12 to $[s^i, t_{i+1}]$ with $K_5 = K_5'$ in these points, we derive:

$$\|Z_1^\star(\underline{t}) - Z^\star(\bar{s}_i)\| = \|Z_1^\star(\underline{t}) - Z^\star(s_2(\underline{t}))\| \leq K_5'\left[\frac{1}{\sqrt{\omega}} + |t_{i+1} - s^i|\left|\int_{s^i}^{t_{i+1}} b(x)\,dx\right|\right],$$

hence, by (8.305):

$$\|Z_1^\star(\underline{t}) - Z^\star(\bar{s}_i)\| \leq K_5'\left[\frac{1}{\sqrt{\omega}} + [t_{i+1} - t_i]M_4\right]$$

Similarly:

$$\|Z_1^\star(\underline{\underline{t}}) - Z^\star(\underline{s}_i)\| \leq K_5'\left[\frac{1}{\sqrt{\omega}} + [t_{i+1} - t_i]M_4\right],$$

which imply that:

$$\left| \frac{\dot{b}(\underline{t})}{\omega a} - c^\star(\bar{s}_i) \right| \le K'_5 \left[\frac{1}{\sqrt{\omega}} + [t_{i-1} - t_i] M_4 \right] ;$$

$$\left| \frac{\dot{b}(\underline{\underline{t}})}{\omega a} - c^\star(\underline{s}_i) \right| \le K'_5 \left[\frac{1}{\sqrt{\omega}} + [t_{i+1} - t_i] M_4 \right] .$$

Using then (8.303) and the fact that $K_7 K_9 > K'_5$ (by choice of K_9), $\dot{b}(\underline{t})$ and $\dot{b}(\underline{\underline{t}})$ have opposite signs. Hence $\dot{b}$ vanishes somewhere in $]s^i, t_{i+1}[$. This is impossible and (8.307) is thereby proven.

From (8.306)-(8.307) and from Lemma 10' applied to $]s^i, t_i[$ and to $]s^i, t_{i+1}[$, we deduce (8.267). Indeed, $|\int_{s^i}^{t_i} b(x)\,dx| \le M_4$ and $|\int_{s^i}^{t_{i+1}} b(x)\,dx| \le M_4$. Hence, if $|s^i - t_i| < \varepsilon_0(M_4)$ and $|s^i - t_{i+1}| < \varepsilon_0(M_4)$ ($\varepsilon_0(f)$ defined in Lemma 10'), we can apply Lemma 10'. These two conditions are realized as soon as $|t_{i+1} - t_i| < \varepsilon_0(M_4)$ which yields a choice of ω_6 with $\omega \ge \omega_6$. This ω_6 is again uniform.

In particular, (8.106)' holds with $t^1 = s^i$, $t^2 = t_i$ and with $t^1 = s^i$, $t^2 = t_{i+1}$; applying it to t_i and t_{i+1} respectively, and remembering that $\bar{z}(t_i) = x_{s_1(t_i)} = x_i^-$ ((8.281)) and $\bar{z}(t_{i+1}) = x_{s_2(t_{i+1})} = x_i^+$, we derive:

$$d(x(t_i), x_i^-) \le \bar{K}[|s^i - t_i| + |s^i - t_i|^2] \le 2\bar{K}|s^i - t_i| \le 2\bar{K}[t_{i+1} - t_i] \tag{8.308}$$

$$d(x(t_{i+1}), x_i^+) \le \bar{K}[|s^i - t_{i+1}| + |s^i - t_{i+1}|^2] \tag{8.309}$$

$$\le 2\bar{K}|s^i - t_{i+1}| \le 2\bar{K}[t_{i+1} - t_i].$$

From (8.306), (8.307) and from the fact that $x_i^- = x_{s_1(t_i)}$ and $x_i^+ = x_{s_2(t_{i+1})}$, we deduce:

$$d(x_i^-, x_i^+) \le (\operatorname*{Sup}_{x \in M} \|v\|) K_9 \left[\frac{1}{\sqrt{\omega}} + [t_{i+1} - t_i] M_4 \right] \tag{8.310}$$

as $\dot{x}_s = v(x_s)$ goes from x_i^- to x_i^+ during a time smaller than $K_9 \left[\frac{1}{\sqrt{\omega}} + [t_{i+1} - t_i] M_4 \right]$. Inequalities (8.308), (8.309), (8.310) imply then (8.267) with $K_8 = 4\bar{K} + K_9 \operatorname*{Sup}_{x \in M} \|v\|$. The two inequalities (8.266),

$\left|\int_{t_{i+1}}^{s^i} b(x)\,dx\right| \le M_4$ and $\left|\int_{s^i}^{t_{i+1}} b(x)\,dx\right| \le M_4$ have already been proven. To lowerbound these two quantities, we proceed as follows: we apply Lemma 12 to $[t_i, s^i]$ with $t_1 = s^i$ and $t_2 = t_i$. We obtain:

$$\bar{\delta}'\left\|Z_1^\star(s^i) - Z_1^\star(t_i)\right\| \le \left|\int_{t_i}^{s^i} b(x)\,dx\right| + \bar{K}_5'\left[\frac{1}{\sqrt{\omega}} + |t_i - s^i|M_4\right] \tag{8.311}$$

hence

$$\left|\int_{t_i}^{s^i} b(x)\,dx\right| \ge \bar{\delta}'\left|\frac{\dot{b}(s^i)}{\omega a} - \frac{\dot{b}(t_i)}{\omega a}\right| - \bar{K}_5'\left[\frac{1}{\sqrt{\omega}} + |t_i - s^i|M_4\right]. \tag{8.312}$$

But $\dot{b}(t_i) = 0$ and $\frac{|\dot{b}(s^i)|}{\omega a} \ge \varepsilon_2 a^{-1}$ ((8.264)).

Thus we have:

$$\left|\int_{t_i}^{s^i} b(x)\,dx\right| \ge \frac{\bar{\delta}'\varepsilon_2}{a} - \bar{K}_5'\left[\frac{1}{\sqrt{\omega}} + |t_i - s^i|M_4\right]. \tag{8.313}$$

Then we choose ω_6 large enough and uniform on all type (II) intervals such that $\bar{K}_5'\left[\frac{1}{\sqrt{\omega}} + [t_{i+1} - t_i]M_4\right] < \frac{\bar{\delta}'\varepsilon_2}{2a_0}$. This implies:

$$\left|\int_{t_i}^{s^i} b(x)\,dx\right| \ge \frac{\bar{\delta}'\varepsilon_2}{2a_1} = \delta_3\varepsilon_2 . \tag{8.314}$$

We proceed in the same way with $[s^i, t_{i+1}]$, (8.266) is thus proved. It remains to prove (8.268).

For this, we use (8.140) of Lemma 12 on $[s^i, t_i]$ at t_i and on $[s^i, t_{i+1}]$ at t_{i+1}. We have then:

$$\left\|Z^\star(s_1(t_i)) - Z_1^\star(t_i)\right\| \le K_5'\left[\frac{1}{\sqrt{\omega}} + |s^i - t_i|M_4\right] \tag{8.315}$$

$$\left\|Z^\star(s_2(t_{i+1})) - Z_1^\star(t_{i+1})\right\| \le K_5'\left[\frac{1}{\sqrt{\omega}} + |s^i - t_{i+1}|M_4\right].$$

We estimate:

$$\left\|Z^\star(s_1(t_i)) - Z^\star(s_2(t_{i+1}))\right\| . \tag{8.316}$$

We have:

$$\dot{Z}^{\star} = -{}^t\Gamma(x_s)\, Z^{\star}. \tag{8.317}$$

On the other hand, by (8.139) of Lemma 12 applied to $[s^i, t_{i+1}]$, we have:

$$\| Z^{\star}(s) \| \leq \overline{K}_1' \quad \forall\, s \in [0, s_2(t_{i+1})]. \tag{8.318}$$

In particular, as $0 \leq s_1(t_i) \leq s_2(t_{i+1})$, (8.318) holds on $[s_1(t_i), s_2(t_{i+1})]$. Thus, using (8.317), we have:

$$\| Z^{\star}(s_1(t_i)) - Z^{\star}(s_2(t_{i+1})) \| \leq (\sup_{x \in M} \| {}^t\Gamma \|)\, \overline{K}_1'(s_2(t_{i+1}) - s_1(t_i)). \tag{8.319}$$

By (8.306), (8.307), we have:

$$| s_2(t_{i+1}) - s_1(t_i) | \leq K_9 \left[\frac{1}{\sqrt{\omega}} + [t_{i+1} - t_i] M_4 \right]. \tag{8.320}$$

Inequalities (8.315) and (8.320) then imply:

$$\| Z_1^{\star}(t_i) - Z_1^{\star}(t_{i+1}) \| \leq [2 K_5' + \sup_{x \in M} \| {}^t\Gamma \| \overline{K}_1' K_9] \left[\frac{1}{\sqrt{\omega}} + [t_{i+1} - t_i] M_4 \right], \tag{8.321}$$

which yields (8.268) with a good choice of $\overline{K}_8$.

To sum up, we have on the type (II) intervals:

<u>if</u> t_i <u>satisfies (H)</u>

$\delta\varepsilon_2 \leq$ length along $v \leq M_4$

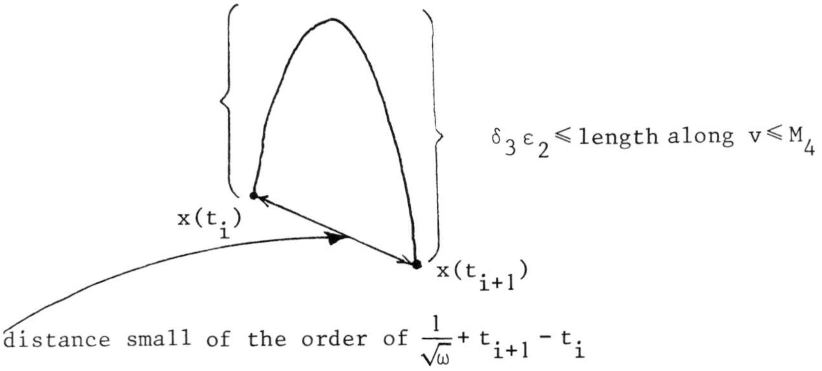

difference of the transport equations small of the order of $\frac{1}{\sqrt{\omega}} + t_{i+1} - t_i$

Analysis of initial and final intervals

We now analyse the situation on a type $[t^+(s_1), t_1]$ or $[t_k, t^-(s_2)]$ interval (see (8.96)), where t_1 (resp. t_k) is the first (resp. the last) zero of $\dot{b}$ after $t^+(s_1)$ (resp. before $t^-(s_2)$) in the interval $(t^+(s_1), t^-(s_2))$.

Such a zero exists as will be seen.

Initial and final intervals are of symmetric type. We will thus reduce to the case of an initial interval $[t^+(s_1), t_1]$, where t_1 might be equal to $t^-(s_2)$.

Let $I_1(s_1)$ be the concentration interval, at given ε_1, having $t^+(s_1)$ as upper boundary. Let $t_1(s_1)$ be the unique point, given by Lemma 8, in $I_1(s_1)$ where $b\dot{b}$ vanishes (see 8.43) of Lemma 8).

In Lemma 9, we proved that, if we transformed the curve $x(t)$ between $t^-(s_1)$ and $t^+(s_1)$ through $\phi(u(t), .)$ where $u(t) = -[\int_{t_1(s_1)}^{t} b(x)\,dx + \bar{c}(t_1(s_1))]$ ((8.40)) and if $|\bar{c}(t_1(s_1))| \leq c\sqrt{\alpha_1}$, then the curve obtained $y(t) = \phi(u(t), x(t))$ remains close at order $K\sqrt{\alpha_1}(t^+(s_1) - t^-(s_1))$ to the curve $z(t)$ starting at $y(t_-(s_1))$ and tangent to $a\xi$ during the time $t^+(s_1) - t^-(s_1)$ (see (8.53)). We first fix the choice of $\bar{c}(t_1(s_1))$ which we left open.

For this, we consider the curve $x(.)$ for t in $[t^-(s_1), t^+(s_1)]$.

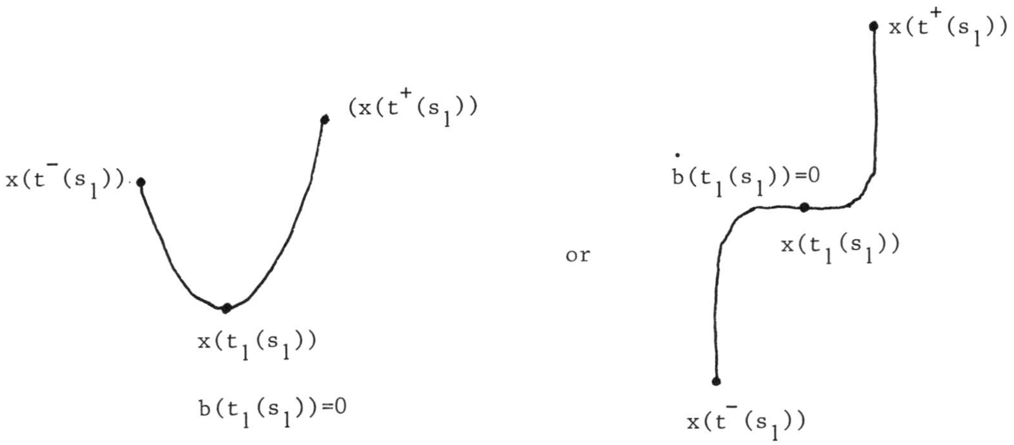

We consider a v-orbit x'_s, starting at $x(t_1(s_1))$, and we consider the solution of the transport equation (8.132) with initial data

$$Z'^{\star}_0 = \begin{bmatrix} 1 - \dfrac{b^2(t_1(s_1))}{2\omega a} + \dfrac{\int_0^1 b^2}{2\omega a} \\ 0 \\ \dfrac{\dot{b}(t_1(s_1))}{\omega a} \end{bmatrix},$$

i.e.

$$\dot{x}'_s = v(x'_s) \; ; \quad x'_0 = x(t_1(s_1)) \tag{8.322}$$

$$\dot{Z}'^{\star} = -{}^t\Gamma(x'_s)\, Z'^{\star} \; ; \quad Z'^{\star}(0) = Z'^{\star}_0 = \begin{bmatrix} 1 - \dfrac{b^2(t_1(s_1))}{2\omega a} + \dfrac{\int_0^1 b^2}{2\omega a} \\ 0 \\ \dfrac{\dot{b}(t_1(s_1))}{\omega a} \end{bmatrix}. \tag{8.323}$$

By (8.42) of Lemma 8, we have:

$$\frac{|\dot{b}|}{\omega}(t_1(s_1)) < \sqrt{\alpha_1} \; ; \quad \frac{b^2(t_1(s_1))}{\omega} < \varepsilon_1 . \tag{8.324}$$

We choose $\bar{\varepsilon}_1 > 0$ and ω_7, $\bar{\varepsilon}_1$ small enough and ω_7 large enough such that:

$$1 + \frac{\int_0^1 b^2}{2\omega a} - \frac{\varepsilon_1}{a} \geq \tfrac{1}{2} \; ; \quad 1 + \frac{\int_0^1 b^2}{2\omega a} + \frac{\varepsilon_1}{a} + \frac{1}{a}\sqrt{\alpha_1} < 2; \quad \frac{1}{a}\sqrt{\alpha_1} < \chi(2, \tfrac{1}{2}), \tag{8.325}$$

(χ is given in Lemma 13) for $0 < \varepsilon_1 < \bar{\varepsilon}_1$ and $\omega \geq \omega_7$.
We may find such an $\bar{\varepsilon}_1$ and such an ω_7, as $\alpha_1(\varepsilon_1) \xrightarrow[\varepsilon_1 \to 0]{} 0$ and as $(\int_0^1 b^2/\omega) \to 0$ when ω goes to $+\infty$.
Under (8.325), we may apply Lemma 13 to $Z^{\star\prime}(s)$; indeed, in $s = 0$, we have:

$$Z^{\star\prime}(0) = \begin{bmatrix} a^{\star\prime}(0) \\ 0 \\ c^{\star\prime}(0) \end{bmatrix} ; \quad |c^{\star\prime}(0)| = \frac{|\dot{b}(t_1(s_1))|}{\omega a} < \frac{1}{a}\sqrt{\alpha_1} < \chi(2, \tfrac{1}{2}) \; ; \tag{8.326}$$

$$|a^{\star\prime}(0)| + |c^{\star\prime}(0)| = \left|1 - \frac{b^2(t_1(s_1))}{2\omega a} + \frac{\int_0^1 b^2}{2\omega a}\right| + \frac{|\dot{b}(t_1(s_1))|}{\omega a}$$

$$< 1 + \frac{\int_0^1 b^2}{2\omega a} + \frac{\varepsilon_1}{a} + \frac{1}{a}\sqrt{\alpha_1} < 2,$$

$$|a^{\star\prime}(0)| = \left|1 - \frac{b^2(t_1(s_1))}{2\omega a} + \frac{\int_0^1 b^2}{2\omega a}\right| \geq 1 + \frac{\int_0^1 b^2}{2\omega a} - \frac{b^2(t_1(s_1))}{2\omega a}$$

$$\geq 1 + \frac{\int_0^1 b^2}{2\omega a} - \frac{\varepsilon_1}{a} \geq \tfrac{1}{2}.$$

Lemma 13 applies with $s_1 = 0$ (notations of this lemma), $M = 2$, $m = \frac{1}{2}$. We deduce that $c^{\star\prime}(s)$ vanishes in a neighbourhood of $s = 0$ of order $K_6(2, \frac{1}{2}) \times c^{\star\prime}(0)$. Let $\tilde{s}_2'$ be the point where $c^{\star\prime}(s)$ vanishes. Thus:

$$c^{\star\prime}(\tilde{s}_2') = 0; \quad |\tilde{s}_2'| \leq K_6(2, \tfrac{1}{2})\ |c^{\star\prime}(0)| = K_6(2, \tfrac{1}{2}) \frac{|\dot{b}|}{\omega a}(t_1(s_1)) \tag{8.327}$$

$$< \frac{1}{a} K_6 \sqrt{\alpha_1}.$$

We set:

$$\bar{c}(t_1(s_1)) = -\tilde{s}_2'. \tag{8.328}$$

It is clear that $\bar{c}(t_1(s_1))$ satisfies (8.50).

Once the choice of $\bar{c}(t_1(s_1))$ is settled, we consider another wing x_s, tangent to v, this time starting at $x(\overset{+}{t}(s_1))$, i.e.

$$\dot{x}_s = v(x_s)\ ; \quad x_0 = x(\overset{+}{t}(s_1)). \tag{8.329}$$

We consider also the solution of the transport equation (8.132) along x_s' with initial data

$$Z_0^{\star} = \begin{bmatrix} 1 - \dfrac{b^2(\overset{+}{t}(s_1))}{2\omega a} + \dfrac{\int_0^1 b^2}{2\omega a} \\ 0 \\ \dfrac{\dot{b}(t^+(s_1))}{\omega a} \end{bmatrix}$$

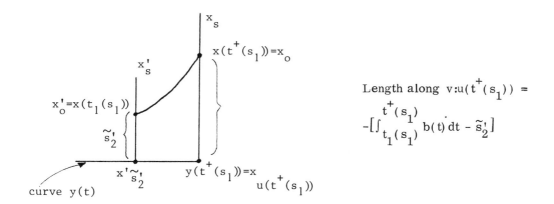

i.e.

$$\dot{Z}^\star = -{}^t\Gamma(x_s)\, Z^\star; \quad Z^\star(0) = Z_0^\star \,. \qquad (8.330)$$

Notice that we have:

$$y(t^+(s_1)) = \phi(u(t^+(s_1)), x(t^+(s_1))) = x_{u(t^+(s_1))} \qquad (8.331)$$

where $u(t^+(s_1)) = -[\int_{t_1(s_1)}^{t^+(s_1)} b(x)\,dx - \tilde{s}_2']$.

Indeed, $y(t^+(s_1))$ is obtained by integrating (8.329) during the time $u(t^+(s_1))$ as $\phi(s_1)$ is the one-parameter group generated by v.

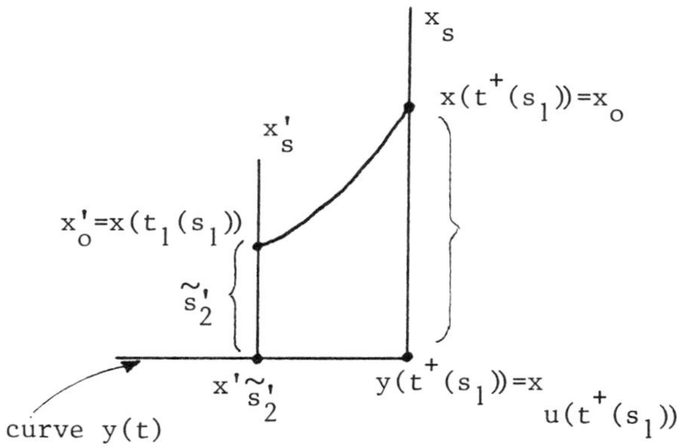

Length along $v: u(t^+(s_1)) = -[\int_{t_1(s_1)}^{t^+(s_1)} \dot{b}(t)\,dt - \tilde{s}_2']$

We may also apply Lemma 13 to x_s and $Z^\star(s)$.

Indeed, in $s = 0$, we have:

by (8.42) and (8.325):

$$Z^\star(0) = \begin{bmatrix} a^\star(0) \\ 0 \\ c^\star(0) \end{bmatrix}, \text{ with } |c^\star(0)| = \frac{|\dot{b}(t^+(s_1))|}{\omega a} < \frac{1}{a}\sqrt{\alpha_1} < \chi(2, \tfrac{1}{2});$$

$$|a^\star(0)| = \left|1 - \frac{b^2(t^+(s_1))}{2\omega a} + \frac{\int_0^1 b^2}{2\omega a}\right| \geq 1 + \frac{\int_0^1 b^2}{2\omega a} - \frac{b^2(t^+(s_1))}{2\omega a}$$

$$\geq 1 + \frac{\int_0^1 b^2}{2\omega a} - \frac{\varepsilon_1}{a} \geq \tfrac{1}{2};$$

$$|a^\star(0)| + |c^\star(0)| = \left|1 - \frac{b^2(t^+(s_1))}{2\omega a} + \frac{\int_0^1 b^2}{2\omega a}\right| + \frac{|\dot{b}(t^+(s_1))|}{\omega a}$$

$$< \frac{1}{a}\sqrt{\alpha_1} + 1 + \frac{\int_0^1 b^2}{2\omega a} + \frac{\varepsilon_1}{a} < 2 ;$$

which allows us to apply Lemma 13 with $s_1 = 0$ (same notation as in this lemma, $M = 2$, $m = \frac{1}{2}$).

We then deduce the existence of $\tilde{s}_2^+$ such that:

$$c^\star(\tilde{s}_2^+) = 0; \quad |\tilde{s}_2^+| < K_6(2, \tfrac{1}{2})\,|c^\star(0)| < \frac{1}{a} K_6\sqrt{\alpha_1}. \tag{8.332}$$

We then have:

Lemma 15: There exist ω_6 and $\bar{\bar{\varepsilon}}_1$; ω_8 and $\bar{\bar{\varepsilon}}_1$ uniform on all concentration intervals such that:

$$d(x_{\tilde{s}_2^+}, y(t^+(s_1))) < K_{12}\left[\frac{1}{\sqrt{\omega}} + \sqrt{\alpha_1}(t^+(s_1) - t_1(s_1))\right] ; \tag{8.333}$$

$$Z^\star(\tilde{s}_2^+) = \begin{bmatrix} a^\star(\tilde{s}_2^+) \\ 0 \\ 0 \end{bmatrix} ; \tag{8.334}$$

$$|a^\star(\tilde{s}_2^+) - a'^\star(\tilde{s}_2')| < K_{12}\left[\frac{1}{\sqrt{\omega}} + \sqrt{\alpha_1}(t^+(s_1) - t_1(s_1))\right] \tag{8.334'}$$

$$\left\|Z^\star(\tilde{s}_2^+) - \begin{bmatrix} 1 \\ 0 \\ 0 \end{bmatrix}\right\| = |a^\star(\tilde{s}_2^+) - 1| < K_{12}\left[\frac{\int_0^1 b^2}{\omega} + \varepsilon_1 + \sqrt{\alpha_1}\right], \tag{8.335}$$

for $\omega \geq \omega_8$; $0 < \varepsilon_1 < \bar{\bar{\varepsilon}}_1$.

Here K_{12} is a uniform geometric constant.

Proof of Lemma 15: The proof relies on arguments which have already been used. We thus only outline here the main steps of this proof, without going into all the details.

We consider the interval $[t^+(s_1), t_1(s_1)]$. On the interval $]t^+(s_1), t_1(s_1)[$, b and $\dot{b}$ do not vanish.

The point $t^+(s_1)$ satisfies the hypothesis $(H_{\delta_0}^{\beta_0})$ with $\delta_0 = \frac{1}{2}$ and $\beta_0 = 2$ as, by (8.325):

$$\frac{1}{2} \leq \left| 1 - \frac{b^2(t^+(s_1))}{2\omega a} + \frac{\int_0^1 b^2}{2\omega a} \right| + \left| \frac{\dot{b}(t^+(s_1))}{\omega a} \right| \leq 2, \tag{8.336}$$

if $0 < \varepsilon_1 < \bar{\varepsilon}_1$ and $\omega > \omega_7$.

Using Lemma 9, (8.51) in particular, with $\bar{c}(t_1(s_1)) = 0$, which is allowed by the hypotheses of this lemma ($\bar{c}$ is indeed arbitrary in this lemma, if it satisfies $|\bar{c}| < c\sqrt{\alpha_1}$). We deduce:

$$\left| \int_{t^+(s_1)}^{t_1(s_1)} b(x)\,dx \right| < K_1\sqrt{\alpha_1} \leq 1 \text{ for } 0 < \varepsilon_1 \text{ small enough.} \tag{8.337}$$

Inequalities (8.336) and (8.337) would allow us to apply Lemmas 12 and 10' to the interval $[t^+(s_1), t_1(s_1)]$, with $t_1 = t^+(s_1)$, $t_2 = t_1(s_1)$ if:

$$|t^+(s_1) - t_1(s_1)| < \nu_0 = \inf(\bar{\varepsilon}_0, \varepsilon_0(1)); \tag{8.338}$$

$\bar{\varepsilon}_0$ as in Lemma 12 with $\beta_0 = 2$ and $\delta_0 = \frac{1}{2}$; $\varepsilon_0(1)$ as in Lemma 10'.

We may always assume that this is the situation.

Indeed, if (8.338) does not hold, we have:

$$|t^+(s_1) - t_1(s_1)| \geq \nu_0 > 0. \tag{8.339}$$

But $d(x_{\tilde{s}_2^+}, y(t^+(s_1))) \leq d(x_{\tilde{s}_2^+}, x_0) + d(x_0, y(t^+(s_1))) = d(x_{\tilde{s}_2^+}, x_0) + d(x_0, x_{u(t^+(s_1))})$ as we have by (8.331) $y(t^+(s_1)) = x_{u(t^+(s_1))}$.

Thus:

$$d(x_{\tilde{s}_2^+}, y(t^+(s_1))) \leq [|\tilde{s}_2^+| + |u(t^+(s_1))|] \sup_{x \in M} \|v\|$$

$$\leq [K_6\sqrt{\alpha_1} + \left| \int_{t_1(s_1)}^{t^+(s_1)} b(x)\,dx - \tilde{s}_2' \right|] \sup_{x \in M} \|v\| \text{ by (8.332)}$$

$$\leq \left[K_6\sqrt{\alpha_1} + \left|\int_{t_1(s_1)}^{t^+(s_1)} b(x)\,dx\right| + |\tilde{s}_2'|\right] \sup_{x\in M} \|v\|$$

$$\leq \left[2K_6\sqrt{\alpha_1} + K_1\sqrt{\alpha_1}\right] \sup_{x\in M} \|v\| \quad \text{by (8.327) and (8.337),}$$

hence:

$$d(x_{\tilde{s}_2^+}, y(t^+(s))) \leq \overline{K}_{12}\sqrt{\alpha_1} = \frac{\overline{K}_{12}}{\nu_0}\sqrt{\alpha_1}\,\nu_0 \leq \frac{\overline{K}_{12}}{\nu_0}\sqrt{\alpha_1}(t^+(s_1) - t_1(s_1)) \quad (8.340)$$

by (8.339), which yields (8.333).

Equation (8.334) is immediate, as by definition of $\tilde{s}_2$, $c^\star(\tilde{s}_2) = 0$.
Finally, (8.335) is obtained as follows:

Given that

$$|Z^\star(0)| = \left|1 - \frac{b^2(t^+(s_1))}{2\omega a} + \frac{\int_0^1 b^2}{2\omega a}\right| + \left|\frac{\dot{b}(t^+(s_1))}{\omega a}\right| < 2,$$

by choice of $\overline{\varepsilon}_1$ and ω_7, and given that $|\tilde{s}_2^+|$ is small, we have

$$\|Z^\star(\tilde{s}_2^+) - Z^\star(0)\| \leq K|\tilde{s}_2^+| \leq K\,K_6\sqrt{\alpha_1}, \quad (8.341)$$

hence $\left|a^\star(\tilde{s}_2^+) - 1 + \dfrac{b^2(t^+(s_1))}{2\omega a} + \dfrac{\int_0^1 b^2}{2\omega a}\right| \leq K\,K_6\sqrt{\alpha_1}$. As, by (8.42)

$$\frac{b^2(t^+(s_1))}{2\omega a} = \frac{\varepsilon_1}{a} \leq \frac{\varepsilon_1}{a_0},$$

we deduce:

$$|a^\star(s_2^+) - 1| \leq K\,K_6\sqrt{\alpha_1} + \frac{\int_0^1 b^2}{2\omega a} + \frac{\varepsilon_1}{a_0}, \quad (8.342)$$

which yields (8.335).

In order to derive (8.334)' we also remark that (8.341) holds with $Z^\star$ replaced by $Z'^\star$ giving:

$$\| Z'^{\star}(\tilde{s}_2') - Z'^{\star}(0) \| \leq K |\tilde{s}_2'| \qquad (8.341)'$$

$$\leq K K_6 \sqrt{\alpha_1} \quad (\text{via } (8.327)),$$

hence:

$$\left| a'^{\star}(\tilde{s}_2') - \left(1 - \frac{b^2(t_1(s_1))}{2\omega a} + \frac{\int_0^1 b^2}{2\omega a} \right) \right| \leq K K_6 \sqrt{\alpha_1} .$$

This implies:

$$|a^{\star}(\tilde{s}_2^+) - a'^{\star}(\tilde{s}_2')| \leq 2K K_6 \sqrt{\alpha_1} + \frac{b^2(t^+(s_1))}{2\omega a} \leq \frac{2\varepsilon_1}{a_0} + 2K K_6 \sqrt{\alpha_1}.$$

We have $\alpha_1 = \varphi^{-1}(M\varepsilon_1)$, with $\varphi(x) \leq Cx^2$. Thus $\varepsilon_1 \leq C_1 \alpha_1^2$.

With ε_1 and hence α_1 small enough, we ensure $\frac{2\varepsilon_1}{a} + 2K K_6 \sqrt{\alpha_1} < K'\sqrt{\alpha_1}$.

Hence

$$|a^{\star}(\tilde{s}_2^+) - a'^{\star}(\tilde{s}_2')| < K'\sqrt{\alpha_1} = \frac{K'\sqrt{\alpha_1}}{\nu_0} . \nu_0 \leq K'\sqrt{\alpha_1}(t^+(s_1) - t_1(s_1)),$$

which yields (8.334)'.

Under (8.339), Lemma 15 is thus proven.

Hence, we may assume that (8.338) is satisfied for the remainder of the proof; which allows us to apply Lemmas 12 and 10'. In what follows, K is a uniform constant.

Lemma 10' tells us that the curve $x(.)$, when t increases from $t^+(s_1)$ to $t_1(s_1)$ remains close up to the order $K\left[\frac{1}{\sqrt{\omega}} + |t^+(s_1) - t_1(s_1)| \left|\int_{t_1(s_1)}^{t^+(s_1)} b(x)\,dx\right|\right]$ to the curve $x_{s(t)}$ where

$$s(t) = \int_{t^+(s_1)}^{t} b(x)\,dx \qquad (8.343)$$

is the change of variables as in (8.120).

Applying Lemma 12, we have:

$$\| Z_1^{\star}(t_1(s_1)) - Z^{\star}(s(t_1(s_1))) \| \leq K\left[\frac{1}{\sqrt{\omega}} + |t^+(s_1) - t_1(s_1)| \left|\int_{t_1(s_1)}^{t^+(s_1)} b(x)\,dx\right|\right], \qquad (8.344)$$

where $Z_1^\star$ is defined by (8.133) as usual.

Now

$$\|Z_1^\star(t_1(s_1)) - Z'^\star(0)\| = \left\| \begin{bmatrix} 1 - \dfrac{b^2(t_1(s_1))}{2\omega a} + \dfrac{\int_0^1 b^2}{2\omega a} \\ \dfrac{b(t_1(s_1))}{\omega} \\ \dfrac{\dot b(t_1(s_1))}{\omega a} \end{bmatrix} - \begin{bmatrix} 1 - \dfrac{b^2(t_1(s_1))}{2\omega a} + \dfrac{\int_0^1 b^2}{2\omega a} \\ 0 \\ \dfrac{\dot b(t_1(s_1))}{\omega a} \end{bmatrix} \right\| \quad (8.345)$$

$$= \frac{|b(t_1(s_1))|}{\omega} < \sqrt{\frac{2\varepsilon_1}{\omega}} \qquad \text{by } (8.42),$$

hence, taking $\bar{\bar{\varepsilon}}_1 < 1$,

$$\|Z'^\star(0) - Z^\star(s(t_1(s_1)))\| \le K\left[\frac{1}{\sqrt{\omega}} + |t^+(s_1) - t_1(s_1)|\,\left|\int_{t_1(s_1)}^{t^+(s_1)} b(x)\,dx\right|\right] + \sqrt{\frac{2}{\omega}}$$

$$= K\left[\frac{1}{\sqrt{\omega}} + |t^+(s_1) - t_1(s_1)|\,\left|\int_{t_1(s_1)}^{t^+(s_1)} b(x)\,dx\right|\right]. \qquad (8.346)$$

Finally, applying Lemmas 10' and 13, we obtain:

$$d(x(t_1(s_1)), x_{s(t_1(s_1))}) \le K\left[\frac{1}{\sqrt{\omega}} + |t^+(s_1) - t_1(s_1)|\,\left|\int_{t_1(s_1)}^{t^+(s_1)} b(x)\,dx\right|\right], \quad (8.347)$$

(by Lemma 10').

$$\|Z'^\star(0) - Z^\star(s(t_1(s_1)))\| \le K\left[\frac{1}{\sqrt{\omega}} + |t^+(s_1) - t_1(s_1)|\,\left|\int_{t_1(s_1)}^{t^+(s_1)} b(x)\,dx\right|\right]. \quad (8.348)$$

Having derived these estimates, we consider the x's pieces defined by (8.322), starting at $x(t_1(s_1))$, and $\tilde{x}_s = x_{s+s(t_1(s_1))}$, obtained by a $s(t_1, s_1)$ time translation in x_s, which is defined by (8.329).

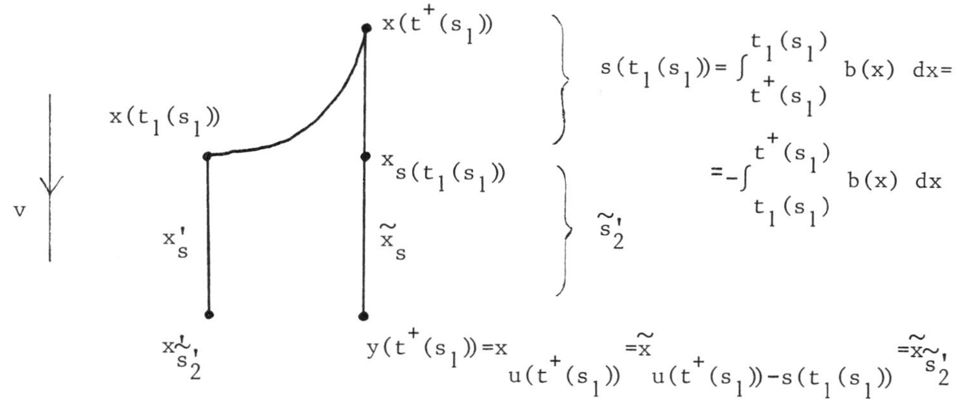

As shown in the figure, we have:

$$s(t_1(s_1)) = \int_{t^+(s_1)}^{t_1(s_1)} b(x)\,dx \text{ by } (8.343) \tag{8.349}$$

$$y(t^+(s_1)) = x_{u(t^+(s_1))} = \tilde{x}_{u(t^+(s_1)) - s(t_1(s_1))} = \tilde{x}_{\tilde{s}'_2}\,, \tag{8.350}$$

as, by (8.331), $y(t^+(s_1)) = x_{u(t^+(s_1))}$, $\tilde{x}_s = x_{s+s(t_1(s_1))}$, and

$u(t^+(s_1)) = -[\int_{t_1(s_1)}^{t^+(s_1)} b(x)\,dx - \tilde{s}'_2] = \tilde{s}'_2 + s(t_1(s_1))$.

By (8.347), the distance between the starting points of x'_s and $\tilde{x}_s$ is upper-bounded by:

$$\begin{aligned} d(x'_0, \tilde{x}_0) &= d(x(t_1(s_1)), x_{s(t_1(s_1))}) \\ &\leq \left[K \frac{1}{\sqrt{\omega}} + |t^+(s_1) - t_1(s_1)|\,\left|\int_{t_1(s_1)}^{t^+(s_1)} b(x)\,dx\right|\right], \end{aligned} \tag{8.351}$$

x'_s and $\tilde{x}_s$ being tangent to v, the distance of x'_s to $\tilde{x}_s$ remains of this order on $[0, \tilde{s}'_2]$, as by (8.327), $|\tilde{s}'_2| < K_6\sqrt{\alpha_1} < K_6$, for $\bar{\bar{\varepsilon}}_1$ small enough and $\varepsilon_1 < \bar{\bar{\varepsilon}}_1$ ($\alpha_1 \to 0$ when $\varepsilon_1 \to 0$),

$$d(x'_s, \tilde{x}_s) \leq K\left[\frac{1}{\sqrt{\omega}} + |t^+(s_1) - t_1(s_1)| \left|\int_{t_1(s_1)}^{t^+(s_1)} b(x)\,dx\right|\right], \quad s \in [0, \tilde{s}'_2]. \quad (8.352)$$

We consider now the transport equations along these pieces.

For x'_s, we consider $Z'^{\star}_s$, solution of $\dot{Z}'^{\star} = -{}^t\Gamma(x'_s) Z'^{\star}$; $Z'^{\star}_0 = Z'^{\star}(0)$ defined by (8.323).

For $\tilde{x}_s$, we consider $\tilde{Z}^{\star}_s$, solution of $\dot{\tilde{Z}}^{\star} = -{}^t\Gamma(\tilde{x}_s)\tilde{Z}^{\star}$; $\tilde{Z}^{\star}(0) = Z^{\star}(s(t_1(s_1)))$, where $Z^{\star}$ is defined in (8.330).

As $\tilde{x}_s$ is obtained from x_s by time translation $s(t_1(s_1))$ and as $\tilde{Z}^{\star}(0) = Z^{\star}(s(t_1(s_1)))$, we have:

$$\tilde{Z}^{\star}(s) = Z^{\star}(s + s(t_1(s_1))), \quad (8.353)$$

indeed, as $\tilde{x}_s$ is an $s(t_1(s_1))$-time translated of x_s, $\tilde{Z}^{\star}$ is an $s(t_1(s_1))$-time translated of $Z^{\star}$.

By (8.348), we have:

$$\|Z'^{\star}(0) - \tilde{Z}^{\star}(0)\| \leq K\left[\frac{1}{\sqrt{\omega}} + |t^+(s_1) - t_1(s_1)| \left|\int_{t_1(s_1)}^{t^+(s_1)} b(x)\,dx\right|\right]. \quad (8.354)$$

From (8.352) and (8.354), we deduce:

$$\|\tilde{Z}'^{\star}(\tilde{s}'_2) - \tilde{Z}^{\star}(\tilde{s}'_2)\| \leq K\left[\frac{1}{\sqrt{\omega}} + |t^+(s_1) - t_1(s_1)| \left|\int_{t_1(s_1)}^{t^+(s_1)} b(x)\,dx\right|\right]. \quad (8.355)$$

Indeed, we have:

$$\begin{cases} \dot{Z}'^{\star} = -{}^t\Gamma(x'_s) Z'^{\star} \\ \dot{\tilde{Z}}^{\star} = -{}^t\Gamma(\tilde{x}_s)\tilde{Z}^{\star} \end{cases}$$

hence

$$\dot{\overbrace{Z'^{\star} - \tilde{Z}^{\star}}} = -({}^t\Gamma(x'_s) - {}^t\Gamma(\tilde{x}_s)) Z'^{\star} + {}^t\Gamma(\tilde{x}_s)(\tilde{Z}^{\star} - Z'^{\star}),$$

hence:

$$\dot{\overbrace{\|Z'^{\star} - \tilde{Z}^{\star}\|}} \leq k\, d(x'_s, \tilde{x}_s)\, \|Z'^{\star}\| + \sup_{x \in M} \|{}^t\Gamma\|\, \|\tilde{Z}^{\star} - Z'^{\star}\|, \quad (8.356)$$

where k is a Lipschitz constant of ${}^t\Gamma$.

$\|Z'\star\|$ remains bounded on $[s, \tilde{s}'_2]$. Indeed,

$$\|Z'\star\|(0) = \left|1 - \frac{b^2(t_1(s_1))}{2\omega a} + \frac{\int_0^1 b^2}{2\omega a}\right| + \left|\frac{\dot{b}(t_1(s_1))}{\omega a}\right| < 2$$

is bounded by 2; and $|\tilde{s}'_2|$ is small if $\bar{\bar{\varepsilon}}_1$ is small. Hence $\|Z'\star\|$ will be bounded by 3, for instance, on $[s, s'_2]$ for $\bar{\bar{\varepsilon}}_1$ small enough.

Inequality (8.356) then implies via (8.354) and (8.352) and the fact that $|\tilde{s}'_2|$ is small:

$$\|Z'\star - \tilde{Z}\star|(\tilde{s}'_2) \tag{8.357}$$

$$\leq e^{(\operatorname{Sup}\|{}^t\Gamma\|)\,|\tilde{s}'_2|}\,\|Z'\star(0) - \tilde{Z}\star(0)\| +$$

$$+ 3k\,K\left[\frac{1}{\sqrt{\omega}} + |t^+(s_1) - t_1(s_1)|\,\left|\int_{t_1(s_1)}^{t^+(s_1)} b(x)\,dx\right|\,|\tilde{s}_2|\right]$$

$$\leq K\left[\frac{1}{\sqrt{\omega}} + |t^+(s_1) - t_1(s_1)|\,\left|\int_{t_1(s_1)}^{t^+(s_1)} b(x)\,dx\right|\right],$$

which yields (8.355).

Then using (8.337) we derive:

$$\|Z'\star - \tilde{Z}\star\|(\tilde{s}'_2) \leq K\left[\frac{1}{\sqrt{\omega}} + |t^+(s_1) - t_1(s_1)|\sqrt{\alpha_1}\right] \tag{8.358}$$

or else:

$$\|Z'\star(\tilde{s}'_2) - Z\star(\tilde{s}'_2 + s(t_1(s_1)))\| \leq K\left[\frac{1}{\sqrt{\omega}} + |t^+(s_1) - t_1(s_1)|\sqrt{\alpha_1}\right]. \tag{8.359}$$

At $\tilde{s}'_2$, the third component $c'\star(\tilde{s}'_2)$ of $Z'\star$ is zero, by definition of $\tilde{s}'_2$. Thus, the third component $c\star(\tilde{s}'_2 + s(t_1(s_1)))$ of $Z\star$ is upperbounded by:

$$|c\star(\tilde{s}'_2 + s(t_1(s_1)))| \leq K\left[\frac{1}{\sqrt{\omega}} + |t^+(s_1) - t_1(s_1))|\sqrt{\alpha_1}\right]. \tag{8.360}$$

Inequality (8.360) allows us to apply Lemma 13. Indeed, as $|\tilde{s}'_2|$ is small, we may ensure, if $\bar{\bar{\varepsilon}}_1$ is small enough, that:

$$\frac{1}{3} \leq \| Z'^{\star}(\tilde{s}_2') \| \leq 3 \tag{8.361}$$

as $\frac{1}{2} \leq \| Z'^{\star}(0) \| \leq 2$.

Thus, by (8.360), we have:

$$\frac{1}{3} - K\left[\frac{1}{\sqrt{\omega}} + |t^{+}(s_1) - t_1(s_1)|\sqrt{\alpha_1}\right] \leq \| Z^{\star}(\tilde{s}_2' + s(t_1(s_1))) \| \tag{8.362}$$

$$\leq 3 + K\left[\frac{1}{\sqrt{\omega}} + |t^{+}(s_1) - t_1(s_1)|\sqrt{\alpha_1}\right] ;$$

if ω'_8 is large enough and $\bar{\bar{\varepsilon}}_1$ small enough, we have:

$$\frac{1}{4} \leq \| Z^{\star}(\tilde{s}_2' + s(t_1(s_1))) \| = |a^{\star}| + |c^{\star}|(\tilde{s}_2' + s(t_1(s_1))) \leq 4, \tag{8.363}$$

hence, taking (8.360) into account, and taking ω_8 large enough and $\bar{\bar{\varepsilon}}_1$ small enough, we have:

$$\frac{1}{5} \leq |a^{\star}|(\tilde{s}_2' + s(t_1(s_1))) ; \quad |a^{\star}| + |c^{\star}|(\tilde{s}_2' + s(t_1(s_1))) \leq 4 \tag{8.364}$$

$$|c^{\star}|(\tilde{s}_2' + s(t_1(s_1))) \leq K\left[\frac{1}{\sqrt{\omega}} + |t^{+}(s_1) - t_1(s_1)|\sqrt{\alpha_1}\right] < \chi(4, \tfrac{1}{5}). \tag{8.365}$$

Lemma 13 thus applies with a value of s_1 equal to $\tilde{s}_2' + s(t_1(s_1))$ along with the notation we are using now.

We then deduce that $c^{\star}$ vanishes once and only once in the same interval $]\tilde{s}_2' + s(t_1(s_1)) - \theta(4, \frac{1}{5}), \tilde{s}_2' + s(t_1(s_1)) + \theta(4, \frac{1}{5})[$, at a point $\tilde{s}_3$ such that:

$$|\tilde{s}_3 - \tilde{s}_2' - s(t_1(s_1))| < K_6(4, \tfrac{1}{5})\, |c^{\star}|(\tilde{s}_2' + s(t_1(s_1))) \tag{8.366}$$

$$< K_6 K\left[\frac{1}{\sqrt{\omega}} + \sqrt{\alpha_1}(t^{+}(s_1) - t_1(s_1))\right].$$

For $\bar{\bar{\varepsilon}}_1$ small enough, and $\varepsilon_1 < \bar{\bar{\varepsilon}}_1$, we necessarily have $\tilde{s}_3 = \tilde{s}_2^{+}$.

Indeed, $c^{\star}(\tilde{s}_2^{+}) = 0$ by definition, and $|\tilde{s}_2^{+}| < K_6\sqrt{\alpha_1}$ by (8.332).

On the other hand,

$$|\tilde{s}_2' + s(t_1(s_1))| \leq |\tilde{s}_2'| + |s(t_1(s_1))| = |\tilde{s}_2'| + \left|\int_{t_1(s_1)}^{t^{+}(s_1)} b(x)\,dx\right|$$

$$\leq \left(\frac{K_6}{a} + K_1\right)\sqrt{\alpha_1} \quad \text{by } (8.327) \text{ and } (8.337).$$

Consequently,

$$\tilde{s}_2' + s(t_1(s_1)) - \theta(4,\tfrac{1}{5}) \leq -\theta(4,\tfrac{1}{5}) + (\frac{K_6}{a} + K_1)\sqrt{\alpha_1}$$

and

$$\tilde{s}_2' + s(t_1(s_1)) + \theta(4,\tfrac{1}{5}) \geq \theta(4,\tfrac{1}{5}) - (\frac{K_6}{a} + K_1)\sqrt{\alpha_1} .$$

If $\bar{\bar{\varepsilon}}_1$ is small enough so that $(\frac{2K_6}{a} + K_1)\sqrt{\alpha_1} < \theta(4,\frac{1}{5})$, then $\tilde{s}_2^+ \in \,]\tilde{s}_2' + s(t_1, s_1) - \theta(4,\frac{1}{5}),\ \tilde{s}_2' + s(t_1(s_1)) + \theta(4,\frac{1}{5})[$. Hence $\tilde{s}_2^+ = \tilde{s}_3$, and we have, by (8.366):

$$|s_2^+ - \tilde{s}_2' - s(t_1(s_1))| < K_6 K\left[\frac{1}{\sqrt{\omega}} + \frac{1}{a}\sqrt{\alpha_1}\,(t^+(s_1) - t_1(s_1))\right] \tag{8.367}$$

$$= K\left[\frac{1}{\sqrt{\omega}} + \sqrt{\alpha_1}\,(t^+(s_1) - t_1(s_1))\right],$$

hence

$$d(x_{\tilde{s}_2^+}, x_{\tilde{s}_2'+s(t_1(s_1))}) \leq (\sup_{x\in M} \|v\|)\, K\left[\frac{1}{\sqrt{\omega}} + \sqrt{\alpha_1}(t^+(s_1) - t_1(s_1))\right] \tag{8.368}$$

$$\leq K\left[\frac{1}{\sqrt{\omega}} + \sqrt{\alpha_1}\,(t^+(s_1) - t_1(s_1))\right],$$

or else:

$$d(x_{\tilde{s}_2^+}, x_{u(t^+(s_1))}) = d(x_{\tilde{s}_2^+}, y(t^+(s_1))) \leq K\left[\frac{1}{\sqrt{\omega}} + \sqrt{\alpha_1}(t^+(s_1) - t_1(s_1))\right]$$

by (8.331) and (8.350) $(x_{\tilde{s}_2'+s(t_1(s_1))} = x_{u(t^+(s_1))} = y(t^+(s_1)))$, which gives (8.333).

Equations (8.334) and (8.335) are established exactly as when we were supposing (8.339) instead of (8.338). (The proof we gave then did not use (8.334)', as can be checked.)

It remains to establish (8.334)'. By (8.359), we have

$$|a'^{\star}(\tilde{s}_2') - a^{\star}(\tilde{s}_2' + s(t_1(s_1)))| \leq K\left[\frac{1}{\sqrt{\omega}} + (t^+(s_1) - t_1(s_1))\sqrt{\alpha_1}\right]. \tag{8.368'}$$

As $\dot{a}^\star = c^\star$ and as $|c^\star|$ is upperbound by M+1 = 4+1 = 5 on the interval $]\tilde{s}'_2 + s(t_1(s_1)) - \theta,\ \tilde{s}'_2 + s(t_1(s_1)) + \theta[$, by (8.145), we have:

$$|a^\star(\tilde{s}_2^+) - a^\star(\tilde{s}'_2 + s(t_1(s_1)))| = \left|\int_{\tilde{s}_2^+}^{\tilde{s}'_2+s(t_1(s_1))} c^\star(s)\,ds\right| \tag{8.368)''}$$

$$\leq 5\,|\tilde{s}_2^+ - \tilde{s}'_2 - s(t_1(s_1))|$$

$$\leq K\left[\frac{1}{\sqrt{\omega}} + \sqrt{\alpha_1}(t^+(s_1) - t_1(s_1))\right] \quad \text{via (8.366)}.$$

Inequalities (8.368)' and (8.368)'' imply:

$$|a^\star(\tilde{s}_2^+) - a'^\star(\tilde{s}'_2)| < K\left[\frac{1}{\sqrt{\omega}} + \sqrt{\alpha_1}(t^+(s_1) - t_1(s_1))\right]$$

which proves (8.364)' and completes the proof of Lemma 15.

In case $t^+(s_1)$ is replaced by $t^-(s_1)$, Lemma 15 has the following analogue: we denote x_s the v-orbit starting at $x(t^-(s_1))$. Let $Z^\star$ be the solution of the transport equation. We denote

$x_{\tilde{s}_2^-}$ the corresponding point to $x_{\tilde{s}_2^+}$ with $t^-(s_1)$ instead of $t^+(s_1)$. (8.369)

We then have:

Lemma 15': There exist ω_8, $\bar{\bar{\varepsilon}}_1$, uniform on all intervals of that type such that:

$$d(x_{\tilde{s}_2^-}, y(t^-(s_1))) < K_{12}\left[\frac{1}{\sqrt{\omega}} + \sqrt{\alpha_1}(t_1(s_1) - t^-(s_1))\right] \tag{8.370}$$

$$Z^\star(\tilde{s}_2^-) = \begin{bmatrix} a^\star(\tilde{s}_2^-) \\ 0 \\ 0 \end{bmatrix}; \tag{8.371}$$

$$|a^\star(\tilde{s}_2^-) - a'^\star(\tilde{s}'_2)| < K_{12}\left[\frac{1}{\sqrt{\alpha}} + \sqrt{\alpha_1}(t_1(s_1) - t^-(s_1))\right] \tag{8.371)'}$$

$$\left\|Z^\star(\tilde{s}_2^-) - \begin{bmatrix} 1 \\ 0 \\ 0 \end{bmatrix}\right\| = |a^\star(\tilde{s}_2^-) - 1| < K_{12}\left(\int_0^1 \frac{b^2}{\omega} + \varepsilon_1 + \sqrt{\alpha_1}\right), \tag{8.372}$$

for $\omega \geq \omega_8$; $0 < \varepsilon_1 < \bar{\bar{\varepsilon}}_1$; K_{12} is a geometric constant.

We start now the final steps in our estimates:

We consider x_s the tangent piece to v starting at $x(t^1(s_1))$; $x_{\tilde{s}_2^+}$ is as before. We assume that:

$$b(t^+(s_1)) > 0; \quad \text{the other case being similar.} \tag{8.373}$$

Let $Z^\star(s)$ be the solution of (8.330). We remember that

$$Z^\star(t) = \begin{bmatrix} 1 - \frac{b^2}{2\omega a}(t) + \frac{\int_0^1 b^2}{2\omega a} \\ \frac{b(t)}{\omega} \\ \frac{\dot{b}(t)}{\omega a} \end{bmatrix}. \text{ Let:}$$

$$\bar{s} \text{ be defined by } \psi(x_{\tilde{s}_2^+}) = x_{\bar{s}}. \tag{8.374}$$

We have:

Lemma 16: Under (8.373), there exist ω_9, M_4, K_{13}, $\bar{\bar{\varepsilon}}_1$, $\gamma_1 > 0$ uniform such that: if the interval we consider is $[t^+(s_1), t_1]$ (i.e. $\dot{b}(t_1) = 0$; t_1 the first zero of $\dot{b}$ after $t^+(s_1)$), we have:

$$\gamma_1 \le \left| \int_{t^+(s_1)}^{t_1} b(x)\,dx \right| \le M_4; \quad Z^\star(\bar{s}) = \begin{bmatrix} a^\star(\bar{s}) \\ 0 \\ 0 \end{bmatrix} \tag{8.375}$$

$$d(\psi(x_{\tilde{s}_2^+}), x(t_1)) = d(x_{\bar{s}}, x(t_1)) < K_{13}\left[\frac{1}{\sqrt{\omega}} + |t_1 - t^+(s_1)|\right] \tag{8.376}$$

$$\|Z^\star(\bar{s}) - Z_1^\star(t_1)\| = \|Z^\star(\psi(x_{\tilde{s}_2^+})) - Z_1^\star(t_1)\| < K_{13}\left[\frac{1}{\sqrt{\omega}} + |t_1 - t^+(s_1)|\right], \tag{8.377}$$

for $\omega \ge \omega_9$ and $\varepsilon_1 < \bar{\bar{\varepsilon}}_1$.

Remark: If, instead of (8.373), we have $b(t^+(s_1)) < 0$, we replace $\psi(x_{\tilde{s}_2^+}) = x_{\bar{s}}$ by $\psi^{-1}(x_{\tilde{s}_2^+}) = x_{\underline{s}}$ in Lemma 16.

Remark: In Lemma 16, we replace, if $b(t^-(s_2)) > 0$ and if the considered interval is $[t_k, t^-(s_2)]$, ψ by ψ^{-1}, $x_{\underset{\sim}{s}_2^+}$ by $x_{\underset{\sim}{s}_2^-}$ and $\bar{s}$ by $\underline{s}$ such that $x_{\underline{s}} = \psi^{-1}(x_{\underset{\sim}{s}_2^-})$. If $b(t^-(s_2)) < 0$, we keep ψ, replace $x_{\underset{\sim}{s}_2^+}$ by $x_{\underset{\sim}{s}_2^-}$, and keep $\bar{s}$ with $x_{\bar{s}} = \psi(x_{\underset{\sim}{s}_2^+})$. In any case, t_1 is replaced by t_k and $t^+(s_1)$ by $t^-(s_2)$.

The following drawings summarize the contents of Lemma 16 and the previous remarks.

a) If $[t^+(s_1), t_1]$; $b(t^+(s_1)) > 0$.

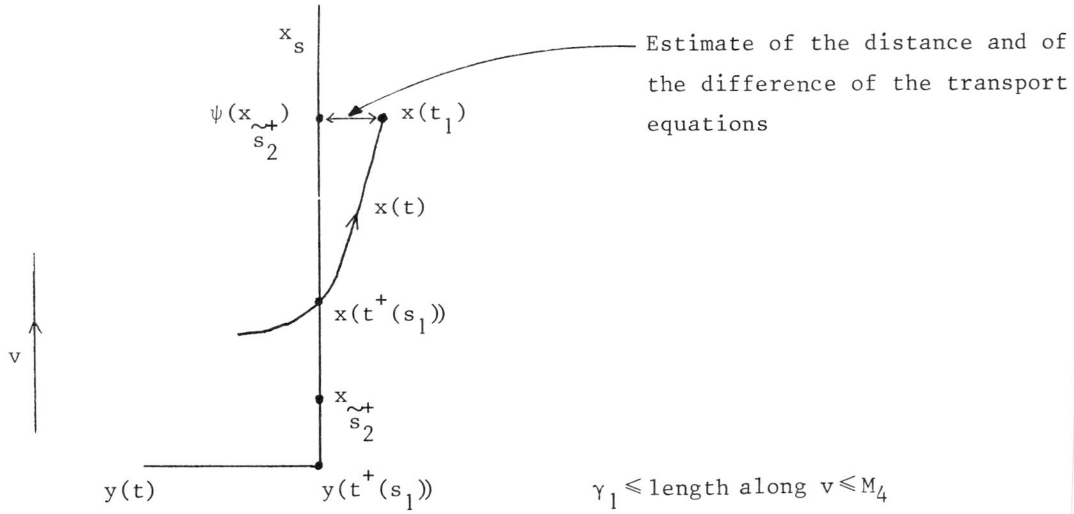

b) If $\left[t^{+}(s_1), t_1\right]$; $b(t^{-}(s_2)) < 0$.

$y(t)$ $\qquad$ $y(t^{+}(s_1))$

$x_{\underset{s_2}{\sim}+}$

$x(t^{+}(s_1))$

$\gamma_1 \leqslant$ length along $v \leqslant M_4$

v

$\psi^{-1}(x_{\underset{s_2}{\sim}+})$ $\qquad$ $x(t_1)$

x_s

Estimate of the distance and of
the difference of the transport
equations

c) if $\left[t_k, t^{-}(s_2)\right]$; $b(t^{-}(s_2)) > 0$.

$y(t^{-}(s_1))$ $\qquad$ $y(t)$

$x_{\underset{s_2}{\sim}-}$

$x(t^{-}(s_2))$

$\gamma_1 \leqslant$ length along $v \leqslant M_4$

v

$x(t_k)$ $\qquad$ $\psi^{-1}(x_{\underset{s_2}{\sim}-})$

x_s

Estimate of the distance and of the difference
of the transport equations

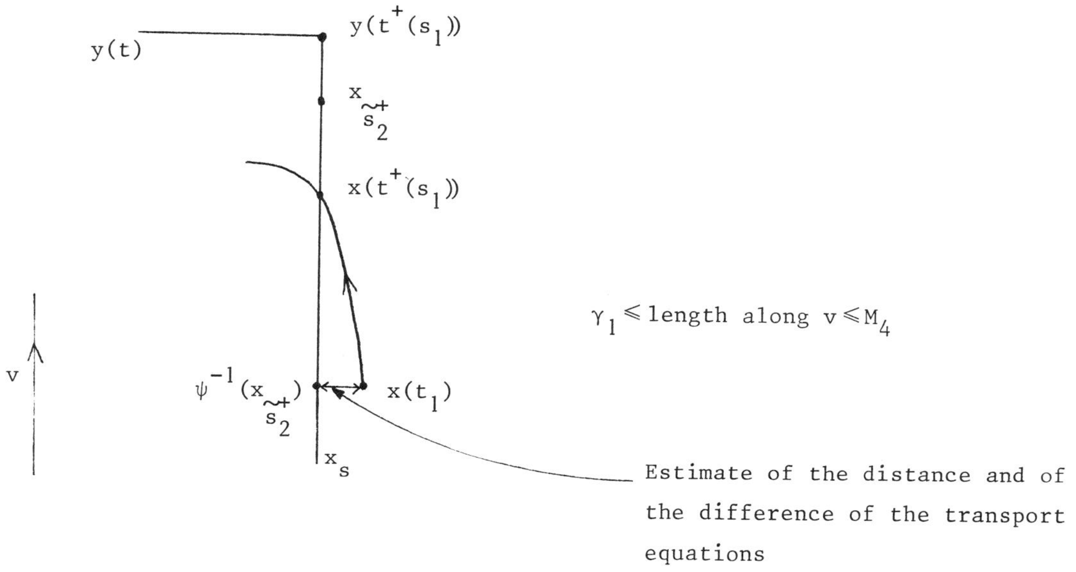

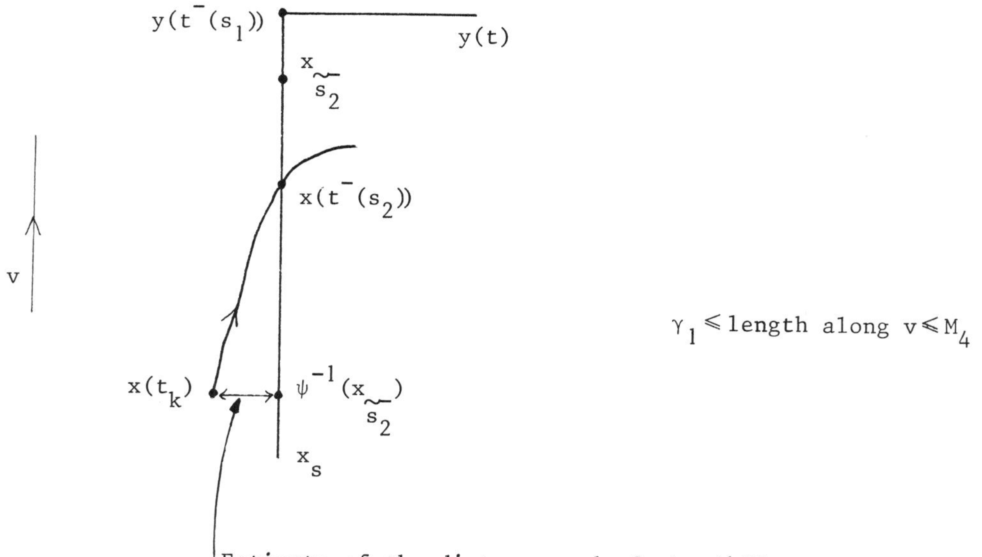

d) If $[t_k, t^-(s_2)]$; $b(t^-(s_2)) < 0$.

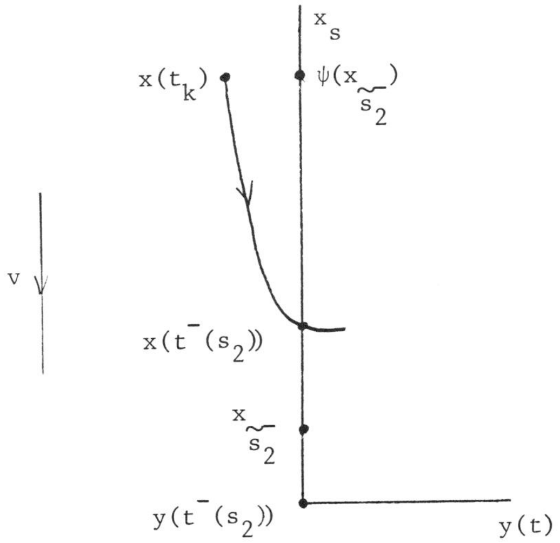

<u>Proof of Lemma 16</u>: The proof relies on Lemmas 10', 12 and 13. Here are the main steps of it.

We first remark that b and $\dot{b}$ are never zero in $]t^+(s_1), t_1)[$. Indeed, by definition $I_1(s_1)$ is the maximal time interval containing s_1 such that $\frac{b^2}{\omega}$ remains less than $2\varepsilon_1$. Hence $\frac{b^2(t^+(s_1))}{\omega} = 2\varepsilon_1$; and $b\dot{b}$ is positive on a right neighbourhood of $t^+(s_1)$. As $b(t^+(s_1))$ is positive by hypothesis, $\dot{b}$ is positive on a right neighbourhood of $t^+(s_1)$. Hence, $\dot{b}$ is positive on all $]t^+(s_1), t_1[$.

We also remark that, by (8.336), the point $t^+(s_1)$ satisfies the hypothesis (H) with $\delta_0 = \frac{1}{2}$ and $\beta_0 = 2$.

Finally, $|t_1 - t^+(s_1)|$ is as small as we wish following an appropriate choice of a uniform ω_9. This allows us to apply Lemmas 10' and 12 to $[t^+(s_1), t_1]$ as follows:

We first choose ω_9 large enough so that, for $\omega \geq \omega_9$, all the non-concentration set has total measure less than $\inf(\bar{\varepsilon}_0, \varepsilon_0(\sup_{x\in M}(\bar{s}(x) + \bar{\alpha})))$, where Lemma 12 gives $\bar{\varepsilon}_0$ when δ_0 is $\frac{1}{2}$ and β_0 is 2; while $\varepsilon_0(\sup_{x\in M}(\bar{s}(x) + \bar{\alpha}))$

is the value of $\varepsilon_0(C)$ in Lemma 10' with $C = \operatorname{Sup}_{x \in M} (\bar{s}(x) + \bar{\alpha})$.

For $\omega \geq \omega_9$, we thus have $|t^+(s_1) - t_1| < \bar{\varepsilon}_0$. Lemma 12 then applies and we have:

$$\left| \int_{t^+(s_1)}^{t_1} b(x)\,dx \right| < \bar{s}(x(t^+(s_1))) + \bar{\alpha} < \operatorname{Sup}_{x \in M} (\bar{s}(x) + \bar{\alpha}) = \bar{C}.$$

As $|t^+(s_1) - t_1| < \varepsilon_0(\bar{C})$, Lemma 10' applies.

We thus have:

$$\left| \int_{t^+(s_1)}^{t_1} b(x)\,dx \right| \leq \bar{s}(x(t^+(s_1))) + \bar{\alpha}. \tag{8.378}$$

By Lemma 10', if

$$s(t) = \int_{t^+(s_1)}^{t} b(x)\,dx \text{ is the change of variables (8.120)}, \tag{8.379}$$

the distance $d(x_{s(t)}, x(t))$, for t in $[t^+(s_1), t_1]$, remains bounded by

$$K\left[\frac{1}{\sqrt{\omega}} + |t_1 - t^+(s_1)| \left| \int_{t^+(s_1)}^{t_1} b(x)\,dx \right| \right] \leq K'\left[\frac{1}{\sqrt{\omega}} + |t_1 - t^+(s_1)|\right].$$

In the following, K is a generic uniform constant.

In particular, we have

$$d(x_{s(t_1)}, x(t_1)) < K\left[\frac{1}{\sqrt{\omega}} + |t_1 - t^+(s_1)|\right]. \tag{8.380}$$

Lemma 12 tells us, via the upperboundedness of $\left| \int_{t^+(s_1)}^{t_1} b(x)\,dx \right|$, that

$$\|Z^\star(s(t)) - Z_1^\star(t)\| \leq K\left[\frac{1}{\sqrt{\omega}} + |t_1 - t^+(s_1)|\right], \quad t \in [t^+(s_1), t_1]. \tag{8.381}$$

As the third component of $Z_1^\star(t)$ is $\frac{\dot{b}(t)}{\omega a}$, this third component vanishes at t_1.
Thus the third component of $Z^\star(s(t))$, $c^\star(s(t))$ is at t_1, upperbounded by:

$$|c^\star(s(t_1))| \leq K\left[\frac{1}{\sqrt{\omega}} + |t_1 - t^+(s_1)|\right]. \tag{8.382}$$

At this point, as we have already done on many earlier occasions, we verify that all the hypotheses of Lemma 14 are satisfied and we apply this lemma. We then deduce the existence of an $\tilde{s} \in \left] s(t_1) - K\left[\frac{1}{\sqrt{\omega}} + |t_1 - t^+(s_1)|\right], s(t_1) + K\left[\frac{1}{\sqrt{\omega}} + |t_1 - t^+|\right]\right[$

such that:

$$c^\star(\tilde{s}) = 0; \quad |\tilde{s} - s(t_1)| < K\left[\frac{1}{\sqrt{\omega}} + |t_1 - t^+(s_1)|\right], \tag{8.383}$$

hence

$$d(x_{\tilde{s}}, x_{s(t_1)}) < K \operatorname{Sup}\|v\| \left[\frac{1}{\sqrt{\omega}} + |t_1 - t^+(s_1)|\right] = K\left[\frac{1}{\sqrt{\omega}} + |t_1 - t^+(s_1)|\right]. \tag{8.384}$$

Equations (8.380) and (8.384) imply:

$$d(x_{\tilde{s}}, x(t_1)) < K\left[\frac{1}{\sqrt{\omega}} + |t_1 - t^+(s_1)|\right]. \tag{8.385}$$

Furthermore, we have:

$$\|Z^\star(\tilde{s}) - Z^\star(s(t_1))\| \leq C|\tilde{s} - s(t_1)| \leq K\left[\frac{1}{\sqrt{\omega}} + |t_1 - t^+(s_1)|\right], \tag{8.386}$$

which implies, via (8.381):

$$\|Z^\star(s) - Z_1^\star(t_1)\| \leq K\left[\frac{1}{\sqrt{\omega}} + |t_1 - t^+(s_1)|\right]. \tag{8.387}$$

Inequalities (8.385) and (8.387) would imply (8.376) and (8.377) if we proved that $x_{\tilde{s}} = (x_{\tilde{s}_2^+})$.

In order to prove that $x_{\tilde{s}} = \psi(x_{\tilde{s}_2^+})$, we proceed as follows:

We consider the curve $\tilde{x}_s$ starting at $x_{\tilde{s}_2^+}$ tangent to v. This curve is deduced from x_s by translation of $\tilde{s}_2^+$:

$$\tilde{x}_s = x_{s+\tilde{s}_2^+}. \tag{8.388}$$

If $\tilde{Z}^\star$ is the solution of the transport equation along $\tilde{x}_s$ with initial data

$\tilde{Z}^\star(0) = Z^\star(\tilde{s}_2^+)$, $\tilde{Z}^\star$ is also deduced from $Z^\star$ by time translation:

$$\tilde{Z}^\star(s) = Z^\star(s + \tilde{s}_2^+). \tag{8.389}$$

As, by definition of $\tilde{s}_2^+$, we have $c^\star(\tilde{s}_2^+) = 0$, we also have:

$$\tilde{Z}^\star(0) = \begin{bmatrix} \tilde{a}^\star(0) \\ 0 \\ 0 \end{bmatrix} = \begin{bmatrix} a^\star(\tilde{s}_2^+) \\ 0 \\ 0 \end{bmatrix} = a^\star(\tilde{s}_2^+) \begin{bmatrix} 1 \\ 0 \\ 0 \end{bmatrix}, \tag{8.390}$$

$a^\star(\tilde{s}_2^+) \neq 0$ since $Z^\star \not\equiv 0$.

If we denote $\tilde{Z}^\star(s) = \begin{bmatrix} \tilde{a}^\star(0) \\ 0 \\ 0 \end{bmatrix}$, the point $\psi(x_{\tilde{s}_2^+})$ corresponds to the first coincidence point of $x_{\tilde{s}_2^+}$, that is the first positive s such that:

$$\tilde{c}^\star(s) = c^\star(s + \tilde{s}_2^+) = 0 \text{ first positive } s. \tag{8.391}$$

Now, at $\tilde{s}$, by (8.383), we have $c^\star(\tilde{s}) = 0$.

Thus $x_{\tilde{s}}$ has to be $\psi(x_{\tilde{s}_2^+})$ if:

$$\tilde{s}_2^+ < \tilde{s}, \tag{8.392}$$

and if $\tilde{s} - \tilde{s}_2^+$ is less than the second positive zero of $\tilde{c}^\star(s)$, i.e. than $\bar{s}(x_{\tilde{s}_2^+}) + \bar{s}(\tilde{x}_{\bar{s}(x_{\tilde{s}_2^+})})$, as $\bar{s} = M \to \mathbb{R}$ defines ψ, i.e.

$$\tilde{s} < \tilde{s}_2^+ + \bar{s}(x_{\tilde{s}_2^+}) + \bar{s}(\tilde{x}_{\bar{s}(x_{\tilde{s}_2^+})}). \tag{8.393}$$

We want to prove (8.393).

From (8.383) we know that:

$$\tilde{s} < s(t_1) + K\left[\frac{1}{\sqrt{\omega}} + \left|t_1 - t^+(s_1)\right|\right] = \left|\int_{t^+(s_1)}^{t_1} b(\mathbf{x})\,dx\right| + K\left[\frac{1}{\sqrt{\omega}} + \left|t_1 - t^+(s_1)\right|\right]. \tag{8.394}$$

Thus, by (8.378)

$$\tilde{s} < \bar{s}(x(t^+(s_1))) + \bar{\alpha} + K\left[\frac{1}{\sqrt{\omega}} + |t_1 - t^+(s_1)|\right]. \tag{8.395}$$

The function $\bar{s}$ is C^∞. Thus

$$|\bar{s}(x(t^+(s_1))) - s(x_{\tilde{s}_2^+})| \leq K\, d(x(t^+(s_1)), x_{\tilde{s}_2^+}). \tag{8.396}$$

Now $x(t^+(s_1)) = x_0$ by definition. Thus, $d(x(t^+(s_1)), x_{\tilde{s}_2^+}) \leq (\operatorname{Sup}_{x\in M} \|v\|)\, |\tilde{s}_2^+|$, hence

$$\bar{s}(x_{\tilde{s}_2^+}) > \bar{s}(x(t^+(s_1))) - \operatorname{Sup}_{x\in M} \|v\|\, |\tilde{s}_2^+|, \tag{8.397}$$

hence, by (8.332),

$$\bar{s}(x_{\tilde{s}_2^+}) > \bar{s}(x(t^+(s_1))) - K\sqrt{\alpha_1}. \tag{8.398}$$

Inequalities (8.394) and (8.398) imply:

$$\tilde{s} < \bar{s}(x_{\tilde{s}_2^+}) + \bar{\alpha} + K\left[\sqrt{\alpha_1} + \frac{1}{\sqrt{\omega}} + |t_1 - t^+(s_1)|\right]. \tag{8.399}$$

By the definition of $\bar{\alpha}$ ((8.137)), we have:

$$0 < \bar{\alpha} = \inf_{x\in M} \left(\frac{\bar{s}(x)}{3}\right).$$

On the other hand, by (8.332), $|\tilde{s}_2^+| < \frac{1}{a} K_6 \sqrt{\alpha_1}$.

We choose $\bar{\varepsilon}_1$ small enough and ω_9 large enough such that, as soon as $\varepsilon_1 < \bar{\bar{\varepsilon}}_1$ and $\omega \geq \omega_9$, we have:

$$\frac{1}{a} K_6 \sqrt{\alpha_1} + K\left[\sqrt{\alpha_1} + \frac{1}{\sqrt{\omega}} + |t_1 - t^+(s_1)|\right] < \bar{\alpha}. \tag{8.400}$$

Inequalities (8.400) and (8.399) then imply:

$$\tilde{s} < \bar{s}(x_{\tilde{s}_2^+}) + 2\bar{\alpha} - \frac{1}{a} K_6 \sqrt{\alpha_1} < \bar{s}(x_{\tilde{s}_2^+}) - |s_2^+| + \frac{2}{3} \inf_{x\in M} (\bar{s}(x)), \tag{8.401}$$

hence:

$$\tilde{s} < \tilde{s}_2^+ + \bar{s}(x_{\tilde{s}_2^+}) + \frac{2}{3}\bar{s}(\tilde{x}_{\bar{s}(x_{\tilde{s}_2^+})}) \text{, which implies (8.393).} \qquad (8.402)$$

We prove now (8.392).

Equation (8.392) is a consequence of the not yet proved inequality (8.375), i.e.

$$\left| \int_{t^+(s_1)}^{t_1} b(x)\, dx \right| \geq \gamma_1 . \qquad (8.403)$$

Indeed, if (8.403) holds, then $s(t_1) = \left| \int_{t^+(s_1)}^{t_1} b(x)\, dx \right| \geq \gamma_1 > 0$. As $b(t^+(s_1)) > 0$ and b does not vanish on $]t^+(s_1), t_1[$, in fact we have $s(t_1) \geq \gamma_1$. Hence, by (8.383),

$$\tilde{s} \geq \gamma_1 - K\left[\frac{1}{\sqrt{\omega}} + \left|t_1 - t^+(s_1)\right|\right] \geq \frac{\gamma_1}{2} > \frac{1}{a}K_6\sqrt{\alpha_1} > \left|s_2^+\right| \quad \text{(via (8.332))}, \qquad (8.404)$$

if ω_9 and $\bar{\bar{\varepsilon}}_1$ are chosen small enough; this proves (8.392).

We now have to prove (8.403).

For this, we proceed as we did to prove $\left| \int_{t_i}^{t_{i+1}} b(x)\, dx \right| \geq \delta_1$ in Lemma 11. We know that $b\dot{b}(t_1(s_1)) = 0$ (by definition of $t_1(s_1)$), we know also that $\dot{b}(t_1) = 0$. Thus $(b\dot{b})^{\cdot}$ vanishes somewhere in the interval $]t_1(s_1), t_1[$. At this point, denoted t^1, we have:

$$(\dot{b}^2 + b\ddot{b})(t^1) = 0, \qquad (8.405)$$

hence:

$$b\ddot{b}(t^1) < 0. \qquad (8.406)$$

b is never zero in $]t_1(s_1), t_1[$. Indeed, as already noticed, b is never zero in $]t^+(s_1), t_1[$. On the other hand, b is never zero in $]t_1(s_1), t^+(s_1)]$ as, by Lemma 6, $t_1(s_1)$ is the unique point in $I_1(s_1)$ where b might possibly take the value zero. Hence $b(t^1)$ has the same sign as $b(t^+(s_1))$, which is positive.

Hence:

$$\ddot{b}(t^1) < 0; \quad b(t^1) > 0. \tag{8.407}$$

On the other hand, by (5.38) :

$$\frac{\ddot{b}(t^1)}{\omega a} = -b(t^1)\left(\frac{b^2(t^1)}{2\omega a} - 1 - \frac{\int_0^1 b^2}{2\omega a} + \frac{a\tau}{\omega} - \frac{b(t^1)}{\omega}\bar{\mu}_\xi + \bar{\mu}\frac{\dot{b}(t^1)}{\omega a}\right)$$

Thus:

$$\delta(t^1) = \frac{b^2(t^1)}{2\omega a} - 1 - \frac{\int_0^1 b^2}{2\omega a} + \frac{a\tau}{\omega} - \frac{b(t^1)}{\omega}\bar{\mu}_\xi + \bar{\mu}\frac{\dot{b}(t^1)}{\omega a} > 0. \tag{8.408}$$

The point t^1 must belong to the interval $]t^+(s_1), t_1[$: if not, it would belong to $I_1(s_1)$. But in $I_1(s_1)$, by (8.42), we have:

$$\frac{b^2(t^1)}{\omega} \leq 2\varepsilon_1 \quad \text{and} \quad \frac{|\dot{b}|(t^1)}{\omega a} < \frac{1}{a}\sqrt{\alpha_1}. \tag{8.409}$$

Equations (8.408) and (8.409) then imply:

$$1 < \frac{2\varepsilon_1}{a_1} + \frac{a_1}{\omega}\operatorname*{Sup}_{x\in M}(|\tau(x)|) + \operatorname*{Sup}_{x\in M}(|\bar{\mu}_\xi(x)|)\frac{\sqrt{2\varepsilon_1}}{\sqrt{\omega}} + \operatorname*{Sup}_{x\in M}|\bar{\mu}(x)| . \frac{\sqrt{\alpha_1}}{a_1}, \tag{8.410}$$

which we can exclude taking $\bar{\bar{\varepsilon}}_1$ small enough.

Thus $t_1 \in]t^+(s_1), t_1[$.

Let:

$$\delta(t^+(s_1)) = \frac{b^2(t^+(s_1))}{2\omega a} - 1 - \frac{\int_0^1 b^2}{2\omega a} - \frac{a\tau(x(t^+(s_1)))}{\omega} \tag{8.411}$$
$$- \frac{b(t^+(s_1))}{\omega}\bar{\mu}_\xi(x(t^+(s_1))) + \bar{\mu}(x(t^+(s_1)))\frac{\dot{b}(t^+(s_1))}{\omega a}.$$

We have:

$$|\delta(t^1) - \delta(t^+(s_1))| \leq \left|\left(1 - \frac{b^2(t^1)}{2\omega a} + \frac{\int_0^1 b^2}{2\omega a}\right) - \left(1 - \frac{b^2(t^+(s_1))}{2\omega a} + \frac{\int_0^1 b^2}{2\omega a}\right)\right| \tag{8.412}$$
$$+ |\bar{\mu}_\xi(x(t^1))|\left|\frac{b(t^1)}{\omega} - \frac{b(t^+(s_1))}{\omega}\right|$$

$$+ \frac{|b(t^+(s_1))|}{\omega} |\bar{\mu}_\xi(x(t^1)) - \bar{\mu}(x(t^+(s_1)))|$$

$$+ |\bar{\mu}(x(t^1))| \left| \frac{\dot{b}(t^1)}{\omega a} - \frac{\dot{b}(t^+(s_1))}{\omega a} \right|$$

$$+ \frac{|b(t^+(s_1))|}{\omega a} |\bar{\mu}(x(t^1)) - \bar{\mu}(x(t^+(s_1)))| + 2\frac{a_1}{\omega} \operatorname*{Sup}_{x \in M} |\tau(x)|,$$

hence, with a constant $C \geq 1 + 2 \operatorname*{Sup}_{x\in M} |\bar{\mu}_\xi| + 2 \operatorname*{Sup}_{x\in M} |\tau(x)| + 2 \operatorname*{Sup}_{x\in M} |\bar{\mu}|$,

$$|\delta(t^1) - \delta(t^+(s_1))| \tag{8.413}$$

$$\leq C\left[\left| \left(1 - \frac{b^2(t^1)}{2\omega a} + \frac{\int_0^1 b^2}{2\omega a}\right) - \left(1 - \frac{b^2(t^+(s_1))}{2\omega a} + \frac{\int_0^1 b^2}{2\omega a}\right) \right| \right.$$

$$+ \left| \frac{b(t^1)}{\omega a} - \frac{b(t^+(s_1))}{\omega} \right| + \left| \frac{\dot{b}(t^1)}{\omega a} - \frac{\dot{b}(t^+(s_1))}{\omega a} \right|$$

$$\left. + \frac{a_1}{\omega} + \frac{|\dot{b}(t^+(s_1))|}{\omega a_0} + \frac{|b(t^+(s_1))|}{\omega} \right]$$

$$\leq C\left[\| Z_1^{\star}(t^1) - Z_1^{\star}(t^+(s_1)) \| + \frac{a_1}{\omega} + \frac{|\dot{b}(t^+(s_1))|}{\omega a_0} + \frac{|b(t^+(s_1))|}{\omega} \right].$$

Finally, by (8.42), $\frac{|\dot{b}(t^+(s_1))|}{\omega} < \sqrt{\alpha_1}$ and $\frac{|b(t^+(s_1))|}{\sqrt{\omega}} = \sqrt{2\varepsilon_1}$. Hence:

$$|\delta(t^1) - \delta(t^+(s_1))| \leq C\left[\| Z_1^{\star}(t^1) - Z_1^{\star}(t^+(s_1)) \| + \frac{a_1}{\omega} + \frac{\sqrt{\alpha_1}}{a_0} + \frac{\sqrt{2\varepsilon_1}}{\sqrt{\omega}} \right]. \tag{8.414}$$

On the other hand, we always have by (8.42):

$$\delta(t^+(s_1)) \leq -1 + \frac{\varepsilon_1}{a_0} + \frac{a_1}{\omega} \operatorname*{Sup}_{x\in M} |\tau(x)| + \frac{\sqrt{2\varepsilon_1}}{\sqrt{\omega}} \operatorname*{Sup}_{x\in M} |\bar{\mu}_\xi| + \operatorname*{Sup}_{x\in M} \bar{\mu} \frac{\sqrt{\alpha_1}}{a_0}, \tag{8.415}$$

For $\bar{\bar{\varepsilon}}_1$ small enough and ω_9 large enough, we thus have:

$$\delta(t^+(s_1)) \leq -\tfrac{1}{2}\,; \tag{8.416}$$

Equations (8.408) and (8.416) then imply:

$$\left| \delta(t^1) - \delta(t^+(s_1)) \right| \geq \tfrac{1}{2}. \tag{8.417}$$

Hence, by (8.414):

$$\left\| Z_1^\star(t^1) - Z_1^\star(t^+(s_1)) \right\| \geq \frac{1}{2C} - \frac{a_1}{\omega} - \frac{\sqrt{\alpha_1}}{a_0} - \frac{\sqrt{2\varepsilon_1}}{\sqrt{\omega}} \geq \frac{1}{4C}, \tag{8.418}$$

for $\bar{\bar{\varepsilon}}_1$ small enough and ω_9 large enough.

We apply then Lemma 12 to the interval $[t^+(s_1), t^1]$; which is possible as this interval, being contained in $[t^+(s_1), t_1]$ satisfies the hypothesis of the lemma. In particular, by (8.141) of this lemma, we have:

$$\frac{\bar{\delta}}{4C} \leq \bar{\delta} \left\| Z_1^\star(t^1) - Z_1^\star(t^+(s_1)) \right\| \tag{8.419}$$

$$\leq \left| \int_{t^+(s_1)}^{t^1} b(\mathbf{x})\, dx \right| + \bar{K}_5 \left[\frac{1}{\sqrt{\omega}} + \left| t^1 - t^+(s_1) \right| \left| \int_{t^+(s_1)}^{t^1} b(x)\, dx \right| \right],$$

hence, b being positive, using the fact that $\left| \int_{t^+(s_1)}^{t^1} b(x)\, dx \right| \leq \left| \int_{t^+(s_1)}^{t_1} b(x)\, dx \right|$ and that $\left| \int_{t^+(s_1)}^{t^1} b(x)\, dx \right| \leq M_4$ (which is a consequence of the preceding inequality and of the estimate we proved on $\left| \int_{t^+(s_1)}^{t_1} b(x)\, dx \right|$), we derive

$$\frac{\bar{\delta}}{4C} \leq \left| \int_{t^+(s_1)}^{t_1} b(x)\, dx \right| + K \left[\frac{1}{\sqrt{\omega}} + \left| t_1 - t^+(s_1) \right| \right]. \tag{8.420}$$

With a good choice of ω_9, (8.420) implies:

$$\frac{\bar{\delta}}{8C} \leq \left| \int_{t^+(s_1)}^{t_1} b(x)\, dx \right|, \tag{8.421}$$

which gives (8.403) or the other inequality of (8.375) and completes the proof of Lemma 16. ■

<u>Lemma 17</u>: Between $t^+(s_1)$ and $t^-(s_2)$, $\dot{b}$ is zero somewhere.

The function b^2 is convex in a neighbourhood of $t^+(s_1)$, by Lemmas 5 and 6, for ε_1 small enough. Furthermore, by Lemma 6, (8.42) in particular, $\frac{b^2}{\omega}(t^+(s_1)) = \frac{b^2}{\omega}(t^-(s_2)) = 2\varepsilon_1$. Hence $\dot{b}$ must take the value zero in $]t^+(s_1), t^-(s_2)[$. Indeed, b^2 is an increasing function in a right neighbourhood of $t^+(s_1)$ (see the beginning of the proof of Lemma 16).

We complete these estimates by the following easy lemma.

Lemma 18: Let ε_1 be given small enough so all the preceding lemmas are valid for $\omega \geq \omega_{10}$.
Let j_1 be the total number of type (I) and (II) intervals satisfying the hypothesis (H) with given δ_0 and β_0 (i.e., at t_i, (H) is satisfied with δ_0 and β_0).
Let j_2 be the total number of initial and final intervals.

Then $\frac{j_1+j_2}{\sqrt{\omega}} \to 0$ when $\omega \to +\infty$.

Proof: We set:

I_1 = type (I) intervals on which (H) is satisfied with given δ_0 and β_0;
I_2 = type (II) intervals on which (H) is satisfied with given δ_0 and β_0;
I_3 = initial intervals ending at t_1;
I_4 = final intervals beginning at t_k;

we have:

$$\left(\int_0^1 b^2\right)^{\frac{1}{2}} \geq \int_0^1 |b|$$

$$\geq \sum_{I_1} \int_{t_i}^{t_{i+1}} |b| + \sum_{I_2} \int_{t_i}^{t_{i+1}} |b| + \sum_{I_3} \int_{t^+(s_i)}^{t_1} |b| + \sum_{I_4} \int_{t_k}^{t^-(s_i)} |b|$$

$$\geq n_1\delta_1 + n_2\delta_3\varepsilon_2 + n_3\gamma_1 + n_4\gamma_1,$$

by Lemmas 11, 14, 16 and 17, where:

n_1 = number of type (I) intervals satisfying (H) with given δ_0 and β_0;

n_2 = number of type (II) intervals satisfying (H) with given δ_0 and β_0;

n_3 = number of initial intervals ending at t_1;

n_4 = number of final intervals beginning at t_k;

δ_1, δ_3, γ_1 are uniform constants; ε_2 depends on ε_1 ($\varepsilon_2 = \inf\, \alpha_0, \frac{\sqrt{\alpha_1}}{3}$)

see Lemma 6.

Now, when ε_1 is given, ε_2 is fixed.

Let $\theta(\varepsilon_1) = \inf(\delta_1, \delta_3, \varepsilon_2, \gamma_1)$.

As $n_1 + n_2 + n_3 + n_4 = j_1 + j_2$, we have:

$$(j_1 + j_2)\ \theta(\varepsilon_1) \leq \int_0^1 |b| \leq \left(\int_0^1 b^2\right)^{\frac{1}{2}} \;;$$

hence:

$$\frac{j_1+j_2}{\sqrt{\omega}} \leq \frac{1}{\theta(\varepsilon_1)}\ \frac{1}{\sqrt{\omega}}\ \left(\int_0^1 b^2\right)^{\frac{1}{2}} \xrightarrow[\omega \to +\infty]{} 0 \;;$$

which proves Lemma 18 and completes a step in the estimates.

9 The convergence theorem

We sum up the content of our estimates.

We first studied the concentration intervals and we straightened up the curve on these intervals. This straightened curve is then close to a curve tangent to $a\xi$ during the time $[t^-(s), t^+(s)]$ at order $K\sqrt{\alpha_1}(t^+(s)-t^-(s))$ (Lemma 9).

We then studied the situation between two concentration intervals and we distinguished three main cases:

- On $[t_i, t_{i+1}]$ type (I) intervals, when (H) is satisfied with given δ_0 and β_0, we approximated $x(t)$ by a curve tangent to v starting at $x(t_i)$ and ending at $\psi(x(t_i))$ or $\psi^{-1}(x(t_i))$.
  This approximation is of order $K[\frac{1}{\sqrt{\omega}} + [t_{i+1} - t_i]]$. The difference between the transport equations $Z^\star$ and $Z_1^\star$ on all this interval, in particular at t_{i+1}, is of the same order (Lemma 11).
- On $[t_i, t_{i+1}]$ type (II) intervals, we showed that, if (H) is satisfied at t_i with fixed δ_0 and β_0, then $x(t_i)$ and $x(t_{i+1})$ are close at order $K[\frac{1}{\sqrt{\omega}} + [t_{i+1} - t_i]]$ and $\| Z_1^\star(t_i) - Z_1^\star(t_{i+1}) \|$ is of the same order (see Lemma 14).
- On initial or final type intervals, we defined distinguished points $x_{\tilde{s}_2^+}$ or $x_{\tilde{s}_2^-}$, which are on v-orbits starting at $x(t^+(s))$ or $x(t^-(s))$, close to the ends of the incomplete straightened curve y at order $K[\frac{1}{\sqrt{\omega}} + \sqrt{\alpha_1}(t^\pm(s) - t_1(s))]$. At these points $Z^\star = a^\star(\tilde{s}_2^\pm)\begin{bmatrix}-1\\0\\0\end{bmatrix}$ is close to $\begin{bmatrix}-1\\0\\0\end{bmatrix}$, at order $\varepsilon_1 + \sqrt{\alpha_1} + \int_0^1 \frac{b^2}{\omega}$.

We then considered $\psi(x_{\tilde{s}_2^\pm})$ or $\psi^{-1}(x_{\tilde{s}_2^\pm})$ depending on the situation.

If the interval is of type $[t^+(s), t_1]$ or $[t_k, t^-(s)]$ we proved that the distance of $\psi(x_{\tilde{s}_2^\pm})$ (or $\psi^{-1}(x_{\tilde{s}_2^\pm})$) to $x(t_1)$ or $x(t_k)$ is of the order $K[\frac{1}{\sqrt{\omega}} + (t_1 - t^+(s))]$ or $K[\frac{1}{\sqrt{\omega}} + |t_k - t^-(s)|]$, while the difference between the transport equations $Z^\star$ and $Z_1^\star$ is of the same order.

We finally proved that $|a^\star(\tilde{s}_2^\pm) - a'^\star(\tilde{s}_2')|$ is less than $K[\frac{1}{\sqrt{\omega}} + \sqrt{\alpha_1}\,|t^\pm(s) - t_1(s)|]$; (Lemmas 15, 15', 16).

All these estimates hold under the sole hypothesis (A_1).

Convention of notations:

Assume that hypothesis (H) is satisfied on the t_i's separating two given concentration intervals.

To avoid any confusion, we denote by $x_{\tilde{s}_2^+}$ the point built on $I_1(s_1)$ by Lemma 15, and by $x_{\tilde{s}_2^-}$ the point built on $I_1(s_2)$ by Lemma 15' where $I_1(s_2)$ is the concentration interval consecutive to $I_1(s_1)$. We will keep the notation $x_{\tilde{s}_2^-}$ for the point constructed on $I_1(s_1)$ by Lemma 15'. Hence to each concentration interval $I_1(s_i)$ we associate two points $\left(x_{\tilde{s}_2^+}\right)_i$ and $\left(x_{\tilde{s}_2^-}\right)_i$. When two intervals $I_1(s_i)$ and $(I_1(s_{i+1})$ are consecutive and we must estimate quantities involving $\left(x_{\tilde{s}_2^+}\right)_i$ and $\left(x_{\tilde{s}_2^-}\right)_{i+1}$, we will denote these $x_{\tilde{s}_2^+}$ and $\overline{x}_{\tilde{s}_2^-}$.

Then we have the figure annotated (9.1).

In (9.1), $[t_3, t_4]$ and $[t_5, t_6]$ are type (II) intervals.

In figure (9.1), we forget the curve $x(t)$ and draw only the wings tangent to v. We obtain figure (9.2) below.

(9.1)

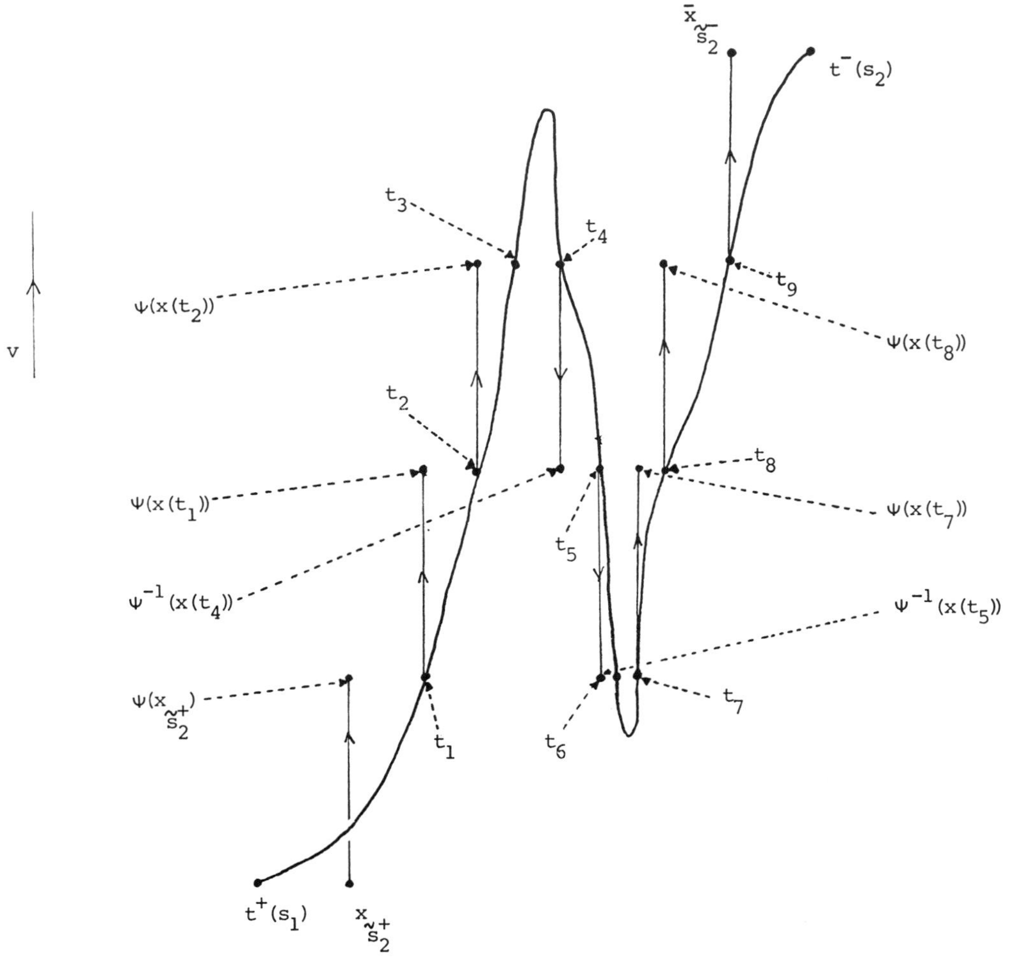

v
$\bar{x}_{\tilde{s}_2^-}$
$t^-(s_2)$
t_3
t_4
$\Psi(x(t_2))$
t_9
$\Psi(x(t_8))$
t_2
t_8
$\Psi(x(t_1))$
$\Psi(x(t_7))$
t_5
$\Psi^{-1}(x(t_4))$
$\Psi^{-1}(x(t_5))$
$\Psi(x_{\tilde{s}_2^+})$
t_7
t_1
t_6
$t^+(s_1)$
$x_{\tilde{s}_2^+}$

(9.2)

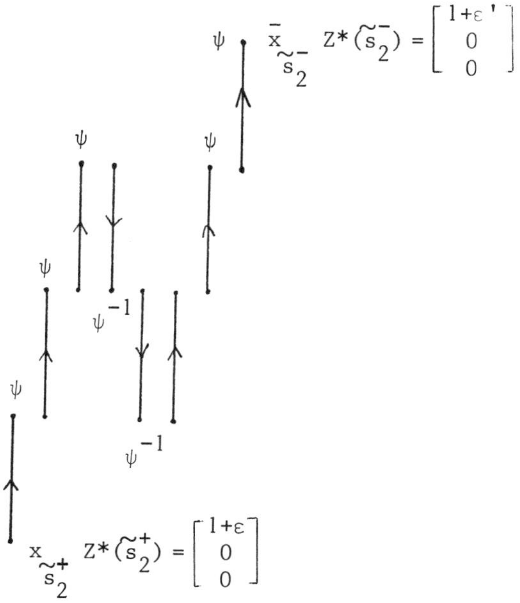

(9.3)

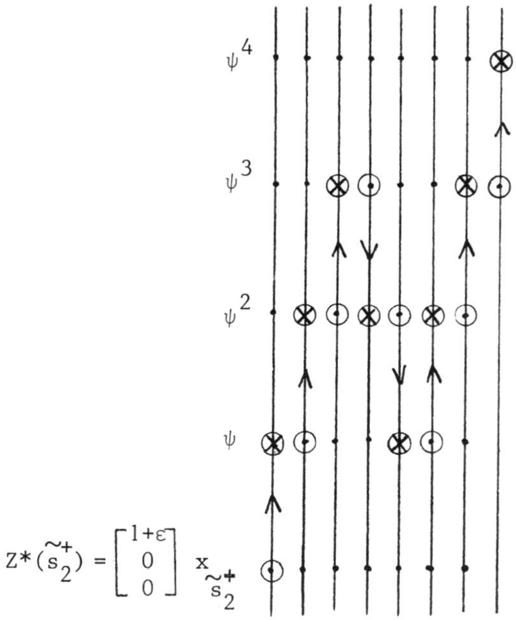

⊙ are the starting points of the pieces

⊗ are the ending points of the pieces; hence, there a jump is made (to another piece)

• are the ψ^i of the starting points

In figure (9.2), we forget the intervals $[t_3, t_4]$ and $[t_5, t_6]$, although their contribution has been indicated by a shift between the wings.

To understand figure (9.2) more clearly, we extend all the pieces in the arrows' direction and iterate ψ along these pieces (figure 9.3)).

Intuitively, each jump occurring at distinguished points $\otimes$ (9.3) amounts to a loss of order $K(\frac{1}{\sqrt{\omega}} + \Delta t)$, where Δt is the time spent running along the previous piece of curve.

We now want to bring back all these movements along v on a unique v-orbit starting at $x_{\tilde{s}_2^+}$.

For this purpose, we need to carry all our estimates back to this v-orbit. This operation requires hypotheses whose essential content is to ensure that estimates holding between two consecutive distinguished points $\otimes$ and $\odot$ do propagate along the pieces through ψ^i, $i \in \mathbb{Z}$.

We provide here two hypotheses which ensure that this operation can be made. These hypotheses can be weakened.

(A_2) $\exists\ \underline{k}$ and $\bar{k} > 0$, $\exists\ \varepsilon_5 > 0$, $\forall\ i \in \mathbb{Z}$, $\underline{k}d(x,y) \leq d(\psi^i(x), \psi^i(y) \leq \bar{k}d(x,y))$ $\forall x, y \in M$ such that $d(x,y) < \varepsilon_5$.

(A_3) Let $\lambda_i : M \to \mathbb{R}$ defined by $\lambda_i(x)[(\psi^i)^\star\alpha]_x = \alpha_x$, $i \in \mathbb{Z}$.
We ask that there exist uniform constants $\alpha_3, \alpha_4, \alpha_5, \varepsilon_6 > 0$ such that:

$$\begin{cases} |\lambda_i(x) - \lambda_i(y)| \leq \alpha_5 d(x,y) & \text{as soon as } d(x,y) < \varepsilon_6 \\ 0 < \alpha_3 \leq |\lambda_i(x)| \leq \alpha_4, & \forall\ i \in \mathbb{Z}, \quad \forall\ x \in M. \end{cases}$$

Hypotheses (A_2)-(A_3) are constraining. They are satisfied in the case of the S^1-fiber bundles. Indeed, in this case, ψ to a certain power is the identity.

We have for instance $\psi^4 = \text{Id}$ in case of S^3. In the case of the S^1-cotangent fiber bundles, we have $\psi^2 = \text{Id}$.

Finally in the case of the lenticular spaces $L(p,q)$, we have $\psi^{2p} = \text{Id}$, ...

It could be interesting to see what happens in case when v is a recurrent vector field.

Now, and for the time being, we leave open the question of the exact meaning of (A_2) and (A_3), having noticed they are satisfied in the case of the S^1-fiber bundles.

These hypotheses are going to allow us to sum up all our estimates in a geometrical result.

Convention of notations:

We now define the notations we need for the following lemma.

We are given two consecutive concentration intervals $I_1(s_1)$ and $\bar{I}_1(s_1)$. We denote $x_{\tilde{s}_2^+}$ the point defined in Lemma 15 on $I_1(s_1)$ and $\bar{x}_{\tilde{s}_2^-}$ the point defined in Lemma 15' on $I_1(s_2)$. We denote $a^\star(\tilde{s}_2^+)$ the coefficient occurring in (8.334) of Lemma 15. We thus have:

$$Z^\star(\tilde{s}_2^+) = a^\star(\tilde{s}_2^+)\begin{bmatrix}1\\0\\0\end{bmatrix},$$

where $Z^\star$ satisfies (8.330) along the wing x_s tangent to v and defined by (8.329).

We denote $\bar{a}^\star(\tilde{s}_2^-)$ the coefficient occurring in (8.371), but here with $t^-(s_2)$ instead of $t^-(s_1)$, i.e. we have:

$$\bar{Z}^\star(\tilde{s}_2^-) = \bar{a}(\tilde{s}_2^-)\begin{bmatrix}1\\0\\0\end{bmatrix},$$

where $\bar{Z}^\star$ satisfies (8.330) too along the wing $\bar{x}_s$ starting now from $x(t^-(s_2))$ (instead of $x(t^-(s_1))$ in Lemma 15').

It is clear that $a^\star(\tilde{s}_2^-)$ satisfies some inequalities identical to (8.371)' and (8.372). If we replace the notations relative to $I_1(s_1)$ with these conventions in notations, we have:

Lemma 19: Under (A_2) and (A_3), hypothesis (H) is satisfied with $\delta_0' = \frac{2\alpha_3}{3}$ and $\beta_0' = \frac{4\alpha_4}{3}$ uniformly at the t_i's for ω large enough and ε_1 small enough.

Furthermore, let $I_1(s_1)$ and $I_1(s_2)$ be two consecutive concentration intervals. Let j be the total number of intervals between $t^+(s_1)$ and $t^-(s_2)$. Let k_1 be the number of wings oriented by ψ in the representation (9.2) and k_2 the humber of wings oriented by ψ^{-1}.

Then $k_2 - k_1$ is even.

Moreover we have:

$$d(\psi^{k_1-k_2}(x_{\tilde{s}_2^+}), \bar{x}_{\tilde{s}_2^-}) < K\left[\frac{j}{\sqrt{\omega}} + t^-(s_2) - t^+(s_1)\right] \tag{9.4}$$

$$|\lambda_{k_1-k_2}(x_{\tilde{s}_2^+}) - 1| < \varepsilon(\varepsilon_1, \omega) \quad \text{where } \varepsilon \xrightarrow[\varepsilon_1 \to 0]{} 0 \tag{9.5}$$

$$\left|\lambda_{k_1-k_2}(x_{\tilde{s}_2^+}) - \frac{\bar{a}\star(\tilde{s}_2^-)}{a\star(\tilde{s}_2^+)}\right| < K\left[\frac{j}{\sqrt{\omega}} + t^-(s_2) - t^+(s_1)\right]. \tag{9.5'}$$

<u>Proof of Lemma 19</u>: Let $\delta'_0 = \frac{2}{3}\alpha_3$, $\beta'_0 = \frac{4}{3}\alpha_4$.

Let $i_0 = \text{Sup}\{i \mid t_j$ satisfies (H) with δ'_0 and β'_0 for $1 \leq j \leq i\}$.

First of all, $i_0 \geq 1$.

Indeed, at t_1, we have by Lemma 16 and particularly by (8.377):

$$\|z\star(\psi(x_{\tilde{s}_2^+})) - z_1^\star(t_1)\| < K_{13}\left[\frac{1}{\sqrt{\omega}} + t_1 - t^+(s_1)\right] \quad \text{if } b(t^+(s_1)) > 0 \tag{9.6}$$

$$\|z\star(\psi^{-1}(x_{\tilde{s}_2^+})) - z_1^\star(t_1)\| < K_{13}\left[\frac{1}{\sqrt{\omega}} + t_1 - t^+(s_1)\right] \quad \text{if } b(t^+(s_1)) < 0. \tag{9.7}$$

On another hand, by (8.334) of Lemma 15, we have:

$$z\star(x_{\tilde{s}_2^+}) = z\star(\tilde{s}_2^+) = a\star(\tilde{s}_2^+)\begin{bmatrix}1\\0\\0\end{bmatrix}. \tag{9.8}$$

As ψ (or ψ^{-1}) sends α on $\lambda_1\alpha$ (or $\lambda_{-1}\alpha$) (notations as in (A_3)) and as $z\star(s)$ satisfies the transport equations of the forms, with $z\star(\tilde{s}_2^+) = a\star(\tilde{s}_2^+)\begin{bmatrix}1\\0\\0\end{bmatrix}$, we necessarily have:

$$z\star(\psi(x_{\tilde{s}_2^+})) = \lambda_1(x_{\tilde{s}_2^+})\, a\star(\tilde{s}_2^+)\begin{bmatrix}1\\0\\0\end{bmatrix} \tag{9.9}$$

or:

$$Z\star(\psi^{-1}(x_{\tilde{s}_2^+})) = \lambda_{-1}(x_{\tilde{s}_2^+})\, a\star(\tilde{s}_2^+) \begin{bmatrix} 1 \\ 0 \\ 0 \end{bmatrix}, \tag{9.10}$$

hence, via (9.6) and (9.7):

$$\left\| Z\star(\psi^{\pm 1}(x_{\tilde{s}_2^+})) - Z_1^\star(t_1) \right\| < K_{13}\left[\frac{1}{\sqrt{\omega}} + t_1 - t^+(s_1)\right]. \tag{9.11}$$

By definition, we have:

$$Z_1^\star(t_1) = \begin{bmatrix} 1 - \dfrac{b^2(t_1)}{2\omega a} + \dfrac{\int_0^1 b^2}{2\omega a} \\ \dfrac{b(t_1)}{\omega} \\ 0 \end{bmatrix} \tag{9.12}$$

From (9.8), (9.11) and (9.12) we deduce:

$$\left| 1 - \frac{b^2(t_1)}{2\omega a} + \frac{\int_0^1 b^2}{2\omega a} - \lambda_{\pm}(x_{\tilde{s}_2^+})\, a\star(\tilde{s}_2^+) \right| < K_{13}\left[\frac{1}{\sqrt{\omega}} + t_1 - t^+(s_1)\right], \tag{9.13}$$

hence

$$\left|\lambda_{\pm}(x_{\tilde{s}_2^+})\, a\star(\tilde{s}_2^+)\right| - K_{13}\left[\frac{1}{\sqrt{\omega}} + t_1 - t^+(s_1)\right] \tag{9.14}$$

$$\leq \left| 1 - \frac{b^2(t_1)}{2\omega a} + \frac{\int_0^1 b^2}{2\omega a} \right| \leq \left|\lambda_{\pm}(x_{\tilde{s}_2^+})\, a\star(\tilde{s}_2^+)\right| + K_{13}\left[\frac{1}{\sqrt{\omega}} + t_1 - t^+(s_1)\right].$$

From (A_3) we derive:

$$\alpha_3 \left|a\star(\tilde{s}_2^+)\right| - K_{13}\left[\frac{1}{\sqrt{\omega}} + t_1 - t^+(s_1)\right] \tag{9.15}$$

$$\leq \left| 1 - \frac{b^2(t_1)}{2\omega a} + \frac{\int_0^1 b^2}{2\omega a} \right| < \alpha_4 \left|a\star(\tilde{s}_2^+)\right| + K_{13}\left[\frac{1}{\sqrt{\omega}} + t_1 - t^+(s_1)\right].$$

By (8.335), we then have:

$$\alpha_3\left(1 - K_{12}\left[\frac{\int_0^1 b^2}{\omega} + \varepsilon_1 + \sqrt{\alpha_1}\right]\right) - K_{13}\left[\frac{1}{\sqrt{\omega}} + t_1 - t^+(s_1)\right] \tag{9.16}$$

$$< \left|1 - \frac{b^2(t_1)}{2\omega a} + \frac{\int_0^1 b^2}{2\omega a}\right| < \alpha_4\left(1 + K_{12}\left[\frac{\int_0^1 b^2}{\omega} + \varepsilon_1 + \sqrt{\alpha_1}\right]\right)$$

$$+ K_{13}\left[\frac{1}{\sqrt{\omega}} + t_1 - t^+(s_1)\right]$$

hence, if ω is large enough and ε_1 small enough:

$$\frac{2\alpha_3}{3} < \left|1 - \frac{b^2(t_1)}{2\omega a} + \frac{\int_0^1 b^2}{2\omega a}\right| < \frac{4}{3}\alpha_4, \tag{9.17}$$

which proves that $i_0 \geq 1$.

Now let i_k and i_{k+1} be two indices such that:

$$t_{i_k} \leq t_{i_{k+1}} \leq t_{i_0}. \tag{9.18}$$

Between t_{i_k} and $t_{i_{k+1}}$, there are only type (II) intervals. (9.19)

We estimate:

$$\| Z_1^\star(t_{i_k}) - Z_1^\star(t_{i_{k+1}}) \| \tag{9.20}$$

$$d(x(t_{i_k}), \quad x(t_{i_{k+1}})). \tag{9.21}$$

By hypothesis, all intervals $[t_i, t_{i+1}]$ with $i_k \leq i \leq i_{k+1}-1$ are of type (II). Moreover, in all these t_i's, as $i \leq i_{k+1} \leq i_0$, hypothesis (H) is verified with δ'_0 and β'_0.

We may then apply Lemma 14. For ω large enough depending on δ'_0 and β'_0, we have, by (8.267) and (8.268):

$$d(x(t_{i_k}), x(t_{i_{k+1}})) \tag{9.22}$$

$$\leq d(x(t_{i_k}), x(t_{i_k+1})) + d(x(t_{i_k+1}), x(t_{i_k+2})) + \ldots + d(x(t_{i_{k+1}-1}), x(t_{i_{k+1}}))$$

$$\leq \overline{K}_8(\delta'_0,\beta'_0)\left[\frac{1}{\sqrt{\omega}} + t_{i_k+1} - t_{i_k}\right] + \ldots + \overline{K}_8\left[\frac{1}{\sqrt{\omega}} + t_{i_{k+1}} - t_{i_{k+1}-1}\right],$$

hence:

$$d(x(t_{i_k}), x(t_{i_{k+1}})) \leq \overline{K}_8\left[\frac{\alpha_k}{\sqrt{\omega}} + t_{i_{k+1}} - t_{i_k}\right], \tag{9.23}$$

where:

$$\alpha_k = \text{number of type (II) intervals between } i_k \text{ and } i_{k+1}. \tag{9.24}$$

In the same way, we derive:

$$\|Z^\star_1(t_{i_k}) - Z^\star_1(t_{i_{k+1}})\| \leq \overline{K}_8\left[\frac{\alpha_k}{\sqrt{\omega}} + t_{i_{k+1}} - t_{i_k}\right]. \tag{9.25}$$

Now let:

$$T_1 < T_2 < \ldots < T_l \tag{9.26}$$

be the lowerbounds of type (I) intervals for $i \leq i_0$. We add T_0 which represents the point $x_{\tilde{s}_2^+}$.

Let:

$$\tau_2 < \ldots < \tau_{l+1} \tag{9.27}$$

be the upperbounds of those type (I) intervals. We add $\tau_1 = t_1$, first zero of b. So the $[T_{i-1}, \tau_i]$, for $1 \leq i \leq l+1$, represent all the other intervals than the type (II) intervals starting at a point $T_k \leq t_{i_0}$.

To each interval $[T_{i-1}, \tau_i]$, we associate an integer:

$$\begin{aligned} n([T_{i-1}, \tau_i]) &= n(T_{i-1}) = +1 \text{ if } \psi \text{ is acting on } [T_{i-1}, \tau_1] \\ &= n(T_{i-1}) = -1 \text{ if } \psi^{-1} \text{ is acting on } [T_{i-1}, \tau_i]. \end{aligned} \tag{9.28}$$

This definition is justified, as hypothesis (H) being satisfied with δ'_0 and β'_0 on these intervals $(T_k \leq t_{i_0})$, all the previous estimates, in particular (8.123) and (8.124) of Lemma 11 and (8.376)-(8.377) of Lemma 16 are valid.

Let:

$$n_i = \sum_{i=0}^{i-1} n(T_k) \; ; \quad n_0 = 0. \tag{9.29}$$

Once we have constructed these intervals $[T_{i-1}, \tau_i]$, we remark that we have:

$$T_{i-1} < \tau_i \leq T_i < \tau_{i+1} . \tag{9.30}$$

Between τ_i and T_i, there are no type (I) intervals nor initial and final intervals. Thus (τ_i, T_i) satisfies (9.18)-(9.19) and the estimates (9.23)-(9.25) hold.

If we set:

$$\bar{\alpha}_i = \text{number of type (II) intervals between } \tau_i \text{ and } T_i. \tag{9.31}$$

These estimates yield:

$$d(x(\tau_i), x(T_i)) \leq \bar{K}_8(\delta_0', \beta_0') \left[\frac{\bar{\alpha}_i}{\sqrt{\omega}} + T_i - \tau_i\right] \tag{9.32}$$

$$\| Z_1^\star(\tau_i) - Z_1^\star(T_i) \| \leq \bar{K}_8(\delta_0', \beta_0') \left[\frac{\bar{\alpha}_i}{\sqrt{\omega}} + T_i - \tau_i\right] . \tag{9.33}$$

Let now:

$$\begin{cases} Z_i^\star = \begin{bmatrix} 1 - \dfrac{b^2(T_i)}{2\omega a} + \dfrac{\int_0^1 b^2}{2\omega a} \\ 0 \\ 0 \end{bmatrix} = a_i^\star \begin{bmatrix} 1 \\ 0 \\ 0 \end{bmatrix} & \text{if } l \geq i \geq 1 \\ Z_0^\star = a^\star(\tilde{s}_2^+) \begin{bmatrix} 1 \\ 0 \\ 0 \end{bmatrix} = a_0^\star \begin{bmatrix} 1 \\ 0 \\ 0 \end{bmatrix} \end{cases} \tag{9.34}$$

$$\begin{cases} \tilde{Z}_i^{\star} = Z^{\star}(\psi^{n(T_{i-1})}(x(T_{i-1}))) = \tilde{a}_i^{\star}\begin{bmatrix}1\\0\\0\end{bmatrix} & \text{if } l+1 \geq i \geq 2 \\ \tilde{Z}_1^{\star} = Z^{\star}(\psi^{n(T_0)}(x_{\tilde{s}_2^+})) = \tilde{a}_1^{\star}\begin{bmatrix}1\\0\\0\end{bmatrix}. \end{cases} \tag{9.35}$$

We remember that $Z^{\star}(\psi^{\pm}(x(t_i))) = Z^{\star}(\bar{s}(x(t_i)))$ is the value of $Z^{\star}(s)$ at the coincidence point $s = \bar{s}(x(t_i))$ or $\underline{s}(x(t_i))$; (where $Z^{\star}(s)$ satisfies the transport equation with $Z^{\star}(0) = \begin{bmatrix} 1 - \frac{b^2(t_i)}{2\omega a} + \frac{\int_0^1 b^2}{2\omega a} \\ 0 \\ 0 \end{bmatrix}$ (With the necessary changes for $Z^{\star}(\psi^{\pm}(x_{\tilde{s}_2^+}).))$.

In particular, with the notations of (9.34) and (9.35), we see that $Z_i^{\star}$ corresponds to the set of initial data on the intervals $[t_i, t_{i+1}]$ while $\tilde{Z}_i^{\star}$ corresponds to the final data through $\psi^{n(T_{i-1})}$ when the initial data is $Z_{i-1}^{\star}$.

Once we have understood this change of notations, we may apply Lemmas 11 or 16, depending if we are on a type (I) interval or on an initial or final one, using the new notations and with $[t_i, t_{i+1}] = [T_{i-1}, \tau_i]$.

In particular, applying (8.123) and (8.124), or (8.376)-(8.377), we have:

$$\|Z^{\star}(\psi^{n(T_{i-1})}(x(T_{i-1})) - Z_1^{\star}(\tau_i)\| = \|\tilde{Z}_i^{\star} - Z_1^{\star}(\tau_i)\| < K\left[\frac{1}{\sqrt{\omega}} + \tau_i - T_{i-1}\right], \tag{9.36}$$

for $l+1 \geq i \geq 2$;

$$\|Z^{\star}(\psi^{n(T_0)}(x_{\tilde{s}_2^+})) - Z_1^{\star}(\tau_1)\| = \|\tilde{Z}_1^{\star} - Z_1^{\star}(\tau_1)\| < K\left[\frac{1}{\sqrt{\omega}} + \tau_1 - t^{+}(s_1)\right]; \tag{9.36'}$$

$$d(x(\tau_i), \psi^{n(T_{i-1})}(x(T_{i-1}))) \leq K\left[\frac{1}{\sqrt{\omega}} + \tau_i - T_{i-1}\right], \tag{9.37}$$

$$d(x(\tau_1), \psi^{n(T_0)}(x_{\tilde{s}_2^+})) \le K\left[\frac{1}{\sqrt{\omega}} + \tau_1 - t^+(s_1)\right]. \qquad (9.37)'$$

Furthermore, we have

$$\|Z_1^\star(T_i) - Z_i^\star\| = \left\| \begin{bmatrix} 1 - \frac{b^2(T_i)}{2\omega a} + \frac{\int_0^1 b^2}{2\omega a} \\ \frac{b(T_i)}{\omega} \\ 0 \end{bmatrix} - \begin{bmatrix} 1 - \frac{b^2(T_i)}{2\omega a} + \frac{\int_0^1 b^2}{2\omega a} \\ 0 \\ 0 \end{bmatrix} \right\| \qquad (9.38)$$

$$= \frac{|b(T_i)|}{\omega}.$$

As by hypothesis $\left|1 - \frac{b^2(T_i)}{2\omega a} + \frac{\int_0^1 b^2}{2\omega a}\right| \le \beta_0' \quad (T_i \le t_{i_0})$, we have:

$$\|Z_1^\star(T_i) - Z_i^\star\| \le \frac{C(\beta_0')}{\sqrt{\omega}}, \qquad (9.39)$$

where $C(\beta_0')$ is an appropriate constant.

Estimates (9.32), (9.33), (9.36), (9.36)', (9.37), (9.37)' and (9.39) then imply:

$$\begin{cases} \|Z_i^\star - \tilde{Z}_i^\star\| = \|a_i^\star - \tilde{a}_i^\star\| \le \|Z_i^\star - Z_1^\star(T_i)\| + \|Z_1^\star(T_i) - Z_1^\star(\tau_i)\| + \\ + \|Z_1^\star(\tau_i) - \tilde{Z}_i^\star\| \le \overline{K}_8\left[\frac{\overline{\alpha}_i}{\sqrt{\omega}} + T_i - \tau_i\right] + \frac{C(\beta_0')}{\sqrt{\omega}} + K\left[\frac{1}{\sqrt{\omega}} + \tau_i - T_{i-1}\right] \\ \text{hence} \\ \|Z_i^\star - \tilde{Z}_i^\star\| = |a_i^\star - \tilde{a}_i^\star| \le K_{16}\left[\frac{\overline{\alpha}_i + 1}{\sqrt{\omega}} + T_i - T_{i-1}\right], \quad i \ge 2 \end{cases} \qquad (9.40)$$

$$d(x(T_i), \psi^{n(T_{i-1})}(x(T_{i-1}))) \le K_{16}\left[\frac{\overline{\alpha}_i + 1}{\sqrt{\omega}} + T_i - T_{i-1}\right]; \quad i \ge 2 \qquad (9.41)$$

and also:

$$\|Z_1^\star - \tilde{Z}_1^\star\| \le K_{16}\left[\frac{1}{\sqrt{\omega}} + T_1 - t^+(s_1)\right] \qquad (9.40)'$$

$$d(x(T_1), \psi^{n(T_0)}(x_{\tilde{s}_2^+})) \leq K_{16}\left[\frac{1}{\sqrt{\omega}} + T_1 - t^+(s_1)\right] \tag{9.41'}$$

Equations (9.40)-(9.41), (9.40)'-(9.41)' are the estimates we need.

We also remark that $\tilde{Z}_i^\star$, being equal to $Z^\star(\psi^{n(T_{i-1})}(x(T_{i-1})))$, gives us the components at $\psi^{n(T_{i-1})}(x(T_{i-1}))$ of the transported form along v which is equal to $Z_{i-1}^\star$ at $x(T_{i-1})$.

As $Z_{i-1}^\star = a_{i-1}^\star \begin{bmatrix}1\\0\\0\end{bmatrix}$, we necessarily have:

$$\tilde{a}_i^\star = \lambda_{n(T_{i-1})}(x(T_{i-1}))a_{i-1}^\star, \qquad 2 \leq i \leq l+1. \tag{9.42}$$

Indeed, $Z_{i-1}^\star$ gives the components of $a_{i-1}^\star\alpha$ at $x(T_{i-1})$. By definition of $\psi^{\pm 1}$, the transformed form of $Z_{i-1}^\star$ at $\psi^{\pm 1}(x(T_{i-1}))$ will be collinear to $a_{i-1}^\star\alpha$ ($\psi^{\pm 1}$ sends α on α) and the collinearity coefficient is given by (A_5) equal to $\lambda_{\pm 1}$.

For the forthcoming estimates, we introduce the following notations:

$$x_i = x(T_i); \quad l \geq i \geq 1; \quad x_0 = x_{\tilde{s}_2^+}; \tag{9.43}$$

$$y_i = \psi^{n(T_{i-1})}(x(T_{i-1})); \quad l+1 \geq i \geq 2; \quad y_1 = \psi^{n(T_0)}(x_{\tilde{s}_2^+}) = \psi^{n(T_0)}(x_0);$$

$$\hat{x}_i = \psi^{-n_i}(x_i); \quad \hat{y}_i = \psi^{-n_i}(y_i). \tag{9.44}$$

We have

$$\hat{y}_i = \hat{x}_{i-1} \tag{9.45}$$

as

$$\hat{y}_i = \psi^{-n_i}(y_i) = \psi^{-n_i}(\psi^{n(T_{i-1})}(x(T_{i-1})))$$

$$= \psi^{-n_i} \circ \psi^{n(T_{i-1})}(x_{i-1}) = \psi^{-n_{i-1}}(x_{i-1}) = \hat{x}_{i-1}.$$

We denote also:

$$\hat{a}_i^{\star} = \lambda_{-n_i}(x_i)\, a_i^{\star}; \quad l \geq i \geq 0; \tag{9.46}$$

$$\hat{\tilde{a}}_i^{\star} = \lambda_{-n_i}(y_i)\, \tilde{a}_i^{\star}; \quad l+1 \geq i \geq 1.$$

We have:

$$\hat{\tilde{a}}_i^{\star} = \hat{a}_{i-1}^{\star}. \tag{9.47}$$

Indeed:

$$\hat{\tilde{a}}_i^{\star} = \lambda_{-n_i}(y_i)\, \tilde{a}_i^{\star} \underset{(\text{via } 9.42)}{=\!=} \lambda_{-n_i}(y_i)\, \lambda_{n(T_{i-1})}(x(T_{i-1}))\, a_{i-1}^{\star}$$

$$\hat{a}_{i-1}^{\star} = \lambda_{-n_{i-1}}(x_{i-1})\, a_{i-1}^{\star} \qquad (\text{via } (9.46)).$$

Then (9.47) is equivalent to:

$$\lambda_{-n_{i-1}}(x_i) = \lambda_{-n_i}(y_i)\, \lambda_{n(T_{i-1})}(x(T_{i-1})), \tag{9.48}$$

i.e.

$$\lambda_{-n_{i-1}}(x_{i-1}) = \lambda_{-n_i}(\psi^{n(T_{i-1})}(x_{i-1}))\, \lambda_{n(T_{i-1})}(x_{i-1}). \tag{9.49}$$

To prove (9.49), we start from:

$$((\psi^{-n_i}) \star \alpha)_y = \lambda_{-n_i}(y)\, \alpha_y. \tag{9.50}$$

We take $y = \psi^{n(T_{i-1})}(x_{i-1})$.

$$((\psi^{-n_i}) \star \alpha)_{\psi^{n(T_{i-1})}(x_{i-1})} = \lambda_{-n_i}(\psi^{n(T_{i-1})}(x_{i-1}))\, \alpha_{\psi^{n(T_{i-1})}(x_{i-1})}. \tag{9.51}$$

We have $n_i = n_{i+1} + n(T_{i-1})$.

Then:

$$\alpha_{\psi^{-n_i} \circ \psi^{n(T_{i-1})}(x_{i-1})}(D\psi^{-n_i}(.)) = \alpha_{\psi^{-n_{i-1}}(x_{i-1})}(D\psi^{-n_i}(.)) \tag{9.52}$$

$$= \lambda_{-n_i}(y_i)\,\alpha_{\psi^{n(T_{i-1})}(x_{i-1})}(.) .$$

Equation (9.52) then implies:

$$\alpha_{\psi^{-n_{i-1}}(x_{i-1})}(D\psi^{-n_i} \circ D\psi^{n(T_{i-1})}) = \lambda_{-n_i}(y_i)\,\alpha_{\psi^{n(T_i)}(x_{i-1})}(D\psi^{n(T_{i-1})}) \tag{9.53}$$

$$= \lambda_{-n_i}(t_i)\,((\psi^{n(T_{i-1})})\star\alpha)_{x_{i-1}} ,$$

hence:

$$\alpha_{\psi^{-n_{i-1}}(x_{i-1})}(D\psi^{-n_{i-1}}(.)) = ((\psi^{-n_{i-1}})\star\alpha)_{x_{i-1}} \tag{9.54}$$

$$= \lambda_{-n_i}(y_i)\,(\psi^{\star n(T_{i-1})}\alpha)_{x_{i-1}} ;$$

hence:

$$\lambda_{-n_{i-1}}(x_{i-1}) = \lambda_{-n_i}(y_i)\,\lambda_{n(T_{i-1})}(x_{i-1}) .$$

We now sum up the estimates and notations:

$$|a_i^\star - \tilde{a}_i^\star| \le K_{16}\left[\frac{\bar{\alpha}_i + 1}{\sqrt{\omega}} + T_i - T_{i-1}\right] \quad ((9.40) \text{ and } (9.40)') \tag{9.55}$$

$$d(x_i, y_i) \le K_{16}\left[\frac{\bar{\alpha}_i + 1}{\sqrt{\omega}} + T_i - T_{i-1}\right] \quad ((9.41) \text{ and } (9.41)') \tag{9.56}$$

new notations

$$\hat{x}_i = \psi^{-n_i}(x_i) \; ; \quad \hat{y}_i = \psi^{-n_i}(y_i) \quad ((9.44)) \tag{9.57}$$

$$\hat{y}_i = \hat{x}_{i-1} \qquad ((9.45)) \qquad (9.58)$$

$$\hat{a}^\star_i = \lambda_{-n_i}(x_i)\, a^\star_i ; \quad \hat{\tilde{a}}^\star_i = \lambda_{-n_i}(y_i)\, \tilde{a}^\star_i \qquad ((9.46)) \qquad (9.59)$$

$$\hat{\tilde{a}}^\star_i = \hat{a}^\star_{i-1} \qquad ((9.47)) \qquad (9.60)$$

We want to estimate:

$$d(\hat{x}_0, \hat{x}_l) = d(x_0, \hat{x}_l) \qquad (9.61)$$

($\hat{x}_0 = \psi^{-n_0}(x_0) = x_0$ as n_0 is zero).

$$|\hat{a}^\star_l - \hat{a}^\star_0| . \qquad (9.62)$$

For this, we choose ω large enough such that:

$$\sum_{i=0}^{l} K_{16}\left[\frac{\bar{\alpha}_i+1}{\sqrt{\omega}} + T_i - T_{i-1}\right] < \left[\inf(\varepsilon_5, \varepsilon_6)\right] \frac{1}{\operatorname{Sup}(1, \bar{k})} . \qquad (9.63)$$

Such a choice is possible as $\Sigma\, T_i - T_{i-1} < t^-(s_2) - t^+(s_1)$ and as, by Lemma 18, $\sum_i \frac{\bar{\alpha}_i+1}{\sqrt{\omega}} \to 0$.

We also remark that this choice can be made uniform on all non-concentration intervals, with all possible time intervals inside these non-concentration intervals.

Under (9.63), by (9.56) we have:

$$d(x_i, y_i) < \inf(\varepsilon_5, \varepsilon_6) . \qquad (9.64)$$

Thus (A_2) and (A_3) apply and give:

$$d(\psi^{-n_i}(x_i), \psi^{-n_i}(y_i)) = d(\hat{x}_i, \hat{y}_i) \leq \bar{k}\, d(x_i, y_i) \qquad (9.65)$$

$$\leq \bar{k}\, K_{16}\left[\frac{\bar{\alpha}_i+1}{\sqrt{\omega}} + T_i - T_{i-1}\right] \quad \text{(via (9.56))}.$$

Thus:

$$d(x_0,\hat{x}_l)=d(\hat{x}_0,\hat{x}_l)\leq d(\hat{x}_0,\hat{x}_1)+d(\hat{x}_1,\hat{x}_2)\ \dots + d(\hat{x}_{l-1},\hat{x}_l). \tag{9.66}$$

By (9.58), we have:

$$d(\hat{x}_i,\hat{x}_{i+1}) = d(\hat{y}_{i+1},\hat{x}_{i+1}). \tag{9.67}$$

Equations (9.66) and (9.67) give:

$$d(x_0,\hat{x}_l) \leq d(\hat{y}_1,\hat{x}_1) + d(\hat{y}_2,\hat{x}_2)\ \dots + d(\hat{y}_l,\hat{x}_l) \tag{9.68}$$

hence by (9.65)

$$d(x_0,\hat{x}_l) \leq \bar{k}\, K_{16}\left(\sum_0^l \left[\frac{\bar{\alpha}_i+1}{\sqrt{\omega}} + T_i - T_{i-1}\right]\right). \tag{9.69}$$

As $\hat{x}_l = \hat{y}_{l+1}$, we have also:

$$d(x_0,\hat{y}_{l+1}) \leq \bar{k}\, K_{16}\left(\sum_0^l \left[\frac{\bar{\alpha}_i+1}{\sqrt{\omega}} + T_i - T_{i-1}\right]\right). \tag{9.70}$$

We estimate now $\hat{a}^\star - \hat{a}_0^\star$.

As (8.21) holds, we deduce from (A_3):

$$|\lambda_{-n_i}(x_i) - \lambda_{-n_i}(y_i)| \leq \alpha_5\, d(x_i,y_i) \tag{9.71}$$

$$\leq \alpha_5 K_{16}\left[\frac{\bar{\alpha}_i+1}{\sqrt{\omega}} + T_i - T_{i-1}\right]$$

$$0<\alpha_3 \leq |(\lambda_{-n_i}(x_i)| \leq \alpha_4\ ;\ 0<\alpha_3 \leq |\lambda_{-n_i}(y_i)| \leq \alpha_4 \tag{9.72}$$

hence we deduce:

$$|\hat{a}_i^\star - \hat{\tilde{a}}_i^\star| = |\lambda_{-n_i}(x_i)\, a_i^\star - \lambda_{-n_i}(y_i)\,\tilde{a}_i^\star| \tag{9.73}$$

$$\leq |\lambda_{-n_i}(x_i) - \lambda_{-n_i}(y_i)|\,|a_i^\star| + |\lambda_{-n_i}(y_i)|\,|a_i^\star - \tilde{a}_i^\star|,$$

hence, by (9.71), (9.72) and (9.55):

$$|\hat{a}_i^\star - \hat{\hat{a}}_i^\star| \leq K_{16}\left[\frac{\bar{\alpha}_i+1}{\sqrt{\omega}} + T_i - T_{i-1}\right] \quad (\alpha_5|a_i^\star| + \alpha_4). \tag{9.74}$$

We remark that by hypothesis we have:

$$\delta_0' = \frac{2\alpha_3}{3} \leq |a_i^\star| = \left|1 - \frac{b^2(T_i)}{2\omega a} + \frac{\int_0^1 b^2}{2\omega a}\right| \leq \beta_0' = \frac{4\alpha_4}{3} \; . \tag{9.75}$$

Indeed, T_i being before t_{i_0} satisfies hypothesis (H) with δ_0' and β_0'.
Inequalities (9.74) and (9.75) then imply:

$$|\hat{a}_i^\star - \hat{\hat{a}}_i^\star| \leq K_{16}\left[\frac{4\alpha_5\alpha_4}{3} + \alpha_4\right]\left[\frac{\bar{\alpha}_i+1}{\sqrt{\omega}} + T_i - T_{i-1}\right]. \tag{9.76}$$

From (9.60), inequality (9.76) can also be read:

$$|\hat{a}_i^\star - \hat{a}_{i-1}^\star| \leq K_{16}\,\alpha_4\left[1 + \frac{4\alpha_5}{3}\right]\left[\frac{\bar{\alpha}_i+1}{\sqrt{\omega}} + T_i - T_{i-1}\right] \tag{9.77}$$

as $\hat{\hat{a}}_i^\star = \hat{a}_{i-1}^\star$.
Hence we deduce:

$$|\hat{a}_l^\star - \hat{a}_0^\star| \leq K_{16}\,\alpha_4\left[1 + \frac{4\alpha_5}{3}\right]\left[\sum_0^l\left(\frac{\bar{\alpha}_i+1}{\sqrt{\omega}} + T_i - T_{i-1}\right)\right]. \tag{9.78}$$

As $\hat{a}_0^\star = \lambda_{-n_0}(x_0)\, a_0^\star = a_0^\star$ (n_0 is zero), we have:

$$|\hat{a}_l^\star - a_0^\star| \leq K_{16}\,\alpha_4\left[1 + \frac{4\alpha_5}{3}\right]\left[\sum_0^l\left(\frac{\bar{\alpha}_i+1}{\sqrt{\omega}} + T_i - T_{i-1}\right)\right], \tag{9.79}$$

which can also be read, as $\hat{a}_l^\star = \hat{\hat{a}}_{l+1}^\star$,

$$|\hat{\hat{a}}_{l+1}^\star - a_0^\star| \leq K_{16}\,\alpha_4\left[1 + \frac{4\alpha_5}{3}\right]\left[\sum_0^l\left(\frac{\bar{\alpha}_i+1}{\sqrt{\omega}} + T_i - T_{i-1}\right)\right]. \tag{9.80}$$

As

$$\begin{cases} \hat{a}_l^\star = \lambda_{-n_l}(x_l)\, a_l^\star & ((9.59)) \\ \hat{\hat{a}}_{l+1}^\star = \lambda_{-n_{l+1}}(y_{l+1})\, \tilde{a}_{l+1}^\star \, , \end{cases}$$

we deduce from (9.79) and (9.80) :

$$|\lambda_{-n_l}(x_l)\, a^\star_l - a^\star_0| \le K_{16}\, \alpha_4 \left[1 + \frac{4\alpha_5}{3}\right] \left[\sum_0^l \frac{\bar{\alpha}_i + 1}{\sqrt{\omega}} + T_i - T_{i-1}\right] \tag{9.81}$$

$$|\lambda_{-n_{l+1}}(y_{l+1})\, \tilde{a}^\star_{l+1} - a^\star_0| \le K_{16}\, \alpha_4 \left[1 + \frac{4\alpha_5}{3}\right] \left[\sum_0^l \frac{\bar{\alpha}_i + 1}{\sqrt{\omega}} + T_i - T_{i-1}\right]. \tag{9.82}$$

We notice that:

$$\lambda_i(x) = \frac{1}{\lambda_{-i}(\psi^i(x))} \quad \text{as if} \quad (\psi^{i\star}\alpha)_x = \lambda_i(x)\, \alpha_x \tag{9.83}$$

then $\alpha_{\psi^i(x)}(D\psi^i(.)) = \lambda_i(x)\, \alpha_x$ hence $\alpha_x(D\psi^{-i}(.)) = \frac{1}{\lambda_i(x)} \alpha_{\psi^i(x)}$; thus

$$((\psi^{-i})\star\alpha)_{\psi^i(x)} = \frac{1}{\lambda_i(x)}\alpha_{\psi^i(x)} = \lambda_{-i}(\psi^i(x))\, \alpha_{\psi^i(x)}.$$

Equations (9.83) and (A_5) imply:

$$\alpha_3 \le \left|\frac{1}{\lambda_{-n_l}(x_l)}\right| = |\lambda_{n_l}(\psi^{-n_l}(x_l))| \le \alpha_4 \tag{9.84}$$

and

$$\alpha_3 \le \left|\frac{1}{\lambda_{-n_{l+1}}(y_{l+1})}\right| = |\lambda_{n_{l+1}}(\psi^{-n_{l+1}}(y_{l+1}))| \le \alpha_4 . \tag{9.85}$$

From (9.81)-(9.82), (9.84) and (9.85), we deduce:

$$\alpha_4 |a^\star_0| + ⍟ \ge |a^\star_l| \ge \alpha_3 |a^\star_0| - K_{16} \frac{\alpha_4}{\alpha_3} \left[1 + \frac{4\alpha_5}{3}\right] \left[\sum_0^l \frac{\bar{\alpha}_i + 1}{\sqrt{\omega}} + T_i - T_{i-1}\right] \tag{9.86}$$

$$\alpha_4 |a^\star_0| + ⍟ \ge |\tilde{a}^\star_{l+1}| \ge \alpha_3 |a^\star_0| - K_{16} \frac{\alpha_4}{\alpha_3} \left[1 + \frac{4\alpha_5}{3}\right] \left[\sum_0^l \frac{\bar{\alpha}_i + 1}{\sqrt{\omega}} + T_i - T_{i-1}\right] \tag{9.87}$$

$$\left(⍟ = K_{16} \frac{\alpha_4}{\alpha_3} \left[1 + \frac{4\alpha_5}{3}\right] \left[\sum_0^l \frac{\bar{\alpha}_i + 1}{\sqrt{\omega}} + T_i - T_{i-1}\right] \right).$$

By definition ((9.34)), $a^\star_0 = a^\star(\tilde{s}_2^+)$.

By (8.335), we have:

$$|a^\star(\tilde{s}_2^+) - 1| < K_{12}\left[\frac{\int_0^1 b^2}{\omega} + \varepsilon_1 + \sqrt{\alpha_1}\right]. \tag{9.88}$$

We choose then ω large enough and ε_1 small enough so that:

$$\frac{15}{16} \le |a^\star(\tilde{s}_2^+)| = |a_0^\star| \le \frac{17}{16} \qquad (9.89)$$

$$\frac{1}{16}\alpha_4 \ge \frac{1}{16}\alpha_3 \ge K_{16}\frac{\alpha_4}{\alpha_3}\left[1+\frac{4\alpha_5}{3}\right]\left[\sum_0^{l}\frac{\bar{\alpha}_i+1}{\sqrt{\omega}} + T_i - T_{i-1}\right], \qquad (9.90)$$

Equations (9.86) and (9.87) then imply:

$$\frac{2\alpha_3}{3} < \frac{7}{8}\alpha_3 \le |a_l^\star| \le \frac{18}{16}\alpha_4 = \frac{9}{8}\alpha_4 < \frac{4\alpha_4}{3} \qquad (9.91)$$

$$\frac{7}{8}\alpha_3 \le |\tilde{a}_{l+1}^\star| \le \frac{9}{8}\alpha_4. \qquad (9.92)$$

Equations (9.91) and (9.92) imply that hypothesis (H) is satisfied with $\delta_0' = \frac{\alpha_3}{2}$, $\beta_0' = 2\alpha_4$ on all t_i's between $t^+(s_1)$ and $t^-(s_2)$.

Indeed, suppose that $t_{i_0} < t_{i_0+1} < t^-(s_2)$.

Two cases are possible:

1 The first case where $[t_{i_0}, t_{i_0+1}]$ is a type (II) interval.

In this case, by (8.268) of Lemma 14, we have:

$$\|Z_1^\star(t_{i_0+1}) - Z_1^\star(t_{i_0})\| \le \bar{K}_8\left[t_{i_0+1} - t_{i_0} + \frac{1}{\sqrt{\omega}}\right]. \qquad (9.93)$$

On the other hand, in this case $T_l < t_{i_0}$ and thus

$$\tau_{l+1} \le t_{i_0}. \qquad (9.94)$$

The interval $[\tau_{l+1}, t_{i_0}]$ can then contain only type (II) intervals. Equations (9.18) and (9.19) hold with $t_{i_k} = \tau_{l+1}$ and $t_{i_{k+1}} = t_{i_0}$. Thus, by (9.25) we have:

$$\|Z_1^\star(\tau_{l+1}) - Z_1^\star(t_{i_0})\| \le \bar{K}_8\left[\frac{\gamma_{i_0}}{\sqrt{\omega}} + t_{i_0} - \tau_{l+1}\right], \qquad (9.95)$$

where γ_{i_0} is the number of intervals between τ_{l+1} and t_{i_0}.

Finally, by (9.36) applied with $i = l+1$, we have:

$$\|\tilde{Z}^{\star}_{l+1} - Z^{\star}_{1}(\tau_{l+1})\| < K\left[\frac{1}{\sqrt{\omega}} + \tau_{l+1} - T_l\right]. \qquad (9.96)$$

Equation (9.93), (9.95) and (9.96) imply:

$$\|\tilde{Z}^{\star}_{l+1} - Z^{\star}_{1}(t_{i_0+1})\| \le \operatorname{Sup}(\overline{K}_8, K)\left[\frac{\gamma_{i_0+1}}{\sqrt{\omega}} + t_{i_0+1} - T_l\right]. \qquad (9.97)$$

By definition, $\tilde{Z}^{\star}_{l+1} = \begin{bmatrix} \tilde{a}^{\star}_{l+1} \\ 0 \\ 0 \end{bmatrix}$.

The first component of $Z^{\star}_{1}(t_{i_0+1})$ is $1 - \frac{b^2(t_{i_0+1})}{2\omega a} + \frac{\int_0^1 b^2}{2\omega a}$. Thus:

$$\left|\tilde{a}^{\star}_{l+1} - \left[1 - \frac{b^2(t_{i_0+1})}{2\omega a} + \frac{\int_0^1 b^2}{2\omega a}\right]\right| \le \operatorname{Sup}(\overline{K}_8, K)\left[\frac{\gamma_{i_0+1}}{\sqrt{\omega}} + t_{i_0+1} - T_l\right], \qquad (9.98)$$

hence, via (9.92):

$$\frac{7}{8}\alpha_3 - \operatorname{Sup}(\overline{K}_8, K)\left[\frac{\gamma_{i_0+1}}{\sqrt{\omega}} + t_{i_0+1} - T_l\right] \le \left|1 - \frac{b^2(t_{i_0+1})}{2\omega a} + \frac{\int_0^1 b^2}{2\omega a}\right| \qquad (9.99)$$

$$\le \frac{9}{8}\alpha_4 + \operatorname{Sup}(\overline{K}_8, K)\left[\frac{\gamma_{i_0+1}}{\sqrt{\omega}} + t_{i_0+1} - T_l\right].$$

We then choose ω uniform so that:

$$\operatorname{Sup}(\overline{K}_8, K)\left[\frac{j_1+j_2}{\sqrt{\omega}} + \Sigma[t^-(s_{i+1}) - t^+(s_i)]\right] < \frac{\alpha_3}{100}, \qquad (9.100)$$

where $j_1 + j_2$ is the total number of type (I), type (II), or initial and final intervals.

This is possible as, by Lemma 18, $\frac{j_1+j_2}{\sqrt{\omega}} \to 0$ and $\Sigma[t^-(s_{i+1}) - t^+(s_i)] \to 0$ by Lemma 5.

Under (9.100), we have:

$$\frac{2}{3}\alpha_3 < \left|1 - \frac{b^2(t_{i_0+1})}{2\omega a} + \frac{\int_0^1 b^2}{2\omega a}\right| < \frac{4\alpha_4}{3} \tag{9.101}$$

as $\quad \frac{9}{8}\alpha_4 + \frac{\alpha_3}{100} < \frac{9}{8}\alpha_4 + \frac{\alpha_4}{100} < \frac{4}{3}\alpha_4$

and $\quad \frac{7}{8}\alpha_3 - \frac{\alpha_3}{100} > \frac{2}{3}\alpha_3$.

Now (9.101) means that t_{i_0+1} satisfies hypothesis (H) also with δ'_0 and β'_0; which contradicts the definition of i_0.

[2] The second case where $[t_{i_0}, t_{i_0+1}]$ is a type (I) interval.

Thus $T_l = t_{i_0}$ and $\tau_{l+1} = t_{i_0+1}$.

In this case, we apply (9.36) with $i = l+1$.

We have:

$$\|\tilde{Z}^\star_{l+1} - Z^\star_1(\tau_{l+1})\| = \|\tilde{Z}^\star_{l+1} - Z^\star_1(t_{i_0+1})\| < K\left[\frac{1}{\sqrt{\omega}} + \tau_{l+1} - T_l\right] \tag{9.102}$$

hence

$$\|\tilde{Z}^\star_{l+1} - Z^\star_1(t_{i_0+1})\| < K\left[\frac{1}{\sqrt{\omega}} + t_{i_0+1} - t_{i_0}\right],$$

hence:

$$\left|\tilde{a}^\star_{l+1} - \left[1 - \frac{b^2(t_{i_0+1})}{2\omega a} + \frac{\int_0^1 b^2}{2\omega a}\right]\right| < K\left[\frac{1}{\sqrt{\omega}} + t_{i_0+1} - t_{i_0}\right]. \tag{9.103}$$

Equations (9.92) and (9.103) then allow us to conclude that t_{i_0+1} satisfies (H) with δ'_0 and β'_0 for ω large enough; which contradicts the definition of i_0. Thus we see that hypothesis (H) is globally satisfied on all the t_i's between two consecutive concentration intervals.

Then let t_{k_0} be the last zero of $\dot{b}$ before $t^-(s_2)$. Let T_l defined by (9.26) with $i_0 = k_0$; τ_{l+1} by (9.27). As t_{k_0} bounds the final interval and thus does not bound a type (I) interval, we have:

$$T_l < t_{k_0}\,; \quad \tau_{l+1} \leq t_{k_0} \,. \tag{9.104}$$

By (9.70), we have:

$$d(x_0, \hat{y}_{l+1}) \leq \bar{k}\, K_{16}\left[\sum_0^l \left(\frac{\bar{\alpha}_i + 1}{\sqrt{\omega}} + T_i - T_{i-1}\right)\right], \tag{9.105}$$

and, by (9.80),

$$|\hat{\tilde{a}}^{\star}_{l+1} - a^{\star}_0| \leq K_{16}\,\alpha_4\left(1 + \frac{4\alpha_5}{3}\right)\left[\sum_0^l\left(\frac{\bar{\alpha}_i + 1}{\sqrt{\omega}} + T_i - T_{i-1}\right)\right]. \tag{9.105)'$$

Equation (9.63) implies that $d(x_0, \hat{y}_{l+1}) < \inf(\varepsilon_5, \varepsilon_6)$.

Thus (A_2) and (A_3) can be applied with $x_0 = x$, $y = \hat{y}_{l+1}$. We then deduce:

$$\begin{cases} d(\psi^{n_{l+1}}(x_0), \psi^{n_{l+1}}(\hat{y}_{l+1})) = d(\psi^{n_{l+1}}(x_0), y_{l+1}) \leq \bar{k}\, d(x_0, \hat{y}_{l+1}) & (9.106) \\ \leq \bar{k}^2 K_{16}\left[\sum_0^l\left(\frac{\bar{\alpha}_i + 1}{\sqrt{\omega}} + T_i - T_{i-1}\right)\right] \quad (\text{via } (9.105)) & \\ |\lambda_{n_{l+1}}(x_0) - \lambda_{n_{l+1}}(\hat{y}_{l+1})| \leq \alpha_5\, d(x_0, \hat{y}_{l+1}) & (9.107) \\ \leq \alpha_5 K_{16}\left[\sum_0^l\left(\frac{\bar{\alpha}_i + 1}{\sqrt{\omega}} + T_i - T_{i-1}\right)\right] \quad (\text{via } (9.105)). & \end{cases}$$

By (9.37), we have on the other hand:

$$d(x(\tau_{l+1}), \psi^{n(T_l)}(x(T_l))) \leq K\left[\frac{1}{\sqrt{\omega}} + \tau_{l+1} - T_l\right] \tag{9.108}$$

but $y_{l+1} = \psi^{n(T_l)}(x(T_l))$ ((9.43)), hence:

$$d(x(\tau_{l+1}), y_{l+1}) \leq K\left[\frac{1}{\sqrt{\omega}} + \tau_{l+1} - T_l\right]. \tag{9.109}$$

In the same way, by (9.36):

$$\|Z^{\star}(\psi^{n(T_l)}(x(T_l))) - Z^{\star}_1(\tau_{l+1})\| = \|\tilde{Z}^{\star}_{l+1} - Z^{\star}_1(\tau_{l+1})\| \tag{9.110}$$

$$< K\left[\frac{1}{\sqrt{\omega}} + \tau_{l+1} - T_l\right];$$

($\tilde{Z}^{\star}_{l+1} = Z^{\star}(\psi^{n(T_l)}(x(T_l)))$ by definition (9.35)).

Moreover, as the interval $[\tau_{l+1}, t_{k_0}]$ contains only type (II) intervals, (9.23) and (9.25) hold with $t_{i_k} = \tau_{l+1}$ and $t_{k_0} = t_{i_{k+1}}$,

$$d(x(\tau_{l+1}), x(t_{k_0})) \leq \overline{K}_8\left[\frac{\theta_{k_0}}{\sqrt{\omega}} + t_{k_0} - \tau_{l+1}\right] \tag{9.111}$$

$$\|Z_1(\tau_{l+1}) - Z_1(t_{k_0})\| \leq \overline{K}_8\left[\frac{\theta_{k_0}}{\sqrt{\omega}} + t_{k_0} - \tau_{l+1}\right], \tag{9.112}$$

where θ_{k_0} is the number of type (II) intervals in $[\tau_{l+1}, t_{k_0}]$.

Inequalities (9.109) and (9.111) imply:

$$d(y_{l+1}, x(t_{k_0})) \leq \mathrm{Sup}(K, \overline{K}_8)\left[\frac{\theta_{k_0}+1}{\sqrt{\omega}} + t_{k_0} - T_l\right]. \tag{9.113}$$

Equations (9.110) and (9.112) imply:

$$\|\tilde{Z}^{\star}_{l+1} - Z^{\star}_1(t_{k_0})\| \leq \mathrm{Sup}(\overline{K}_8, K)\left[\frac{\theta_{k_0}+1}{\sqrt{\omega}} + t_{k_0} - T_l\right]. \tag{9.114}$$

Finally, we apply Lemma 16 backwards, i.e. to the final interval $[t_{k_0}, t^-(s_2)]$. Let:

$n(t_{k_0}) = 1$ if this interval is oriented by ψ in the approximation of Lemma 16;

$= -1$ if it is oriented by ψ^{-1}.

From Lemma 16 and the two remarks following this lemma, we have:

$$d\left(\psi^{-n(t_{k_0})}(\bar{x}_{\tilde{\bar{s}}_2}), x(t_{k_0})\right) < K_{13}\left[\frac{1}{\sqrt{\omega}} + t^-(s_2) - t_{k_0}\right] \tag{9.115}$$

$$\left\| Z^\star\left(\psi^{-n(t_{k_0})}(\bar{x}_{\tilde{\bar{s}}_2})\right) - Z_1^\star(t_{k_0}) \right\| < K_{18}\left[\frac{1}{\sqrt{\omega}} + t^-(s_2) - t_{k_0}\right]. \tag{9.116}$$

Now

$$Z^\star\left(\psi^{-n(t_{k_0})}(\bar{x}_{\tilde{\bar{s}}_2})\right) = \lambda_{-n(t_{k_0})}(\bar{x}_{\tilde{\bar{s}}_2})\, \bar{a}^\star(\tilde{\bar{s}}_2). \tag{9.117}$$

Indeed, $Z^\star(s)$ satisfies the transport equations of the forms and $Z^\star(\tilde{\bar{s}}_2) = \bar{a}^\star(\tilde{\bar{s}}_2)\begin{bmatrix}1\\0\\0\end{bmatrix}$ by Lemma 15 applied backwards (i.e. to $[t_{k_0}, t^-(s_2)]$ instead of $[t^+(s_1), t^1]$).

As $\psi^{-n(t_{k_0})}$ sends α on α with a collinearity coefficient equal to $\lambda_{-n(t_{k_0})}$ and as $Z^\star$ at $\tilde{\bar{s}}_2$ is collinear to $\begin{bmatrix}1\\0\\0\end{bmatrix}$ and provides us at this time with the coordinates of α, (9.117) holds.

From (9.113) and (9.115), we deduce:

$$d\left(y_{l+1}, \psi^{-n(t_{k_0})}(\bar{x}_{\tilde{\bar{s}}_2})\right) < K_{18}\left[\frac{\theta_{k_0}+2}{\sqrt{\omega}} + t^-(s_2) - T_l\right] \tag{9.118}$$

From (9.114) and (9.116), we deduce:

$$\left\| Z_{l+1}^\star - Z^\star\left(\psi^{-n(t_{k_0})}(\bar{x}_{\tilde{\bar{s}}_2})\right) \right\| = \left| \tilde{a}_{l+1}^\star - \lambda_{-n(t_{k_0})}(\bar{x}_{\tilde{\bar{s}}_2})\, a^\star(\tilde{\bar{s}}_2) \right| \tag{9.119}$$

$$< K_{18}\left[\frac{\theta_{k_0}+2}{\sqrt{\omega}} + t^-(s_2) - T_l\right].$$

From (9.118) and (9.106), we then deduce:

$$d(\psi^{n_{l+1}}(x_0),\psi^{-n(t_{k_0})}(\bar{x}_{\tilde{s}_2^-})) < K_{19}\left[\frac{\theta_{k_0}+2}{\sqrt{\omega}} + \sum_0^l\left(\frac{\bar{\alpha}_i+1}{\sqrt{\omega}} + T_i - T_{i-1}\right) + t^-(s_2) - T_l\right]$$

$$< K_{20}\left[\frac{j}{\sqrt{\omega}} + t^-(s_2) - t^+(s_1)\right], \tag{9.120}$$

where j represents the total number of intervals between $t^+(s_1)$ and $t^-(s_2)$ and K_{20} is an appropriate constant.

We choose now ω, uniform and large enough so that

$$K_{20}\left[\frac{j_1+j_2}{\sqrt{\omega}} + \Sigma\, t^-(s_{i+1}) - t^+(s_i)\right] < \varepsilon_5, \tag{9.121}$$

(where $j_1 + j_2$ is the total number of intervals on $[0,1]$), which is made possible by Lemmas 18 and 5; we have

$$d(\psi^{n_{l+1}}(x_0),\psi^{-n(t_{k_0})}(\bar{x}_{\tilde{s}_2^-})) < \varepsilon_5. \tag{9.122}$$

Thus (A_2) can be applied with $x = \psi^{n_{l+1}}(x_0)$ and $y = \psi^{-n(t_{k_0})}(\bar{x}_{\tilde{s}_2^-})$ and we have:

$$d(\psi^{n_{l+1}+n(t_{k_0})}(x_0),\bar{x}_{\tilde{s}_2^-}) < \bar{k}\, d(\psi^{n_{l+1}}(x_0),\psi^{-n(t_{k_0})}(\bar{x}_{\tilde{s}_2^-})) \tag{9.123}$$

$$< \bar{k}\, K_{20}\left[\frac{j}{\sqrt{\omega}} + t^-(s_2) - t^+(s_1)\right].$$

Noticing that, by definition, $x_0 = x_{\tilde{s}_2^+}$ ((9.43)), we deduce:

$$d(\psi^{n_{l+1}+n(t_{k_0})}(x_{\tilde{s}_2^+}),\bar{x}_{\tilde{s}_2^-}) < K\left[\frac{j}{\sqrt{\omega}} + t^-(s_2) - t^+(s_1)\right]. \tag{9.124}$$

It then suffices to notice that:

$$n_{l+1} + n(t_{k_0}) = k_1 - k_2 \tag{9.125}$$

to obtain (9.4) of Lemma 19.

We have only to prove now (9.5) and the fact that $k_1 - k_2$ is even. For this, we have the following estimates:

$$|\hat{\tilde{a}}^{\star}_{l+1} - a^{\star}_0| < K\left[\sum_0^{l}\left(\frac{\bar{\alpha}_i+1}{\sqrt{\omega}} + T_i - T_{i-1}\right)\right] \tag{9.105'}$$

$$|\lambda_{n_{l+1}}(x_0) - \lambda_{n_{l+1}}(\hat{y}_{l+1})| < K\left[\sum_0^{l}\left(\frac{\bar{\alpha}_i+1}{\sqrt{\omega}} + T_i - T_{i-1}\right)\right] \tag{9.107}$$

$$|\tilde{a}^{\star}_{l+1} - \lambda_{-n(t_{k_0})}(\bar{x}_{\tilde{s}_2^-})\, a^{\star}(\tilde{s}_2^-)| < K\left[\frac{\theta_{k_0}+2}{\sqrt{\omega}} + t^-(s_2) - T_l\right]. \tag{9.119}$$

Moreover, we have:

$$\hat{\tilde{a}}^{\star}_{l+1} = \lambda_{-n_{l+1}}(y_{l+1})\,\tilde{a}^{\star}_{l+1} . \tag{9.46}$$

Thus:

$$|\tilde{a}^{\star}_{l+1}\lambda_{-n_{l+1}}(y_{l+1}) - a^{\star}_0| < K\left[\sum_0^{l}\left(\frac{\bar{\alpha}_i+1}{\sqrt{\omega}} + T_i - T_{i-1}\right)\right] . \tag{9.126}$$

Taking into account (9.119), we thus have:

$$|\lambda_{-n(t_{k_0})}(\bar{x}_{\tilde{s}_2^-})\,\lambda_{-n_{l+1}}(y_{l+1})\,\bar{a}^{\star}(\tilde{s}_2^-) - a^{\star}_0| < \tag{9.127}$$

$$K\left[\sum_0^{l}\left(\frac{\bar{\alpha}_i+1}{\sqrt{\omega}} + T_i - T_{i-1}\right)\right] + |\lambda_{-n_{l+1}}(y_{l+1})|\,K\left[\frac{\theta_{k_0}+2}{\sqrt{\omega}} + t^-(s_2) - T_l\right]$$

hence

$$|\lambda_{-n(t_{k_0})}(\bar{x}_{\tilde{s}_2^-})\,\lambda_{-n_{l+1}}(y_{l+1})\,\bar{a}^{\star}(\tilde{s}_2^-) - a^{\star}_0| < \bar{K}\left[\frac{j}{\sqrt{\omega}} + t^-(s_2) - t^+(s_1)\right]. \tag{9.128}$$

(We used the fact that $|\lambda_{-n_{l+1}}(y_{l+1})| < \alpha_4$ by (A_3); we upperbounded $K\sum_0^{l}\frac{\bar{\alpha}_i+1}{\sqrt{\omega}} + \alpha_5 K\frac{\theta_{k_0}+2}{\sqrt{\omega}}$ by $\bar{K}\frac{j}{\sqrt{\omega}}$ and we also upperbounded the other terms $\bar{K}(t_i - T_{i-1}) + \bar{K}(t_{i+1} - T_i)$ .... by $K(t^-(s_2) - t^+(s_1))$.)

Notice now that we have:

$$\lambda_{-n_{l+1}}(y_{l+1}) = \frac{1}{\lambda_{n_{l+1}}(\psi^{-n_{l+1}}(y_{l+1}))} = \frac{1}{\lambda_{n_{l+1}}(\hat{y}_{l+1})}, \tag{9.129}$$

by (9. 83) and by the definition of $\hat{y}_{l+1}$ given in (9. 44).

Similarly, by (9. 83)

$$\lambda_{-n(t_{k_0})}(\bar{x}_{\tilde{s}_2}) = \frac{1}{\lambda_{n(t_{k_0})}(\psi^{-n(t_{k_0})}(\bar{x}_{\tilde{s}_2}))}. \tag{9.130}$$

Now, by (9. 120) and with ω large enough, we have:

$$|d(\psi^{n_{l+1}}(x_0), \psi^{-n(t_{k_0})}(\bar{x}_{\tilde{s}_2}))| < K_{20}\left[\frac{j}{\sqrt{\omega}} + t^-(s_2) - t^+(s_1)\right] < \varepsilon_5, \tag{9.131}$$

hence, applying (A_3):

$$|\lambda_{n(t_{k_0})}(\psi^{-n(t_{k_0})}(\bar{x}_{\tilde{s}_2})) - \lambda_{n(t_{k_0})}(\psi^{n_{l+1}}(x_0))| \tag{9.131'}$$

$$< \alpha_5 K_{20}\left[\frac{j}{\sqrt{\omega}} + t^-(s_2) - t^+(s_1)\right].$$

We have now all the necessary estimates for (9. 5).

We replace in (9. 128) $\lambda_{-n_{l+1}}(y_{l+1})$ and $\lambda_{-n(t_0)}(\bar{x}_{\tilde{s}_2})$ by their values given in (9. 129) and (9. 130), which yields:

$$|\bar{a}^\star(\tilde{s}_2) - a_0^\star \lambda_{n_{l+1}}(\hat{y}_{l+1}) \lambda_{n(t_{k_0})}(\psi^{-n(t_{k_0})}(\bar{x}_{\tilde{s}_2}))| \tag{9.132}$$

$$\leq \bar{K} |\lambda_{n_{l+1}}(\hat{y}_{l+1}) \lambda_{n(t_{k_0})}(\psi^{-n(t_{k_0})}(\bar{x}_{\tilde{s}_2}))| \left[\frac{j}{\sqrt{\omega}} + t^-(s_2) - t^+(s_1)\right]$$

$$< \bar{K}\,\alpha_4^2 \left| \frac{j}{\sqrt{\omega}} + t^-(s_2) - t^+(s_1) \right| .$$

We use (9.131)' to derive from (9.132):

$$|\bar{a}^\star(\tilde{s}_2^-) - a_0^\star \lambda_{n_{l+1}}(\hat{y}_{l+1}) \lambda_{n(t_{k_0})}(\psi^{n_{l+1}}(x_0)| \tag{9.133}$$

$$\leq \left[\frac{j}{\sqrt{\omega}} + t^-(s_2) - t^+(s_1)\right] [\bar{K}\,\alpha_4^2 + \alpha_5 K_{20} |a_0^\star \lambda_{n_{l+1}}(\hat{y}_{l+1})||]$$

$$\leq \left[\frac{j}{\sqrt{\omega}} + t^-(s_2) - t^+(s_1)\right] [\bar{K}\,\alpha_4^2 + K_{20}\,\alpha_5\,\alpha_4 |a_0^\star||].$$

We use (9.107) to derive from (9.133):

$$|\bar{a}^\star(\tilde{s}_2^-) - a_0^\star \lambda_{n_{l+1}}(x_0) \lambda_{n(t_{k_0})}(\psi^{n_{l+1}}(x_0))| \tag{9.134}$$

$$\leq \left[\frac{j}{\sqrt{\omega}} + t^-(s_2) - t^+(s_1)\right] [\bar{K}\,\alpha_4^2 + K_{20}\alpha_5\alpha_4 |a_0^\star| + 2K |a_0^\star| |\lambda_{n(t_{k_0})}(\psi^{n_{l+1}}(x_0))||]$$

$$\leq \left[\frac{j}{\sqrt{\omega}} + t^-(s_2) - t^+(s_1)\right] [\bar{K}\,\alpha_4^2 + K_{20}\alpha_5\alpha_4 |a_0^\star| + 2K\,\alpha_4 |a_0^\star||],$$

where we have upperbounded $\sum_0^l \left(\frac{\bar{\alpha}_i + 1}{\sqrt{\omega}}\right) + T_i - T_{i-1}$ by $2\left[\frac{j}{\sqrt{\omega}} + t^-(s_2) - t^+(s_1)\right]$

and $|\lambda_{n(t_{k_0})}(\psi^{n_{l+1}}(x_0)|$ by α_4 as we are allowed by (A_3).

$$|\bar{a}^\star(\tilde{s}_2^-) - a_0^\star \lambda_{n_{l+1}}(x_0) \lambda_{n(t_{k_0})}(\psi^{n_{l+1}}(x_0))| \tag{9.134}$$

$$< [\bar{\bar{K}} + |a_0^\star||] \left[\frac{j}{\sqrt{\omega}} + t^-(s_2) - t^+(s_1)\right].$$

In (9.134), $a_0^\star = a^\star(\tilde{s}_2^+)$.

By Lemmas 15 and 15', we have:

$$|a^\star(\tilde{s}_2^-) - 1| < K_{12}\left[\frac{\int_0^1 b^2}{\omega} + \varepsilon_1 + \sqrt{\alpha_1}\right] \tag{9.135}$$

$$|\bar{a}^\star(\tilde{s}_2^-) - 1| < K_{12}\left[\frac{\int_0^1 b^2}{\omega} + \varepsilon_1 + \sqrt{\alpha_1}\right]. \tag{9.136}$$

We choose ω large enough so that $K_{12}\left[\frac{\int_0^1 b^2}{\omega} + \varepsilon_1 + \sqrt{\alpha_1}\right] < 1$. Then $|a\star(\tilde{s}_2^+)| = |a_0^\star| < 2$. Equation (9.134) thus becomes:

$$|\bar{a}\star(\tilde{s}_2^-) - a_0^\star \lambda_{n_{l+1}}(x_0)\lambda_{n(t_{k_0})}(\psi^{n_{l+1}}(x_0))| < (\bar{\bar{K}}+2)(\frac{j}{\omega}+t^-(s_2)-t^+(s_1))$$

$$< C\left[\frac{j}{\sqrt{\omega}} + t^-(s_2) - t^+(s_1)\right]. \qquad (9.137)$$

Using (9.135) and (9.136) we then deduce:

$$|\lambda_{n_{l+1}}(x_0)\lambda_{n(t_{k_0})}(\psi^{n_{l+1}}(x_0)) - 1| \qquad (9.138)$$

$$< C\left[\frac{j}{\sqrt{\omega}} + t^-(s_2) - t^+(s_1)\right] + K_{12}\left[\frac{\int_0^1 b^2}{\omega} + \varepsilon_1 + \sqrt{\alpha_1}\right](1+\lambda_{n_{l+1}}(x_0)\lambda_{n(t_{k_0})}(\psi^{n_{l+1}}(x_0))).$$

Hence, using (A_3) (upperboundedness of the λ_i's)

$$|\lambda_{n_{l+1}}(x_0)\lambda_{n(t_{k_0})}(\psi^{n_{l+1}}(x_0)) - 1| \qquad (9.139)$$

$$< C\left[\frac{j}{\sqrt{\omega}} + t^-(s_2) - t^+(s_1)\right] + \bar{K}_{12}\left[\frac{\int_0^1 b^2}{\omega} + \varepsilon_1 + \sqrt{\alpha_1}\right].$$

Now, $\lambda_{n_{l+1}}(x_0)\lambda_{n(t_{k_0})}(\psi^{n_{l+1}}(x_0)) = \lambda_{n_{l+1}+n(t_{k_0})}(x_0)$ through a computation similar to the one used to establish (9.48).

On the other hand, $x_0 = x_{\tilde{s}_2^+}$ by definition ((9.43)).

Hence, we have:

$$|\lambda_{n_{l+1}+n(t_{k_0})}(x_{\tilde{s}_2^+}) - 1| \qquad (9.140)$$

$$< C\left[\frac{j}{\sqrt{\omega}} + t^-(s_2) - t^+(s_1)\right] + \bar{K}_{12}\left[\frac{\int_0^1 b^2}{\omega} + \varepsilon_1 + \sqrt{\alpha_1}\right].$$

As by (9.125) $n_{\ell+1} + n(t_{k_0}) = k_1 - k_2$, we deduce:

$$|\lambda_{k_1-k_2}(x_{\tilde{s}_2^+}) - 1| < C\left[\frac{j}{\sqrt{\omega}} + t^-(s_2) - t^+(s_1)\right] + \bar{K}_{12}\left[\frac{\int_0^1 b^2}{\omega} + \varepsilon_1 + \sqrt{\alpha_1}\right] \quad (9.141)$$

$$< \varepsilon(\varepsilon_1, \omega) = C\left[\frac{j_2+j_1}{\sqrt{\omega}} + \Sigma t^-(s_{i+1}) - t^+(s_i)\right] + \bar{K}_{12}\left[\frac{\int_0^1 b^2}{\omega} + \varepsilon_1 + \sqrt{\alpha_1}\right],$$

which proves (9.5).

The evenness of $k_1 - k_2$ follows immediately.

Indeed, for ε_1 and ω small enough, we have $\varepsilon(\varepsilon_1, \omega) < \frac{1}{2}$; hence:

$$\lambda_{k_1-k_2}(x_{\tilde{s}_2^+}) > \tfrac{1}{2} > 0. \quad (9.142)$$

The collinearity coefficient of

$((\psi^{k_1-k_2})\star\alpha)_{x_{\tilde{s}_2^+}}$ on $\alpha_{x_{\tilde{s}_2^+}}$ being positive,

$k_1 - k_2$ is even: indeed ψ corresponds to a comeback of the field of planes α on itself after one and only one rotation. Hence, after an odd iteration of ψ, the transported form and the forms are oriented in opposite directions and the collinearity coefficient cannot possibly be positive.

It remains to prove (9.5)' which is deduced from (9.134) by noticing that $a_0^\star = a^\star(\tilde{s}_2^+)$ and that, as we have already shown:

$$\lambda_{n_{\ell+1}}(x_0)\lambda_{n(t_{k_0})}(\psi^{n_{\ell+1}}(x_0)) = \lambda_{n_{\ell+1}+n(t_{k_0})}(x_0) = \lambda_{k_1-k_2}(x_{\tilde{s}_2^+}). \blacksquare$$

After the proof of Lemma 19, we sum up the situation: between two concentration intervals, we established that we could replace the phenomenon described in (9.3) by a simpler phenomenon:

$$\psi^{k_1 - k_2}(x_{\underset{s_2}{\sim}+})$$

$\bar{x}_{\underset{s_2}{\sim}+}$

v

distance of order $K\left(\dfrac{j}{\sqrt{\omega}} + t^-(s_1) - t^+(s_2) \right)$

j = number of intervals between $t^+(s_1)$ and
$$t^-(s_2)$$

$| \lambda_{k_1 - k_2}(x_{\underset{s_2}{\sim}+}) - 1 |$ small

ψ^2

ψ

$x_{\underset{s_2}{\sim}+}$

Points $x_{\underset{s_2}{\sim}+}$ and $\bar{x}_{\underset{s_2}{\sim}-}$ are, because of Lemma 19, very constrained:

between $x_{\underset{s_2}{\sim}+}$ and $\psi^{k_1 - k_2}(x_{\underset{s_2}{\sim}+})$, the second point being very close to $\bar{x}_{\underset{s_2}{\sim}-}$, the

form a has made an even number of revolutions and has almost come back on

itself <u>as a form</u>. This means that $x_{\underset{s_2}{\sim}+}$ must lie in a neighbourhood of the

conjugate points of the form a.

Given x in M, under (A_1), there is of course a discrete number of

coincidence (and oriented coincidence) points of the form a: these are the

$\psi^i(x)$; $i \in \mathbb{Z}$.

Among these points, the existence of one such that $\lambda_i(x) = 1$ raises a

constraint. For i given in $\mathbb{Z}$, such points are expected to belong to a hyper-

surface Γ_i of M.

Such a phenomenon is generical:

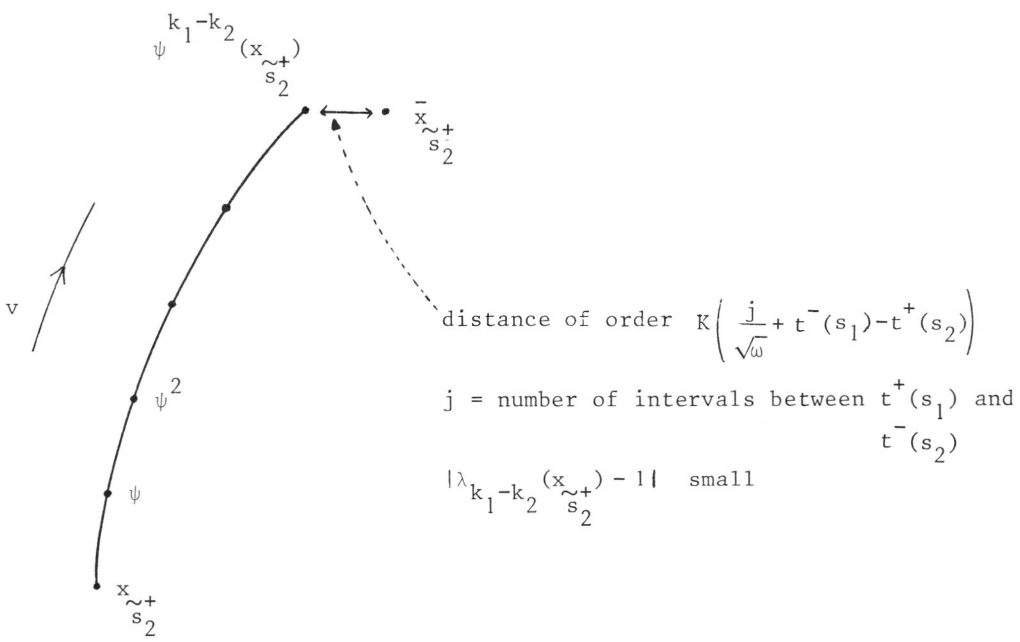

Let us examine the case of S^3 with $\alpha = \lambda\alpha_0$ and $\alpha_0 = \Sigma(x_i dy_i - y_i dx_i)$.

In this case, with v defining a Hopf fibration, α makes one revolution along a v-orbit after having described 1/4 of a v-orbit. The first oriented coincidence point is obtained after 1/2 of a v-orbit. On the remaining half of the v-orbit, there is a coincidence point at 3/4 of the orbit (reverse orientation) and a conjugate point when the orbit is completely described, i.e. we come back to our starting point. Thus:

$$\psi^2(x_0) = -x_0; \quad \psi^4(x_0) = x_0. \tag{9.144}$$

Here, coincidence points which are conjugate points are either, for a given x_0, x_0 itself or $-x_0$, under the conditions

$$\lambda(-x_0) = \lambda(x_0) \tag{9.145}$$

where λ defines α with respect to α_0.

Consequently, the curves x_ε are of the following type:

(9.146)

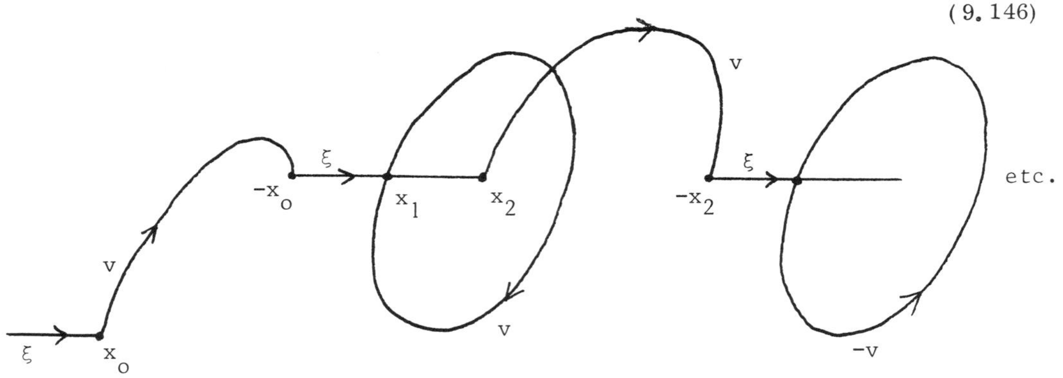

the points x_i satisfy $\lambda(x_i) = \lambda(-x_i)$ for $i = 0$, $i = 2$.

When we are dealing with a convex Hamiltonian H, these points lie on the hypersurface:

$$\Gamma = \{x \mid 1 = H(x) = H(-x)\} \subset \Sigma = \{x \mid H(x) = 1\}.$$

If H is even, $\Gamma = \Sigma$. However, when we slightly perturb this case, we find a generical case where Γ is a hypersurface.

We now turn back to the general situation and sum up our analysis.

Summary of the analysis

[1] Inside the concentration intervals, we defined two end points x_i^+ and x_i^- such that:

$$x_i^- = \psi^{k_i'}(x_{i-1}^+)\ ; \quad x_{i-1}^+ = (x_{\tilde{s}_2^+})_{i-1}\ ; \quad k_i' = (k_1 - k_2)_{i-1}; \tag{9.147}$$

(x_{i-1}^+, x_i^-) corresponds to the pair $(x_{\tilde{s}_2^+}, \psi^{k_1-k_2}(x_{\tilde{s}_2^+}))$ of Lemma 19. k_i' is even and $\lambda_{k_i'}(x_i^+)$ is close to 1.

[2] On a concentration interval, we approximated the curve, at order $\sqrt{\alpha_1}\,\delta t_i$, by a curve z_i, tangent to $a\xi$ during the time δt_i, and starting at $y(t_i^-(s_i))$. This was derived in Lemma 9, with $\delta t_i = t_i^+(s_i) - t_i^-(s_i)$, i being the index of the concentration interval. We proved, in Lemma 15', that $y(t_i^-(s_i))$ is close to $(x_{\tilde{s}_2^-})_i$ at order $K\left[\frac{1}{\sqrt{\omega}} + \sqrt{\alpha_1}\,\delta t_i\right]$; while in Lemma 19, we proved that $(x_{\tilde{s}_2^-})_i$ is close to $x_i^- = \psi^{k_i'}(x_{i-1}^+)$, at order $K\left[\frac{\bar{j}_i}{\sqrt{\omega}} + \delta t_i^-\right]$, where $\bar{j}_i$ is the total number of distinguished intervals between $t_i^+(s_{i-1})$ and $t_i^-(s_i)$; and $\delta t_i^- = t_i^-(s_i) - t_i^+(s_{i-1})$.

Finally, we know, from Lemma 15 that $(x_{\tilde{s}_2^+})_i$ is close to $y(t_i^+(s_i))$ at order $\left[\frac{1}{\sqrt{\omega}} + \sqrt{\alpha_1}\,\delta t_i\right]$.

From this summary we deduce:

Lemma 20: Consider the curve $\tilde{z}_i$ tangent to a ξ during the time δt_i and starting at $x_i^- = \psi^{k_i'}(x_{i-1}^+) = \psi^{k_i'}(x_{(\tilde{s}_2^+)_{i-1}})$. Then the extremity $\bar{x}_i^+$ of this curve is close to $x_i^+ = (x_{\tilde{s}_2^+})_i$ at order $\frac{\bar{j}_i+1}{\sqrt{\omega}} + \delta t_i^- + \sqrt{\alpha_1}\,\delta t_i$.

Proof: The curves $\tilde{z}_i$ and z_i are both tangent to aξ during the same time. They start respectively at $x_i^- = \psi^{k_i}(x_{i-1}^+)$ and at $y(t_i^-(s_i))$.

The distance between these two points is upperbounded by the sum of the distance of $(x_{\tilde{s}_2^-})_i$ to x_i^-, which is of the order $K\left[\frac{\bar{j}_i}{\sqrt{\omega}} + \delta t_i^-\right]$ (Lemma 19), and the distance of $(x_{\tilde{s}_2^-})_i$ to $y(t_i^-(s_i))$, which is of the order $K\left[\frac{1}{\sqrt{\omega}} + \sqrt{\alpha_1}\ \delta t_i\right]$ (Lemma 15').

This distance is thus of the order $K\left[\frac{\bar{j}_i+1}{\sqrt{\omega}} + \delta t_i^- + \sqrt{\alpha_1}\ \delta t_i\right]$. This implies that $\bar{x}_i^+$ is close to the end point of z_i, denoted z_i^+, at the same order:

$$d(\bar{x}_i^+, z_i^+) < K\left[\frac{\bar{j}_i+1}{\sqrt{\omega}} + \delta t_i^- + \sqrt{\alpha_1}\ \delta t_i\right].$$

On the other hand, the curve z_i remains close, through Lemma 9, to the curve $y(t)$ at order $\sqrt{\alpha_1}\ \delta t_i$. In particular, we have:

$$d(z_i^+, y(t_i^+(s_i))) < K\sqrt{\alpha_1}\ \delta t_i .$$

Finally, we have, by Lemma 15:

$$d((x_{\tilde{s}_2^+})_i, y(t_i^+(s_i))) < K\left[\frac{1}{\sqrt{\omega}} + \sqrt{\alpha_1}\ \delta\ t_i\right].$$

We thus deduce:

$$d(\bar{x}_i^+, (x_{\tilde{s}_2^+})_i) = d(\bar{x}_i^+, x_i^+) < 3K\left[\frac{\bar{j}_i+1}{\sqrt{\omega}} + \delta t_i^- + \sqrt{\alpha_1}\ \delta t_i\right].$$

We now forget our original curve and we consider the obtained approximate:

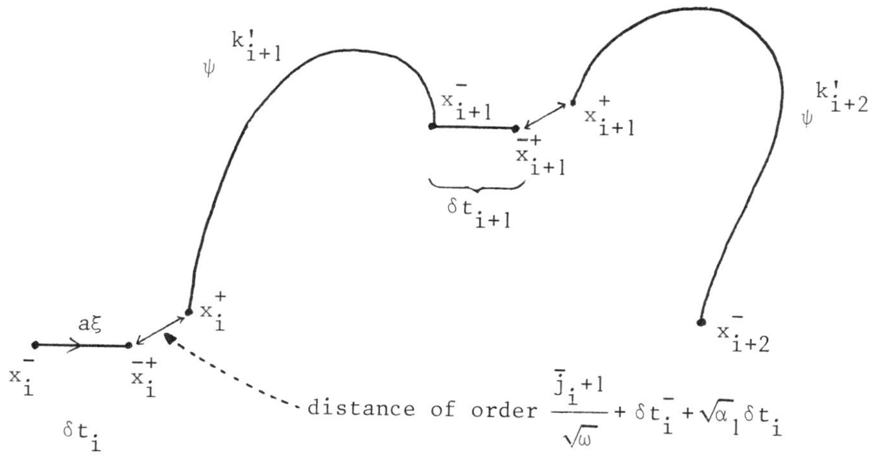

(9.150)

$|\lambda_{k'_{i+1}}(x_i^+) - 1|$ small; k'_i even

We prove the following theorem which shows that α is somewhat globally preserved on (9.150).

<u>Theorem 6</u>: Let n_0 (depending on ε_1 and ω) be the total number of concentration intervals. Let $\lambda_{k'_i}(x_i^+)$ be the characteristic numbers of $\psi^{k'_i}$, at points x_i^+. We have

$$\left|\prod_{i=1}^{m} \lambda_{k'_i}(x_{i-1}^+) - 1\right| \to 0 \text{ uniformly for } m \leq n_0. \qquad (9.150)'$$

The significance of Theorem 6 is that the curve (9.150) preserves the <u>form</u> α more and more. Indeed, it first implies that, for each $\psi^{k'_i}$, $\lambda_{k'_i}(x_{i-1}^+)$ tends to 1. Hence α is sent back on itself from x_{i-1}^+ to $\psi^{k'_i}(x_{i-1}^+)$. On the other hand, pieces tangent to ξ preserve α, as ξ is the characteristic vector field of α (hence $L_\xi \alpha = 0$, when L_ξ is the Lie derivative along ξ). Finally, the <u>global</u> influence of the $\psi^{k'_i}$ on α is small by (9.150)'.

In this sense, we may say that (9.150) is a "geodesic to α relatively to v".

<u>Proof of Theorem 6</u>: We know that $x_{i-1}^+ = (x_{\tilde{s}_2^+})_{i-1}$. We denote $a_i^{\star -} = \bar{a}^\star((\tilde{s}_2^-)_i)$;

$a_i^{\star +} = a^\star((\tilde{s}_2^+)_i)$ (notations preceding Lemma 19).

By (9.5)' of Lemma 19, we have:

$$\left|\lambda_{k_i'}(x_{i-1}^+) - \frac{a_i^{\star -}}{a_{i-1}^{\star +}}\right| = \left|\lambda_{k_i'}((x_{\tilde{s}_2^-})_{i-1}) - \frac{\bar{a}^\star((\tilde{s}_2^-)_i)}{a^\star((\tilde{s}_2^+)_{i-1})}\right| < K\left[\frac{\bar{j}_i}{\sqrt{\omega}} + \delta t_i^-\right].$$

Thus:

$$\prod_{i=1}^m \left[\frac{a_i^{\star -}}{a_{i-1}^{\star +}} - K \frac{\bar{j}_i}{\sqrt{\omega}} + \delta t_i^-\right] < \prod_{i=1}^m \lambda_{k_i'}(x_{i-1}^+) < \prod_{i=1}^m \left[\frac{a_i^{\star -}}{a_{i-1}^{\star +}} + K \frac{\bar{j}_i}{\sqrt{\omega}} + \delta t_i^-\right].$$

On the other hand, by (8.334)' of Lemma 15, we have:

$$\left|a^\star(\tilde{s}_2^+)_i) - a'^\star((\tilde{s}_2')_i)\right| < K_{12}\left[\frac{1}{\sqrt{\omega}} + \sqrt{\alpha_1}\ \delta t_i\right].$$

In the same say, by (8.371)' of Lemma 15', we have:

$$\left|a^\star((\tilde{s}_2^-)_i) - a'^\star((\tilde{s}_2')_i)\right| < K_{12}\left[\frac{1}{\sqrt{\omega}} + \sqrt{\alpha_1}\ \delta t_i\right],$$

hence, we deduce:

$$\left|a^\star((\tilde{s}_2^+)_i) - a^\star((\tilde{s}_2^-)_i)\right| < 2\,K_{12}\left[\frac{1}{\sqrt{\omega}} + \sqrt{\alpha_1}\delta t_i\right].$$

We here remark that $a^\star((\tilde{s}_2^-)_i) = \bar{a}^\star((\tilde{s}_2^-)_i) = a_i^{\star -}$; indeed we decided to draw a bar over a when there could be a confusion, in particular in the proof of Lemma 19, between $a^\star((\tilde{s}_2^-))_{i-1}$ and $a^\star((\tilde{s}_2^-)_i)$. (There was no index before this!)

In the same way, $a^\star((\tilde{s}_2^+)_i) = a_i^{\star +}$.

Thus, we have:

$$\left|a_i^{\star +} - a_i^{\star -}\right| < 2\,K_{12}\left[\frac{1}{\sqrt{\omega}} + \sqrt{\alpha_1}\ \delta t_i\right].$$

As, by Lemmas 15 and 15', the $a_i^{\star +}$'s and $a_i^{\star -}$'s are very close to 1, for ε_1 small enough and ω large enough, we have:

$$\left|\frac{a^{\star -}_i}{a^{\star +}_i} - 1\right| < 4K_{12}\left[\frac{1}{\sqrt{\omega}} + \sqrt{\alpha_1}\delta t_i\right].$$

Consequently, we have:

$$\prod_{i=1}^{m}\left[\frac{a^{\star -}_i}{a^{\star +}_{i-1}} \pm K\left(\frac{\bar{j}_i}{\sqrt{\omega}} + \delta t^-_i\right)\right] = \frac{\prod_{i=1}^{m}\left(a^{\star -}_i \pm \left[K\frac{\bar{j}_i}{\sqrt{\omega}} + \delta t^-_i\right] a^{\star +}_{i-1}\right)}{\prod_{i=0}^{m} a^{\star +}_i}$$

$$= \frac{a^{\star +}_m}{a^{\star +}_0} \prod_{i=1}^{m}\left[\frac{a^{\star -}_i}{a^{\star +}_i} \pm K\left(\frac{\bar{j}_i}{\sqrt{\omega}} + \delta t^-_i\right) a^{\star +}_{i-1}\right]$$

$$= \frac{a^{\star +}_m}{a^{\star +}_0} \prod_{i=1}^{m}\left[1 + \delta_i \pm K\left(\frac{\bar{j}_i}{\sqrt{\omega}} + \delta t^-_i\right) a^{\star +}_{i-1}\right]$$

with $|\delta_i| < 4K_{12}\left[\frac{1}{\sqrt{\omega}} + \sqrt{\alpha_1}\ \delta t_i\right]$.

Now

$$\sum_{i=1}^{m} |\delta_i| + K\sum_{i=1}^{m}\left[\frac{\bar{j}_i}{\sqrt{\omega}} + \delta t^-_i\right] a^{\star +}_{i-1}$$

$$\leq 4K_{12}\sum_{i=1}^{m}\left[\frac{1}{\sqrt{\omega}} + \sqrt{\alpha_1}\ \delta t_i\right] + 2K\left[\sum_{i=1}^{m}\frac{\bar{j}_i}{\sqrt{\omega}} + \delta t^-_i\right],$$

(for ε_1 small enough and ω large enough, we have $|a^{\star -}_{i-1}| < 2$ as $a^{\star +}_{i-1}$ goes to 1 uniformly).

$$\leq (4K_{12} + 2K)\frac{j_1 + j_2}{\sqrt{\omega}} + 2K\,\Sigma\,\delta t^-_i + 4K_{12}\sqrt{\alpha_1} \xrightarrow[\substack{\varepsilon_1 \to 0 \\ \omega \to +\infty}]{} 0.$$

On the other hand $\dfrac{a^{\star +}_m}{a^{\star +}_0}$ goes to 1 uniformly on m.

This yields that, uniformly on m, $\prod_{i=1}^{m} \lambda_{k'_i}(x^+_{i-1})$ goes to 1, as it is upper and lower bounded by two quantities which converge to 1 uniformly on m. ■

Our aim now is to "rearrange" the curve (9.150); this means that we will define, close to (9.150), an almost closed curve, made up with pieces, some of them tangent to ξ, the others tangent to v between endpoints x and $\psi^{k_i'}(x)$ which are almost conjugate (i.e. $|\lambda_{k_i'}(x) - 1| \to 0$). These curves will converge to closed curves as ε_1 goes to zero and ω to $+\infty$.

The device consists in cumulating all losses between points x_i^{-+} and x_i^+ due to concentration intervals on a unique concentration interval. Indeed, if we add all these losses, we arrive at something of the order:

$\Sigma\left(\frac{\bar{j}_i+1}{\sqrt{\omega}} + \delta t_i^- + \sqrt{\alpha_1}\ \delta t_i\right)$; which is less than $\frac{j_1+j_2}{\sqrt{\omega}} + \Sigma\ \delta t_i^- + \sqrt{\alpha_1}$. This last quantity goes to zero as ε_1 goes to zero and ω to $+\infty$, as, by Lemmas 5 and 18, $\frac{j_1+j_2}{\sqrt{\omega}}$ and $\Sigma\ \delta t_i^-$ go then to zero.

Consequently, if we are able to simply cumulate all these losses without multiplying them by factors going to $+\infty$, we will obtain an almost closed curve.

Through this operation, we lose (9.150)'.

This is why, in the following theorems, we will retain the points x_i^+ satisfying (9.150)'. It is probable that, with a better rearrangement process, it is possible to obtain (9.150)' on the almost closed curve.

For our purpose here, we will leave aside this problem. However we show in Theorem 7 that this information, in generical situation, is not lost.

There are two ways to obtain an almost closed curve.

The first way is direct and involves a strong hypothesis on the one parameter groups of the vector fields $D\psi^{2k}(\xi)$; $k \in \mathbb{Z}$. However, there are two arguments in favour of this approach: the phenomenon is thereby clearly seen and this allows us to overcome the case when M fibers in S^1 over a surface.

The second way takes more into account the behaviour of α along the involved v-orbits and its dynamics. It also relies on a generical local situation.

We will, at the end, give a convergence theorem which mixes both ways and we will show that the limit curves live in a certain part of M (section 10).

<u>The first approach is the following:</u>

We assume:

$$(\mathcal{A}_1)\begin{cases}(A_2) \text{ and } (A_3) \text{ are satisfied} \\ \text{Let } \theta_s^{2k} \text{ be the one-parameter groups of } \xi_k = D\psi^{2k}(\xi) \quad (k \in \mathbb{Z}). \\ \text{We assume the following hypothesis of local equicontinuity on the} \\ \theta_s^{2k}\ ; \quad \alpha_6,\ s_0 > 0 \text{ such that } \forall k \in \mathbb{Z} \text{ and } \forall s,\ |s| \le s_0,\ \forall x,y \in M, \\ \qquad d(\theta_s^{2k}(x), \theta_s^{2k}(y)) \le (1 + \alpha_6 |s|)\, d(x,y).\end{cases}$$

We then have:

<u>Theorem</u> $\mathcal{A}_1$: Under $(\mathcal{A}_1)$, the curve (9.150) may be rearranged in a curve $\hat{x}_\varepsilon$, $\varepsilon > 0$ given arbitrary, made up with pieces $[\hat{x}_0^-, \hat{x}_0^+]$, $[\hat{x}_0^+, \hat{x}_1^-]$, $[\hat{x}_1^-, \hat{x}_1^+], \dots,$ $[\hat{x}_{n_0-1}^-, \hat{x}_{n_0-1}^+]$, $[\hat{x}_{n_0-1}^+, \hat{x}_{n_0}^-]$. On $[\hat{x}_i^-, \hat{x}_i^+]$, the rearranged curve is tangent to ξ during the time $a\delta t_i$. On $[\hat{x}_{i-1}^+, \hat{x}_i^-]$, the curve is tangent to v and we have:

$$\hat{x}_i^- = \psi^{k_i'}(\hat{x}_{i-1}^+)\ ; \quad k_i' \text{ even}; \quad |\lambda_{k_i'}(\hat{x}_{i-1}^+) - 1| < \varepsilon$$

$$d(\hat{x}_{n_0}^-, \hat{x}_0^-) < \varepsilon. \tag{9.152}$$

Moreover, we have:

$$d(\hat{x}_i^\pm, x_i^\pm) \le K_2 \left(\sum_{k=0}^{i} \frac{\bar{j}_k + 1}{\sqrt{\omega}} + \delta t_k^- + \sqrt{\alpha_1}\ \delta t_k \right) \to 0 \tag{9.152'}$$

uniformly on i, where $x_i^\pm$, are the points of (9.150) satisfying in particular (9.150)'.

<u>Proof of Theorem</u> $\mathcal{A}_1$: Consider the curve given in (9.150) with the points x_i^-, x_i^+, $\bar{x}_i^+$, the characteristic periods δt_i, δt_i^-, the exponents k_i and the characteristic intervals j_i.

Consider another curve $\hat{x}_\varepsilon$ constructed as follows: θ_s^0 is, according to the notations of $\mathcal{A}_1$, the one parameter group of ξ,

$$\hat{x}_0^- = x_0^- \; ; \quad \hat{x}_0^+ = \theta^0_{a\delta t_0}(\hat{x}_0^-) \; ; \tag{9.153}$$

$$\hat{x}_1^- = \psi^{k'_0}(\hat{x}_0^+), \quad \hat{x}_1^+ = \theta^0_{a\delta t_1}(\hat{x}_1^-) \ldots \hat{x}_i^- = \psi^{k'_{i-1}}(\hat{x}_{i-1}^+),$$

$$\hat{x}_i^+ = \theta^0_{a\delta t_i}(\hat{x}_i^-), \ldots, \hat{x}_{n_0}^- = \psi^{k'_{n_0-1}}(\hat{x}_{n_0-1}^+).$$

The curve so defined is of the specified form.

We prove under $(\mathfrak{A}_1)$, that $d(\hat{x}_{n_0}^-, \hat{x}_0^-) < \varepsilon$ and that $|\lambda_{k'_i}(\hat{x}_{i-1}^+) - 1| < \varepsilon$.

For this, we consider two arbitrary points on M; ${}^0X$, ${}^0Y$. We denote

$${}^0d = d({}^0X, {}^0Y). \tag{9.154}$$

We transform ${}^0X$, ${}^0Y$ by ψ and its iterates and by θ^0_s as follows: we consider a sequence of integers:

$$k'_1, \quad \ldots, \quad k'_m \in \mathbb{Z} \tag{9.155}$$

and a sequence of real numbers

$$s_1, \ \ldots, \ s_m. \tag{9.156}$$

We consider

$${}^0X_1^- = \psi^{k'_1}({}^0X), \quad {}^0X_1^+ = \theta^0_{s_1}({}^0X_1^-), \quad {}^0X_2^- = \psi^{k'_2}({}^0X_1^+), \ldots, {}^0X_m^+ = \theta^0_{s_m}({}^0X_m^-)$$

$$\tag{9.157}$$

$${}^0Y_1^- = \psi^{k'_1}({}^0Y), \quad {}^0Y_1^+ = \theta^0_{s_1}({}^0Y_1^-), \quad {}^0Y_2^- = \psi^{k'_2}({}^0Y_1^+), \ldots, {}^0Y_m^+ = \theta^0_{s_m}({}^0Y_m^-)$$

We denote:

$$\begin{cases} {}^0X_0^+ = {}^0X \\ {}^0Y_0^+ = {}^0Y . \end{cases}$$

We denote:

$$q_i = \text{integer part of } \frac{|s_i|}{s_0}\ ; \quad \delta_i = |s_i - \operatorname{sgn}(s_i)\, q_i s_0| . \tag{9.158}$$

We assume:

$$d_0 < \varepsilon_5 .$$

Then we have:

$$d({}^0X_l^{\pm}, {}^0Y_l^{\pm}) < d_0\, \bar{k}(1+\alpha_6 s_0)^{\sum_{i=1}^{m} q_i} \prod_{i=1}^{m} (1+\alpha_6 \delta_0)\ ; \tag{9.159}$$

$l = 0, 1, \ldots, m.$

Indeed, denoting:

$$\mathcal{J}_r = \sum_{i=r+1}^{l} k_i'\ ;\quad r \le l-1;\quad \mathcal{J}_l = 0, \tag{9.160}$$

we have:

$$\psi^{\mathcal{J}_i}({}^0X_i^+) = \psi^{\mathcal{J}_{i+1}}({}^0X_{i+1}^-). \tag{9.161}$$

Indeed ${}^0X_{i+1}^- = \psi^{k_{i+1}}({}^0X_i^+)$ by (9.157); hence:

$$\psi^{\mathcal{J}_{i+1}}({}^0X_{i+1}^-) = \psi^{\mathcal{J}_{i+1}} \circ \psi^{k_{i+1}}({}^0X_i^+) = \psi^{\mathcal{J}_{i+1}+k_{i+1}}({}^0X_i^+) = \psi^{\mathcal{J}_i}({}^0X_i^+).$$

In the same way:

$$\psi^{\mathcal{J}_i}({}^0Y_i^+) = \psi^{\mathcal{J}_{i+1}}({}^0Y_{i+1}^-). \tag{9.162}$$

Let:

$$X_i = \psi^{\mathcal{J}_i}({}^0X_i^+) = \psi^{\mathcal{J}_{i+1}}({}^0X_{i+1}^-)\ ;\quad Y_i = \psi^{\mathcal{J}_i}({}^0y_i^+) = \psi^{\mathcal{J}_{i+1}}({}^0y_{i+1}^-). \tag{9.163}$$

For $i = 0$, we have:

$$X_0 = \psi^{\mathcal{J}_0}({}^0X_0^+) = \psi^{\mathcal{J}_0}({}^0X)\ ;\quad Y_0 = \psi^{\mathcal{J}_0}({}^0Y_0^+) = \psi^{\mathcal{J}_0}({}^0Y). \tag{9.164}$$

On the other hand, we have:

$$X_{i+1} = \theta_{s_{i+1}}^{\mathcal{J}_{i+1}}(X_i)\,; \quad Y_{i+1} = \theta_{s_{i+1}}^{\mathcal{J}_{i+1}}(Y_i). \tag{9.165}$$

Indeed:

$$X_{i+1} = \psi^{\mathcal{J}_{i+1}}({}^0X_{i+1}^+) \underset{(9.157)}{=\!=\!=} \psi^{\mathcal{J}_{i+1}}(\theta_{s_{i+1}}^0({}^0X_{i+1}^-)). \tag{9.166}$$

By definition of θ_s^k, we have:

$$\psi^r \circ \theta_s^k = \theta_s^{k+r} \circ \psi^r, \tag{9.167}$$

as θ_s^i is the one parameter group of $D\psi^i(\xi)$.

Thus:

$$X_{i+1} = \theta_{s_{i+1}}^{\mathcal{J}_{i+1}} \circ \psi^{\mathcal{J}_{i+1}}({}^0X_{i+1}^-) \underset{(9.163)}{=\!=\!=} \theta_{s_{i+1}}^{\mathcal{J}_{i+1}}(X_i), \tag{9.167'}$$

which proves (9.165).

Finally, we have:

$$X_l = \psi^{\mathcal{J}_l}({}^0X_l^+) = \psi^0({}^0X_l^+) = {}^0X_l^+\,; \quad Y_l = \psi^{\mathcal{J}_l}({}^+Y_0^l) = \psi^0({}^0Y_l^+) = {}^0Y_l^+. \tag{9.168}$$

To obtain (9.159), we estimate:

$$d(X_l, Y_l) = d({}^0X_l^+, {}^0Y_l^+). \tag{9.169}$$

(The case ${}^0X_l^-$, ${}^0Y_l^-$ is equivalent.)

For that, we first estimate $d(X_0, Y_0)$ using (A_2). We have

$$d({}^0X, {}^0Y) < \varepsilon_5. \tag{9.170}$$

Thus (A_2) can be applied and gives:

$$d(\psi^{\mathcal{J}_0}({}^0X), \psi^{\mathcal{J}_0}({}^0Y)) = d(X_0, Y_0) < \bar{k}d({}^0X, {}^0Y) = \bar{k}d_0. \tag{9.171}$$

Starting with (X_0, Y_0), we go to (X_1, Y_1) through $\theta_{s_1}^{\mathcal{J}_1}$, then to (X_2, Y_2) starting at (X_1, Y_1) through $\theta_{s_2}^{\mathcal{J}_2}$, etc. We end up this process with

$(X_l, Y_l) = ({}^0X_l^+, {}^0Y_l^+)$ through $\theta_{s_l}^{\mathcal{J}_l} = \theta_{s_l}^0$.

At each of these steps, we may apply hypothesis ($\mathcal{Q}_1$) as follows. We divide the interval s_i in q_i intervals of length s_0. An interval of size $\delta_i = |s_i - \operatorname{sgn}(s_i) q_i s_0|$ then remains.

Each time an interval of length s_0 is crossed, the initial distance is multiplied by a factor upperbounded by $(1 + \alpha_6 s_0)$. After having crossed the q_i intervals of length s_0, we then have multiplied the initial distance $d(X_{i-1}, Y_{i-1})$ by $(1 + \alpha_6 s_0)^{q_i}$ at most. An interval of length $\delta_i \leq s_0$ then remains; where the distance is multiplied by $(1 + \alpha_6 \delta_i)$.

Summing up, we obtain:

$$d(X_i, Y_i) \leq (1 + \alpha_6 s_0)^{q_i} (1 + \alpha_6 \delta_i) d(X_{i-1}, Y_{i-1}). \tag{9.172}$$

The inequalities (9.172) imply, together with (9.171), the inequalities (9.159).

We now come back to the curve (9.150) and the curve $\hat{x}_\varepsilon$. We consider a sequence of curves defined as follows:

$\hat{x}_\varepsilon^i$ starts at x_i^+ and "follows the movement" thereafter, i.e. starting at x_i^+, we apply $\psi^{k'_i}$, then we take a ξ-orbit of length $a\delta t_{i+1}$; then we apply $\psi^{k'_{i+1}}$ and we take a ξ-orbit of length $a\delta t_{i+2}$; ... The process is stopped after all k'_j for $j \geq i+1$ are used. (9.173)

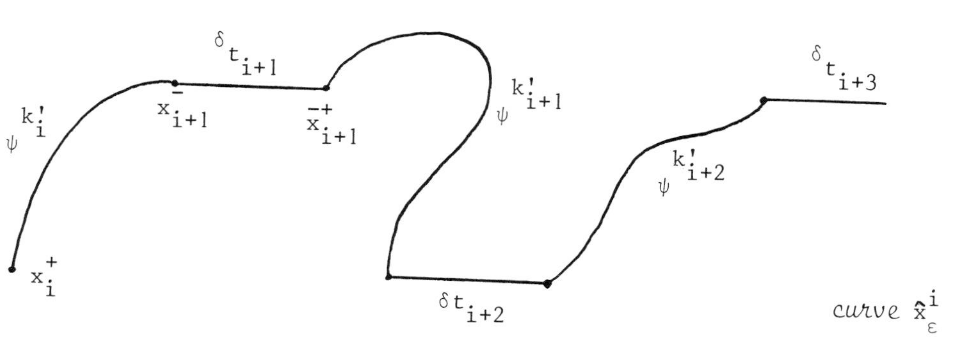

Considering two such curves $\hat{x}_\varepsilon^i$ and $\hat{x}_\varepsilon^{i+1}$, they start at two shifted points: the first one starts at x_i^+; the second one at x_{i+1}^+.

Hence, after having applied $\psi^{k_i'}$ and described a piece of ξ-orbit of length $a\delta t_{i-1}$, we arrive at a point $\bar{x}_{i+1}$ which thus belongs to $\hat{x}_\varepsilon^i$. Starting at $\bar{x}_{i+1}^+$ for $\hat{x}_\varepsilon^i$ and x_{i+1}^+ for $\hat{x}_\varepsilon^{i+1}$, the two curves $\hat{x}_\varepsilon^i$ and $\hat{x}_\varepsilon^{i+1}$ follow the same movement: $\psi^{k'_{i+1}}$; ξ during time $a\delta t_{i+1}, \ldots$, etc.

Consequently, we may apply (9.159) (i.e. we are in the situation of ${}^0X = \bar{x}_{i+1}^+$; ${}^0Y = x_{i+1}^+$ and (9.157)) if the initial distance $d(\bar{x}_{i+1}^+, x_{i+1}^+)$ is such that:

$$d(\bar{x}_{i+1}^+, x_{i+1}^+) < \varepsilon_5 \,. \tag{9.174}$$

We know that this distance is of the order $\frac{\bar{j}_i + 1}{\sqrt{\omega}} + \delta t_i^- + \sqrt{\alpha_1}\, \delta t_i$. As $\Sigma \delta t_i^-$ goes to zero as well as $\Sigma \frac{\bar{j}_i + 1}{\sqrt{\omega}}$ and as α_1 may be chosen small following a choice on ε_1, (9.174) may be made to hold for all i.

Let us denote now by "making an operation" the fact of applying $\psi^{k_r'}$ and then describing a ξ-orbit piece of length $a\delta t_{r+1}$. (9.175)

Let us denote δ_i^{l+} the distance between the point of $\hat{x}_\varepsilon^i$ obtained after $l+1$ operations and the point of $\hat{x}_\varepsilon^{i+1}$ obtained after l operations. (9.176)

Equivalently, δ_i^{l+} is the distance between the point of $\hat{x}_\varepsilon^i$ obtained after having made l operations <u>starting at</u> $\bar{x}_{i+1}^+$ (which is the result after <u>one</u> operation) and the point of x_ε^{i+1} obtained after l operations.

Denote δ_i^{l-} the distance between the point of $\hat{x}_\varepsilon^i$ obtained after having made $l+1$ operations and then having applied $\psi^{k'_{i+l+1}}$ and the point of $\hat{x}_\varepsilon^{i+1}$ obtained after having made l operations and then having applied $\psi^{k'_{i+l+1}}$. (9.177)

Equivalently, $\overset{l}{\delta}{}_i^-$ is the distance between the point of $\hat{x}_\varepsilon^i$ obtained after having made l operations and then having applied $\psi^{k_{i+l+1}}$, starting at $\bar{x}_{i+1}^+$, and the point of $\hat{x}_\varepsilon^{i+1}$ obtained after l operations and then having applied $\psi^{k_{i+l+1}}$.

As (9.174) holds, we may apply (9.159) and we have:

$$\overset{l}{\delta}{}_i^+ < \bar{k}(1 + \alpha_6 s_0)^{\sum_{j=1}^{m_i} q_i} \prod_{j=1}^{m_i} (1 + \alpha_6 \delta_j)\, d(x_{i+1}^+, \bar{x}_{i+1}^+) \tag{9.178}$$

$$\overset{l}{\delta}{}_i^- < \bar{k}(1 + \alpha_6 s_0)^{\sum_{j=1}^{m_i} q_j} \prod_{j=1}^{m_i} (1 + \alpha_6 \delta_j)\, d(x_{i+1}^+, \bar{x}_{i+1}^+).$$

In (9.178), the q_j's are, following (9.158), the integer part of $a\delta t_{i+j+1}/s_0$ and δ_j is, also following (9.158), $a\delta t_{i+j+1} - q_j s_0$. m_i is the number of operations one has to make to use all k_j starting at k_{i+1}.

We have:

$$\sum_{j=1}^{m_i} q_j < K; \quad \prod_{j=1}^{m_i} (1 + \alpha_6 \delta_j) < K. \tag{9.179}$$

Indeed, $\sum_{j=1}^{m_i} q_j \le \frac{a}{s_0} \sum_{j=1}^{m_i} \delta t_{i+j+1} \le \frac{a}{s_0}$, and $\prod_{j=1}^{m_i} (1 + \alpha_6 \delta_j) < K$, as $\sum_{j=1}^{m_i} \delta_j < \sum_{j=1}^{m_i} \delta t_{i+1+j} < a$.

Thus, (9.178) gives with (9.179)

$$\overset{l}{\delta}{}_i^+ < K_1\, d(x_{i+1}^+, \bar{x}_{i+1}^+) \tag{9.180}$$

$$\overset{l}{\delta}{}_i^- < K_1\, d(x_{i+1}^+, \bar{x}_{i+1}^+).$$

We remark now that the curve $\hat{x}_\varepsilon$ defined by (9.153) is in fact the same as $\hat{x}_\varepsilon^0$ defined by (9.173) starting at $\hat{x}_0^+ = x_0^+$ (there is in fact a piece of curve on $\hat{x}_\varepsilon$, which runs along ξ from $\hat{x}_0^- = x_0^-$ to $\hat{x}_0^+ = x_0^+$ during time $a\delta t_0$, which is not on $\hat{x}_\varepsilon^0$).

We want to estimate:

$$\overset{l}{\Delta}{}^{-} = d(\hat{x}_l^-, x_l^-) \quad \text{and} \quad \overset{l}{\Delta}{}^{+} = d(\hat{x}_l^+, x_l^+). \tag{9.181}$$

From (9.153), we know that $\hat{x}_l^+$ is obtained from $\hat{x}_0^+ = \hat{x}_0^+$ by making l operations in the sense of (9.175).

$$\overset{1}{\hat{x}}{}_l^{+} \text{ the point obtained on } \hat{x}_\varepsilon^1 \text{ after } l-1 \text{ operations} \tag{9.182}$$

...

$$\overset{k}{\hat{x}}{}_l^{+} \text{ the point obtained on } \hat{x}_\varepsilon^k \text{ after } l-k \text{ operations}$$

...

$\overset{l}{\hat{x}}{}_l^{+}$ will then be the point obtained on $\hat{x}_\varepsilon^l$ after no operation, i.e. the starting point of $\hat{x}_\varepsilon^l$ which, by definition of $\hat{x}_\varepsilon^l$ (see (9.173)), is x_l^+. Thus we have:

$$\overset{l}{\Delta}{}^{+} = d(\hat{x}_l^+, x_l^+) < d(\hat{x}_l^+, \overset{1}{\hat{x}}{}_l^{+}) + d(\overset{1}{\hat{x}}{}_l^{+}, \overset{2}{\hat{x}}{}_l^{+}) + \dots + d(\overset{l-1}{\hat{x}}{}_l^{+}, x_l^+). \tag{9.183}$$

On another hand, the distance $d(\overset{k}{\hat{x}}{}_l^{+}, \overset{k+1}{\hat{x}}{}_l^{+})$ is, by definition, the distance between the point obtained on $\hat{x}_\varepsilon^k$ after $l-k$ operations and the point obtained on $\hat{x}_\varepsilon^{k+1}$ after $l-k-1$ operations. By (9.176), this distance is $\overset{l-k-1}{\delta}{}_k^{+}$.

$$d(\overset{k}{\hat{x}}{}_l^{+}, \overset{k+1}{\hat{x}}{}_l^{+}) = \overset{l-k-1}{\delta}{}_k^{+}. \tag{9.184}$$

Equalities (9.183) and (9.184) imply:

$$\overset{l}{\Delta}{}^{+} < \sum_{k=0}^{l-1} \overset{l-k-1}{\delta}{}_k^{+}. \tag{9.185}$$

By (9.180), we thus have:

$$\overset{l}{\Delta}{}^{+} < K_1 \sum_{k=0}^{l-1} d(x_{k+1}^+, \bar{x}_{k+1}^+). \tag{9.186}$$

$d(x_{k+1}^+, \bar{x}_{k+1}^+)$ is of order $\dfrac{\bar{j}_k+1}{\sqrt{\omega}} + \delta t_k^- + \sqrt{\alpha_1}\, \delta t_k$. Thus

$$\Delta^{l}_{+} = d(\hat{x}^{+}_{l}, x^{+}_{l}) < K_2\Big(\sum_{k=0}^{l} \frac{\bar{j}_k + 1}{\sqrt{\omega}} + \delta t^{-}_{k} + \sqrt{\alpha_1}\ \delta t_k\Big) \tag{9.187}$$

$$< K_2\Big(\frac{j_1+j_2}{\sqrt{\omega}} + \Sigma\ \delta t^{-}_{i} + \sqrt{\alpha_1}\Big) \rightarrow 0,$$

where j_1+j_2 is the total number of intervals and where we have upperbounded $\Sigma\ \delta t_k$ by 1; K_2 is an appropriate constant. Equation (9.187) proves (9.152)'.

We now prove (9.152).

We thus have:

$$d(\hat{x}^{+}_{l}, x^{+}_{l}) < K_2\left[\frac{j_1+j_2}{\sqrt{\omega}} + \Sigma\,\delta t^{-}_{i} + \sqrt{\alpha_1}\right] \quad \forall\, l = 0, 1, \ldots, n_0. \tag{9.188}$$

Similarly, by a similar argument on Δ^{l}_{-}, we have:

$$d(\hat{x}^{-}_{l}, x^{-}_{l}) < K_2\left[\frac{j_1+j_2}{\sqrt{\omega}} + \Sigma\,\delta t^{-}_{i} + \sqrt{\alpha_1}\right]. \tag{9.189}$$

In particular, after all these transformations are completed, we end up at $x^{-}_{n_0} = x^{-}_{0}$. (The curve (9.150) is closed, up to the $(\bar{x}^{+}_{i}, x^{+}_{i})$ jumps, as it is obtained through a straightening process, through Lemmas 19 and 20, from a closed curve.)

We thus have:

$$d(\hat{x}^{-}_{n_0}, x^{-}_{0}) = d(\hat{x}^{-}_{n_0}, x^{-}_{n_0}) < K_2\left[\frac{j_1+j_2}{\sqrt{\omega}} + \Sigma\ \delta t^{-}_{i} + \sqrt{\alpha_1}\right] \xrightarrow[\substack{\omega \to +\infty \\ \varepsilon_1 \to 0}]{} 0. \tag{9.190}$$

As $\hat{x}^{-}_{0} = x^{-}_{0}$ (by (9.153)), we have:

$$d(\hat{x}^{-}_{n_0}, \hat{x}^{-}_{0}) < K_2\left[\frac{j_1+j_2}{\sqrt{\omega}} + \Sigma\ \delta t^{-}_{i} + \sqrt{\alpha_1}\right] \xrightarrow[\substack{\omega \to +\infty \\ \varepsilon_1 \to 0}]{} 0, \tag{9.191}$$

which gives (9.152).

It remains only to prove that:

$$|\lambda_{k'_i}(\hat{x}^{+}_{i-1}) - 1| < \varepsilon \tag{9.192}$$

(i.e. goes to zero when ω goes to $+\infty$ and ε_1 to zero), which is obtained from (9.188) and from (A_3).

Indeed, by (9.188):

$$d(\hat{x}^+_{i-1}, x^+_{i-1}) < K_2\left[\frac{j_1+j_2}{\sqrt{\omega}} + \Sigma\, \delta t^-_i + \sqrt{\alpha_1}\right].$$

We can choose ω large enough so that $K_2\left[\frac{j_1+j_2}{\sqrt{\omega}} + \Sigma\, \delta t^-_i + \sqrt{\alpha_1}\right] < \varepsilon_6$. We then apply (A_3) with $x = \hat{x}^+_{i-1}$ and $y = x^+_{i-1}$.

We derive:

$$|\lambda_{k'_i}(\hat{x}^+_{i-1}) - \lambda_{k'_i}(x^+_{i-1})| < \alpha_5\, d(\hat{x}^+_{i-1}, x^+_{i-1}) \tag{9.193}$$

$$< K_2\alpha_5\left[\frac{j_1+j_2}{\sqrt{\omega}} + \Sigma\, \delta t^-_i + \sqrt{\alpha_1}\right] \quad \text{(by (9.188))}.$$

On the other hand, by (9.5) of Lemma 19, we have:

$$|\lambda_{k_1-k_2}(x_{\underset{\sim}{s}^+_2}) - 1| < \varepsilon(\varepsilon_1, \omega). \tag{9.5}$$

We have denoted (cf. (9.147)) $(x_{\underset{\sim}{s}^+_2})_{i-1} = x^+_{i-1}$ and $(k_1-k_2)_{i-1} = k'_i$. Thus,

with these notations, we have:

$$|\lambda_{k'_i}(x^+_{i-1}) - 1| < \varepsilon(\varepsilon_1, \omega). \tag{9.194}$$

Equations (9.193) and (9.194) imply:

$$|\lambda_{k'_i}(\hat{x}^+_{i-1}) - 1| < K_2\alpha_5\left[\frac{j_1+j_2}{\sqrt{\omega}} + \Sigma\, \delta t^-_i + \sqrt{\alpha_1}\right] + \varepsilon(\varepsilon_1, \omega) \xrightarrow[\substack{\omega\to+\infty \\ \varepsilon_1\to 0}]{} 0. \tag{9.195}$$

■

<u>The second approach for rearranging the curve is the following:</u>

We assume:

$(\mathfrak{A}_2)(x)$ $\begin{cases} (A_2) \text{ and } (A_3) \text{ are satisfied.} \\ \text{Moreover, we make a local hypothesis: for any sequence of} \\ \text{points of type } x_i^+ \text{ converging to } x \in M, x \text{ satisfies: there exists} \\ \quad \varepsilon_7 > 0 \text{ so that the equation} \\ \qquad |\lambda_{2k}(x) - 1| < \varepsilon_7; \quad k \in \mathbb{Z} \qquad (9.196) \\ \text{has only a finite number of solutions } q_0, \ldots, q_{r_0}. \end{cases}$

This set of hypotheses will be denoted $(\mathfrak{A}_2)(x)$. These results we prove under $(\mathfrak{A}_2)(x)$ are somewhat better than the ones proved under $(\mathfrak{A}_1)$. In particular, we clear up some properties of the characteristic exponents k_i'. So, it seems interesting to extend slightly hypothesis $(\mathfrak{A}_2)(x)$ as follows, so that it covers the fiber cases:

$(\mathfrak{A}_2)(x)$ $\begin{cases} \text{when } v \text{ is the vector field tangent to an } S^1\text{-fiberbundle } M, \\ \text{with } \psi^{2m} = Id, \; m \in \mathbb{N}, \text{ we modify } (9.196): \\ \qquad |\lambda_{2k}(x) - 1| < \varepsilon_7; \quad k \in \mathbb{Z}_m = \mathbb{Z}/m\mathbb{Z} \qquad (9.196)' \\ \text{has only a finite number of solutions } q_0, \ldots, q_{r_0}. \end{cases}$

Given a vector field in $\ker \alpha$, the previous hypothesis is natural: the equation $\lambda_{2k}(x) = 1$ defines, generically on the contact form α (i.e. on the contact forms of type $\lambda\alpha_0$; $\lambda \in C^\infty(M, \mathbb{R}^+)$) a hypersurface Γ_k, if k is not zero. The intersection of four such hypersurfaces is generically empty. Hence, the equation $\lambda_{2k}(x) = 1$, for a given x, will be satisfied for at most four distinct non zero k's.

In case v is not the vector field tangent to the fibers of an S^1-bundle, condition (9.196) is somewhat more constraining. It implies that 1 is not an accumulation point for the λ_{2k} when k runs in $\mathbb{Z}$.

Assume that $(\mathfrak{A}_2)(x)$ is satisfied.

Then we have:

<u>Theorem</u> $(\mathfrak{A}_2)(x)$: Under $(\mathfrak{A}_2)(x)$, there exists δ positive so that for j satisfying:

$$\sum_{r=1}^{j-1} \delta t_{i+r} < \delta , \qquad (9.197)$$

the characteristic summands:

$$k'_{i+1} ; k'_{i+1}+k'_{i+2} ; \dots ; k'_{i+1}+\dots+k'_{i+1} ; \dots ; k'_{i+1}+\dots+k'_{i+j} \qquad (9.198)$$

are in the set $\{2q_0, \dots, 2q_{r_0}\}$ (modulo m in the case of fiber bundles in S^1).

The curve (9.150) can be arranged from x_i^+ to $\bar{x}_{i+j}^+$ according to the process of theorem $(\mathcal{A}_1)$ with similar estimates where the constants depend only on $q_0, \dots, q_{r_0}$.

In particular, if j_0 is the Sup of the j's satisfying (9.197), the curve can be rearranged from x_i^+ to $\bar{x}_{i+j_0}^+$, i.e. it can be rearranged on a time Δt such that:

$$\Delta t \geq \delta t_{i+1} + \dots + \delta t_{i+j_0} \geq \inf(\delta, \tfrac{1}{2}) ; \qquad (9.198)'$$

δ being a constant not depending on x, $q_0, \dots, q_{r_0}$, but only on ε_7.

Proof: The whole proof of the theorem relies on the fact that the characteristic summands (9.198) are in $\{2q_0, \dots, 2q_{r_0}\}$ and that δ is a geometrical constant. Once this is proved, the rearrangement process is conducted exactly in the same way as before.

In order to prove that these characteristic summands are in $\{2q_0, \dots, 2q_{r_0}\}$, we use (9.150)'.

First we draw the piece of curve (9.150) starting from x_i^+.

(9.199)

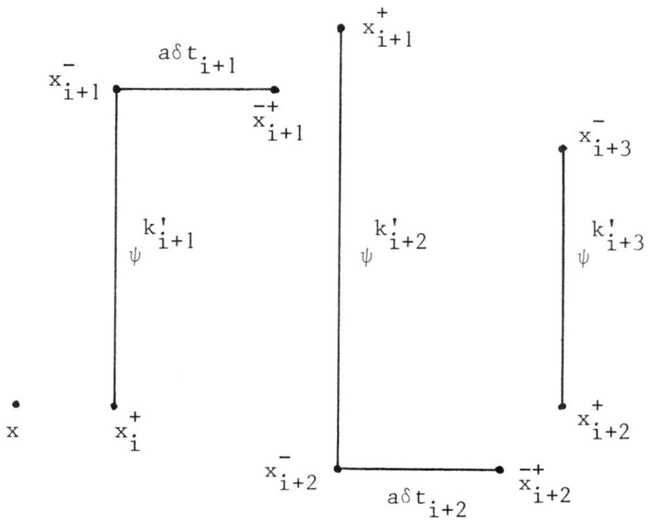

We consider the point $\psi^{\alpha_j}(x_i^+)$ where

$$\alpha_j = \sum_{s=1}^{j} k'_{i+s} \,. \tag{9.200}$$

We want to estimate:

$$\lambda_{\alpha_j}(x_i^+) \,. \tag{9.201}$$

Let:

$$\beta_k^j = \sum_{r=j-k+1}^{j} k'_{i+r} \; ; \quad \beta_0^j = 0. \tag{9.202}$$

Assume that:

$$a\,\delta t_{i+r} < \frac{\inf(\varepsilon_6, \varepsilon_5)}{2 \sup\limits_{x\in M} \| \xi \|} \; ; \; r = 1,\ldots, j-1. \tag{9.203}$$

We choose ε_1 small enough and ω large enough so that:

$$d(\bar{x}^{+}_{i+r}, x^{+}_{i+r}) < K\left[\frac{\bar{j}_{i+r}}{\sqrt{\omega}} + \delta t^{-}_{i+r} + \sqrt{\alpha_1}\, \delta t_{i+r}\right] \text{ (by Lemma 20)} \tag{9.204}$$

$$< \frac{\inf(\varepsilon_6, \varepsilon_5)}{2},$$

for r arbitrary.

With (9.203) and (9.204), we have:

$$\begin{aligned} d(x^{-}_{i+r}, x^{+}_{i+r}) &\leq d(x^{-}_{i+r}, \bar{x}^{+}_{i+r}) + d(\bar{x}^{+}_{i+r}, x^{+}_{i+r}) \\ &\leq a\delta t_{i+r} \operatorname{Sup}\|\xi\| + d(\bar{x}^{+}_{i+r}, x^{+}_{i+r}) \\ &< \inf(\varepsilon_6, \varepsilon_5)\ ; \qquad r = 1, \ldots, j-1. \end{aligned} \tag{9.205}$$

(A_2) and (A_3) may then be applied to $(x^{-}_{i+r}, x^{+}_{i+r})$, $r = 1, \ldots, j-1$.
Hence, for $k = 1, \ldots, j-1$:

$$\begin{aligned} \left|\lambda_{\beta^j_k}(x^{+}_{i+j-k}) - \lambda_{\beta^j_k}(x^{-}_{i+j-k})\right| &\leq \alpha_5 d(x^{+}_{i+j-k}, x^{-}_{i+j-k}) \\ &\leq \alpha_5 [d(x^{-}_{i+j-k}, \bar{x}^{+}_{i+j-k}) + d(\bar{x}_{i+j-k}, x^{+}_{i+j-k})] \\ &\leq \alpha_5\left[\operatorname*{Sup}_{x\in M}\|\xi\| a\delta t_{i+j-k} + K\left[\frac{\bar{j}_{i+j-k}}{\sqrt{\omega}} + \delta t^{-}_{i+j-k} + \sqrt{\alpha_1}\, \delta t_{i+j-k}\right]\right]. \end{aligned} \tag{9.206}$$

We set:

$$\delta^j_k = \alpha_5\left[\operatorname*{Sup}_{x\in M}\|\xi\| a\delta t_{i+j-k} + K\left[\frac{\bar{j}_{i+j-k}}{\sqrt{\omega}} + \delta t^{-}_{i+j-k} + \sqrt{\alpha_1}\, \delta t_{i+j-k}\right]\right]. \tag{9.207}$$

Thus we have:

$$\left|\lambda_{\beta^j_k}(x^{+}_{i+j-k}) - \lambda_{\beta^j_k}(x^{-}_{i+j-k})\right| \leq \delta^j_k . \tag{9.208}$$

On the other hand we have the relation:

$$\lambda_{\beta^j_k}(x^{-}_{i+j-k})\, \lambda_{k'_{i+j-k}}(x^{+}_{i+j-k-1}) = \lambda_{\beta^j_{k+1}}(x^{+}_{i+j-k-1}). \tag{9.209}$$

(9.209) is analogous to (9.49), which we already proved, if we notice that:

$\beta^j_{k+1} = \beta^j_k + k'_{i+j-k}$ and that $x^-_{i+j-k} = \psi^{k'_{i+j-k}}(x^+_{i+j-k-1})$. If we apply (9.209) with $k = k-1$, we obtain:

$$\lambda_{\beta^j_{k-1}}(x^-_{i+j-k+1})\,\lambda_{k'_{i+j-k+1}}(x^+_{i+j-k}) = \lambda_{\beta^j_k}(x^+_{i+j-k}). \tag{9.210}$$

Then, we replace, in (9.208), $\lambda_{\beta^j_k}(x^+_{i+j-k})$ by its value given in (9.210). We find:

$$\left|\lambda_{\beta^j_{k-1}}(x^-_{i+j-k+1}) - \frac{\lambda_{\beta^j_k}(x^-_{i+j-k})}{\lambda_{k'_{i+j-k+1}}(x^+_{i+j-k})}\right| \le \frac{\delta^j_k}{\lambda_{k'_{i+j-k+1}}(x^+_{i+j-k})} \tag{9.211}$$

Hence:

$$\left|\frac{\lambda_{\beta^j_{k-1}}(x^-_{i+j-k+1})}{\prod_{r=2}^{j-k}\lambda_{k'_{i+r}}(x^+_{i+r-1})} - \frac{\lambda_{\beta^j_k}(x^-_{i+j-k})}{\prod_{r=2}^{j-k+1}\lambda_{k'_{i+r}}(x^+_{i+r-1})}\right| \le \frac{\delta^j_k}{\prod_{r=2}^{j-k+1}\lambda_{k'_{i+r}}(x^+_{i+r-1})}\,; \tag{9.212}$$

$k = 1,\dots,j-1.$

Inequalities (9.212), for $k = 1,\dots,j-1$, imply by summing and eliminating:

$$\left|\frac{\lambda_{\beta^j_0}(x^-_{i+j})}{\prod_{r=2}^{j}\lambda_{k'_{i+r}}(x^+_{i+r-1})} - \frac{\lambda_{\beta^j_{j-1}}(x^-_{i+1})}{\lambda_{k'_{i+2}}(x^+_{i+1})}\right| = \left|\frac{\lambda_0(x^-_{i+j})}{\prod_{r=2}^{j}\lambda_{k'_{i+r}}(x^+_{i+r-1})} - \frac{\lambda_{\beta^j_{j-1}}(x^-_{i+1})}{\lambda_{k'_{i+2}}(x^+_{i+1})}\right| =$$

$$= \left|\frac{1}{\prod_{r=2}^{j}\lambda_{k'_{i+r}}(x^+_{i+r-1})} - \frac{\lambda_{\beta^j_{j-1}}(x^-_{i+1})}{\lambda_{k'_{i+2}}(x^+_{i+1})}\right| \le \sum_1^{j-1}\left[\frac{\delta^j_k}{\prod_{r=2}^{j-k+1}\lambda_{k'_{i+r}}(x^+_{i+r-1})}\right]. \tag{9.213}$$

Hence:

$$\left|\lambda_{\beta^j_{j-1}}(x^-_{i+1}) - \frac{1}{\prod_{r=3}^{j}\lambda_{k'_{i+r}}(x^+_{i+r-1})}\right| \le \sum_1^{j-1}\frac{\delta^j_k}{\prod_{r=3}^{j-k+1}\lambda_{k'_{i+r}}(x^+_{i+r-1})}. \tag{9.214}$$

By (9.210), we have:

$$\lambda_{\beta^j_{j-1}}(x^-_{i+1}) = \frac{\lambda_{\beta^j_j}(x^+_i)}{\lambda_{k'_{i+1}}(x^+_i)} . \qquad (9.215)$$

As $\beta^j_j = \sum_1^j k'_{i+r} = \alpha_j$, (9.215) gives:

$$\lambda_{\beta^j_{j-1}}(x^-_{i+1}) = \frac{\lambda_{\alpha_j}(x^+_i)}{\lambda_{k'_{i+1}}(x^+_i)} , \qquad (9.216)$$

hence:

$$\left| \lambda_{\alpha_j}(x^+_i) - \frac{\lambda_{k'_{i+1}}(x^+_i)}{\prod_{r=3}^{j} \lambda_{k'_{i+1}}(x^+_{i+r-1})} \right| \leq \sum_1^{j-1} \left[\frac{\delta^j_k}{\prod_{r=3}^{j-k+1} \lambda_{k'_{i+r}}(x^+_{i+r-1})} \right] \lambda_{k'_{i+1}}(x^+_i) . \qquad (9.217)$$

Equation (9.217) implies, for ε_1 small enough and ω large enough:

$$|\lambda_{\alpha_j}(x^+_i) - 1| \leq \frac{\varepsilon_7}{2} + 2\alpha_5 \sup_{x \in M} \|\xi\| \, a\left(\sum_1^{j-1} \delta t_{i+j-k}\right) . \qquad (9.218)$$

Indeed, by (9.150)', all the products occurring in (9.217) go uniformly to 1, and also $\lambda_{k'_{i+1}}(x^+_i)$.

On the other hand:

$$\sum_1^{j-1} \delta^j_k - \alpha_5 \left[\sup_{x \in M} \|\xi\| \sum_1^{j-1} a \delta t_{i+j-k} \right] \qquad (9.219)$$

$$= K \left[\sum_1^{j-1} \frac{\bar{j}_{i+j-k}}{\sqrt{\omega}} + \delta t^-_{i+j-k} + \sqrt{\alpha_1}\, \delta t_{i+j-k} \right] \to 0$$

as ε_1 goes to zero and ω to infinity.

Inequality (9.218) easily follows.

From (9.218), we deduce that, if:

$$2a\alpha_5 \operatorname*{Sup}_{x\in M} \|\xi\| \sum_1^{j-1} \delta t_{i+j-k} < \frac{\varepsilon_7}{6}, \qquad (9.219)'$$

we have:

$$|\lambda_{\alpha_j}(x_i^+) - 1| \leq \frac{2\varepsilon_7}{3}. \qquad (9.220)$$

We take ε_1 small enough and ω large enough so that:

$$d(x, x_i^+) < \inf\left[\frac{\varepsilon_7}{3\alpha_5}, \varepsilon_6\right]. \qquad (9.221)$$

Under (A_3), we then have:

$$|\lambda_{\alpha_j}(x) - \lambda_{\alpha_j}(x_i^+)| < \alpha_5 d(x, x_i^+) < \frac{\varepsilon_7}{3}. \qquad (9.222)$$

Equations (9.222) and (9.221) imply:

$$|\lambda_{\alpha_j}(x) - 1| < \varepsilon_7. \qquad (9.223)$$

As x satisfies $(\mathfrak{A}_3)(x)$ and as α_j is even because k'_{i+r} is even, we derive:

$$\alpha_j \in \{2q_0, \ldots, 2q_{r_0}\} \qquad (9.224)$$

which is the desired result. We here remark that if (9.219)' and (9.203) are verified with j, they are verified with any $k \leq j$. Thus, (9.219)' and (9.203) imply that $\alpha_k \in \{2q_0, \ldots, 2q_{r_0}\}$ for $k \leq j$.

Two conditions were needed to prove (9.224):

$$a\delta t_{i+r} < \frac{\inf(\varepsilon_6, \varepsilon_5)}{2 \operatorname*{Sup}_{x\in M}\|\xi\|}; \quad r = 1, \ldots, j-1. \qquad (9.202)$$

$$a\sum_1^{j-1} \delta t_{i+j-k} < \frac{\varepsilon_7}{6\alpha_5 \operatorname*{Sup}_{x\in M}\|\xi\|}. \qquad (9.219)'$$

These two conditions (9.203) and (9.219)' can be reduced to the following one:

$$a \sum_{1}^{j-1} \delta t_{i+j-k} < \inf\left(\frac{\varepsilon_6}{2 \operatorname{Sup} \|\xi\|}, \frac{\varepsilon_5}{2 \operatorname{Sup} \|\xi\|}, \frac{\varepsilon_7}{6\alpha_5 \operatorname{Sup} \|\xi\|} \right) . \qquad (9.225)$$

We set:

$$\delta = \frac{1}{a_1} \inf\left(\frac{\varepsilon_6}{2 \operatorname{Sup} \|\xi\|}, \frac{\varepsilon_5}{2 \operatorname{Sup} \|\xi\|}, \frac{\varepsilon_7}{6\alpha_5 \operatorname{Sup} \|\xi\|} \right) , \qquad (9.226)$$

where a is upperbounded by a_1.

As soon as:

$$\sum_{1}^{j-1} \delta t_{i+j-k} < \delta , \qquad (9.227)$$

condition (9.225) is satisfied. Consequently, the characteristic summands $k'_{i+1}, \ldots, k'_{i+1}+\ldots+k'_{i+k}, \ldots, k'_{i+1}+\ldots+k'_{i+j}$ are in $\{2q_0, \ldots, 2q_{r_0}\}$.

This implies that under (9.225), the following quantities:

$$\mathcal{J}_r = \sum_{i+r+1}^{i+j} k'_{i+s} \qquad (9.228)$$

are in the set denoted F

$$\mathcal{J}_r \in F = \{2(q_i - q_j) ; \quad i, j \quad (0, \ldots, r_0)\} . \qquad (9.229)$$

These quantities are similar to those which we looked at in the proof of theorem $(\mathcal{A}_1)$ and which were defined in (9.160). In the proof of theorem $(\mathcal{A}_1)$, we applied the hypothesis of local equicontinuity of the $D\psi^{2k}(\xi)$ to the one-parameter groups corresponding to $2k = \mathcal{J}_r$.

Here, this local equicontinuity hypothesis is automatically satisfied as the $\mathcal{J}_r$ by (9.229) are in finite number; hence the possibility of rearranging the curve up to $\bar{x}^{+}_{i+j}$.

Now, let $j_0 = \operatorname{Sup}\{j$ satisfying (9.227)$\}$.

In $j_0 + 1$, if this index exists in (9.150), we have:

$$\sum_{1}^{j_0} \delta t_{i+r} \geq \delta ; \qquad (9.230)$$

which proves (9.199).

If $j_0 + 1$ does not exist, it means that with the index j_0, we have exhausted all the concentration intervals.

As the complementarity measure of these intervals goes to zero when $\varepsilon_1 \to 0$ and $\omega \to +\infty$, we then have, for ε_1 small enough and ω large enough:

$$1 \leftarrow \sum_{1}^{j_0} \delta t_{i+r} \geq \tfrac{1}{2}. \tag{9.231}$$

which proves (9.199) and completes the proof of theorem $(\mathfrak{A}_2)$. ■

Once the curve has been straightened up, we seek now to understand, under hypothesis $(\mathfrak{A}_2)(x)$, what happens locally. For this purpose, we introduce the following definition:

Definition 8: The situation is said to be generic at x if:

1) the equations $\lambda_{2k}(x) = 1$ have at most one non zero solution k_0;
2) on a neighbourhood of x, the equation $\lambda_{2k_0}(y) = 1$ defines a hypersurface Σ, to which x belongs, if a solution to this equation exists;
3) $\xi(x)$ is non tangent to Σ at x and $\xi(\psi^{2k_0}(x))$ is not tangent to $\psi^{2k_0}(\Sigma)$ at $\psi^{2k_0}(x)$.

It can be proved that such a situation is indeed generical in the following sense: with perturbing α in a neighbourhood of x, in the contact forms $\lambda\alpha$, we may realize the situation of Definition 8, with a λ as close to 1, in the C^∞ sense, as wanted.

We then have the following theorem:

Theorem 7: Let x_i^+, corresponding to a non zero k'_{i+1} and converging to x which satisfies $(\mathfrak{A}_2)(x)$, be such that the situation is generic at x.

Then there exists a non zero k_0 such that $\lambda_{2k_0}(x) = 1$ and

① $\exists \delta_1 > 0$ such that for any j satisfying

$$\sum_{r=1}^{j-1} \delta t_{i+r} < \delta_1, \tag{9.232}$$

we have:

$$k'_{i+1} = 2k_0;\ k'_{i+2} = -2k_0, \dots, k'_{i+k} = (-1)^{k+1} \cdot 2k_0, \dots, k'_{i+j} = (-1)^{j+1} \cdot 2k_0 \tag{9.223}$$

(modulo m in case we are dealing with S^1-fiber bundles).

② Let j_0 be the supremum of the j's satisfying (9.232). We have:

$$\sum_{r=1}^{j_0-1} \delta t_{i+r} \to 0 \quad \text{when } \varepsilon_1 \to 0 \text{ and } \omega \to +\infty . \tag{9.234}$$

③ The straightened curve $\hat{x}$, after having possibly oscillated from $r = 1$ to $j_0 - 1$ around a piece tangent to v of type $[x, \psi^{2k_0}(x)]$ (forwards and backwards), takes the ξ-direction. Thus, in the geometric sense of convergence, the curve $\hat{x}$ converges to one of the following curves, in a neighbourhood of x:

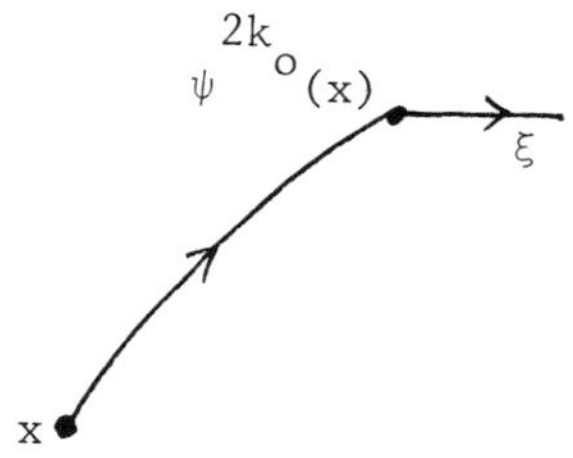

I

limit movement:
the piece $[x, \psi^{2k_0}(x)]$ can be described infinitely many times backwards and forwards "before a final movement" where it is described from x to $\psi^{2k_0}(x)$. Thereafter, the curve is tangent to ξ during a strictly positive time.

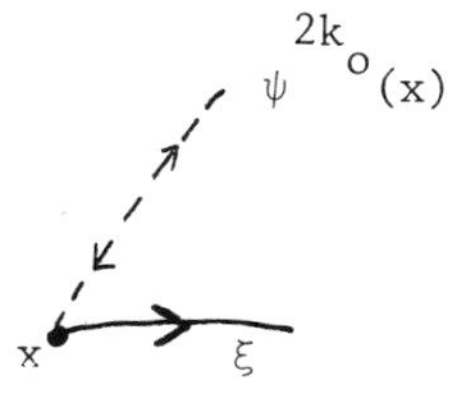

II

limit movement:
the piece $[x, \psi^{2k_0}(x)]$ can be described infinitely many times. The resulting movement is trivial: we came back from x to x.
Thereafter, the curve is tangent to ξ during a strictly positive time.

(See Corollary 2 of Theorem 8 to discriminate between I and II.)

Remark: The hypothesis $k'_{i+1} \neq 0$ is not constraining. Indeed, when $k'_{i+1} = 0$, there are no pointwise oscillations along v, starting at x_i^+; hence the straightened up curve $\hat{x}$ is tangent to ξ in a neighbourhood of x_i^+ and x_i^+ is not a remarkable point.

Proof of Theorem 7: Let $\varepsilon_7 > 0$ be given satisfying $(\mathcal{a}_2)(x)$. Then the equations $|\lambda_{2k}(x) - 1| < \varepsilon_7$ admits at most the solution $q_0, \ldots, q_{r_0}$.

k_0 is one of these integers as $\lambda_{2k_0}(x) = 1$, and also 0 is among these integers.

Consider $\{q_0, \ldots, q_{r_0}\} - \{k_0, 0\} = \{\tilde{q}_0, \ldots, \tilde{q}_{r_0-2}\}$. On this set, as we assume the situation to be generic at x, we have:

$$|\lambda_{2\tilde{q}_i}(x) - 1| > 0.$$

Then let:

$$\varepsilon_8 = \inf\left(\frac{\varepsilon_7}{2}, \tfrac{1}{2}\,|\lambda_{2\tilde{q}_i}(x) - 1|\right) > 0.$$

As soon as $|\lambda_{2k}(x) - 1| < \varepsilon_8$, k has to be equal to k_0 or k has to be zero.

We can even do better, using the continuity of $\lambda_{2q_0}, \ldots, \lambda_{2q_{r_0}}$ at x and find ε_8 such that:

$$|\lambda_{2k}(y) - 1| < \varepsilon_8 \Rightarrow k = k_0 \quad \text{or} \quad k = 0$$

for y in an x-neighbourhood.

With this ε_8, we follow the proof of Theorem $(\mathcal{a}_2)(x)$.

If $\alpha_j = \sum_{r=1}^{j-1} k'_{i+r}$, (9.223) holds with ε_8 instead of ε_7 (we assume (9.203) and (9.219) to hold with ε_8); and also (9.225) will hold with an ε_8.

I. e.

$$a \sum_1^{j-1} \delta t_{i+j-k} < \inf\left(\frac{\varepsilon_6}{2\,\operatorname{Sup}\|\xi\|}, \frac{\varepsilon_5}{2\,\operatorname{Sup}\|\xi\|}, \frac{\varepsilon_8}{6\alpha_5\,\operatorname{Sup}\|\xi\|}\right) = \bar{\delta} \qquad (9.225)'$$

As previously, setting

$$\delta_1 = \frac{\bar{\delta}}{a_1}, \qquad (9.226)'$$

if

$$\sum_{1}^{j-1} \delta t_{i+r} < \delta_1, \qquad (9.227)'$$

equation (9.223) holds with ε_8 instead of ε_7.

Hence

$$|\lambda_{\alpha_j}(x) - 1| < \varepsilon_8, \qquad (9.222)'$$

hence, with the choice of ε_8 we have completed, $\alpha_j = 0$ or $\alpha_j = 2k_0$.

But if $\sum_{1}^{j-1} \delta t_{i+r} < \delta_1$, a fortiori $\sum_{1}^{k-1} \delta t_{i+r} < \delta_1$ for $k \leq j$. Hence $\alpha_k = 0$ or $\alpha_k = 2k_0$ for $k \leq j$.

This implies that $k'_{i+1} = 2k_0$; $k'_{i+2} = -2k_0$, which proves ①.

To prove ②, we use (9.150)'.

By (9.150)', we know that:

$$\left| \prod_{r=0}^{j_0-1} \lambda_{k'_{i+r+1}}(x^+_{i+r}) - 1 \right| \to 0 \qquad (9.235)$$

$$\left| \prod_{r=0}^{j_0-1} \lambda_{k'_{i+r}}(x^+_{i+r}) - 1 \right| \to 0. \qquad (9.236)$$

We will restrict the value of δ_1 found in ① during the proof, so that our curve remains in a neighbourhood of a well-chosen piece $[x, \psi^{2k_0}(x)]$.

First, we simplify (9.235) using ①. Estimate (9.235) then becomes:

$$\left| \prod_{r=1}^{j_0-1} \lambda_{(-1)^{r+2}.2k_0}(x^+_{i+r}) - 1 \right| \to 0. \qquad (9.237)$$

$$\prod_{r-1}^{j_0-1} \lambda_{(-1)^{r+2}.2k_0}(x^+_{i+r})$$

may as well be seen as the product of two by two consecutive terms, one term possibly remaining.

$$\lambda_{-2k_0}(x^+_{i+1})\,\lambda_{2k_0}(x^+_{i+2})\,;\ldots;\lambda_{-2k_0}(x^+_{i+2k+1})\,\lambda_{2k_0}(x^+_{i+2k+2})\,;\ldots \tag{9.238}$$

We want to estimate these products in terms of δt_{i+2k+2}.
A similar process could be done with (9.236) with

$$\lambda_{2k_0}(x^+_i)\,\lambda_{-2k_0}(x^+_{i+1})\,;\ldots;\lambda_{2k_0}(x^+_{i+2k})\,\lambda_{-2k_0}(x^+_{i+2k+1})\,;\ldots \tag{9.239}$$

We consider the case (9.238). The other case is similar.
First, we have:

$$\left|\lambda_{2k_0}(x^+_{i+2k+2}) - \lambda_{2k_0}(\bar{x}^+_{i+2k+2})\right| < K\left[\frac{\bar{j}_{i+2k+2}}{\sqrt{\omega}} + \delta t^-_{i+2k+2} + \sqrt{\alpha_1}\,\delta t_{i+2k+2}\right]. \tag{9.240}$$

For this, we use Lemma 20, which tells us that the distance from $\bar{x}^+_{i+2k+2}$ to x^+_{i+2k+2} is of the order $\frac{\bar{j}_{i+2k+2}}{\sqrt{\omega}} + \delta t^-_{i+2k+2} + \sqrt{\alpha_1}\,\delta t_{i+2k+2}$ and we make use of the differentiability of λ_{2k_0}.

On the other hand, we have:

$$\lambda_{-2k_0}(x^+_{i+2k+1}) = \frac{1}{\lambda_{2k_0}(x^-_{i+2k+2})}. \tag{9.241}$$

Indeed, we know that $x^-_{i+2k+2} = \psi^{k'_{i+2k+2}}(x^+_{i+2k+1}) = \psi^{-2k_0}(x^+_{i+2k+1})$. Thus (9.236) amounts to proving:

$$\lambda_i(x)\;\lambda_{-i}(\psi^i(x)) = 1 = \lambda_0(x) \tag{9.242}$$

which is a relation similar to (9.48), as $i + (-i) = 0$, which we already proved.

Finally, we estimate:

$$\lambda_{2k_0}(\bar{x}^+_{i+2k+2}) - \lambda_{2k_0}(x^-_{i+2k+2}). \tag{9.243}$$

We know that we are going from x^-_{i+2k+2} to $\bar{x}^+_{i+2k+2}$ through ξ during time $a\delta t_{i+2k+2}$. Restricting δ_1 if necessary, we can make an expansion of (9.243) at the second order which gives:

$$|\lambda_{2k_0}(x^+_{i+2k+2})-\lambda_{2k_0}(x^-_{i+2k+2})-(d\lambda_{2k_0}((x^-_{i+2k+2})))\,a\delta t_{i+2k+2}| \tag{9.244}$$

$$\leq C\,a^2(\delta t_{i+2k+2})^2;$$

$i+2k+2 \leq i+j_0-1,$

where C is a uniform constant.

Indeed, by hypothesis we have:

$$\sum_1^{j_0-1} \delta t_{i+r} < \delta_1 \quad \text{hence} \quad \delta t_{i+2k+2} < \delta_1 \tag{9.245}$$

$i+2k+2 \leq i+j_0-1.$

Thus, if δ_1 is small enough, the distance between $\bar{x}^+_{i+2k+1}$ and x^-_{i+2k+2} will be small and upperbounded by $a_1\delta_1$ Sup $\|\xi\|$ and a uniform bounded expansion is possible. We estimate now:

$$d(x, x^-_{i+2k+2}). \tag{9.246}$$

We recall that we have the following relations:

$$x^-_{i+r} = \psi^{k'_{i+r}}(x^+_{i+r-1}) = \psi^{(-1)^{r+1}2k_0}(x^+_{i+r-1}). \tag{9.247}$$

Let:

$$\tilde{\bar{x}}^+_{i+2r+1} = \psi^{-2k_0}(\bar{x}^+_{i+2r+1})\,;\quad \tilde{x}^+_{i+2r+1} = \psi^{-2k_0}(x^+_{i+2r+1}) \tag{9.248}$$

$$\tilde{x}^-_{i+2r+1} = \psi^{-2k_0}(x^-_{i+2r+1}).$$

By (9.247) we have:

$$\tilde{x}^-_{i+2r+1} = x^+_{i+2r} \quad \text{as} \quad \psi^{2k_0}(x^+_{i+2r}) \underset{(9.241)}{=\!=} x^-_{i+2r+1} \tag{9.249}$$

and

$$\tilde{x}^{+}_{i+2r-1} = x^{-}_{i+2r+2} \quad \text{as} \quad x^{-}_{i+2r+2} \underset{(9.241)}{=\!=\!=} \psi^{-2k_0}(x^{+}_{i+2r+1})$$

To make the notations homogeneous, we set:

$$\tilde{\bar{x}}^{+}_{i+2r} = \bar{x}^{+}_{i+2r}; \quad \tilde{x}^{+}_{i+2r} = x^{+}_{i+2r}; \quad \tilde{x}^{-}_{i+2r} = x^{-}_{i+2r} , \tag{9.250}$$

remark that with these notations we have:

$$\tilde{x}^{-}_{i+2r+1} = \tilde{x}^{+}_{i+2r} ; \quad \tilde{x}^{+}_{i+2r+1} = \tilde{x}^{-}_{i+2r+2} . \tag{9.250)'$$

Thus $\tilde{x}^{-}_{i+s+1} = \tilde{x}^{+}_{i+s}$.

We have the following inequalities:

$$d(\tilde{\bar{x}}^{+}_{i+s}, \tilde{x}^{+}_{i+s}) < K\left[\frac{\bar{j}_{i+s}}{\sqrt{\omega}} + \delta t^{-}_{i+s} + \sqrt{\alpha_1}\, \delta t_{i+s}\right] . \tag{9.251}$$

Indeed, if s is even, $\tilde{\bar{x}}^{+}_{i+s} = \bar{x}^{+}_{i+s}$ and $\tilde{x}^{+}_{i+s} = x^{+}_{i+s}$. In this case (9.251) is given by Lemma 20.

If s is odd, we have:

$$d(\tilde{\bar{x}}^{+}_{i+s}, \tilde{x}^{+}_{i+s}) = d(\psi^{-2k_0}(\bar{x}^{+}_{i+s}), \psi^{-2k_0}(x^{+}_{i+s})) \le C\, d(\bar{x}^{+}_{i+s}, x^{+}_{i+s}) \tag{9.252}$$

(for instance C upperbounds $\|D\psi^{-2k_0}\|$) and Lemma 20 gives us (9.245) again.

In the same way, we have:

$$d(\tilde{x}^{-}_{i+s}, \tilde{\bar{x}}^{+}_{i+s}) \le a_1 C_1 \delta t_{i+s} . \tag{9.253}$$

Indeed, if s is even, we have $d(\tilde{x}^{-}_{i+s}, \tilde{\bar{x}}^{+}_{i+s}) = d(x^{-}_{i+s}, \bar{x}^{+}_{i+s})$. As we go from x^{-}_{i+s} to $\bar{x}^{+}_{i+s}$ through ξ during time $a\delta t_{i+1}$, we have an inequality of type (9.253). If s is odd, we have:

$$d(\tilde{x}^{-}_{i+s}, \tilde{\bar{x}}^{+}_{i+s}) = d(\psi^{-2k_0}(\bar{x}^{+}_{i+s}), \psi^{-2k_0}(x^{-}_{i+s})) \le C d(\bar{x}^{+}_{i+s}, x^{-}_{i+s}) \tag{9.254}$$

and we are again in the previous situation.

With these estimates, we have:

$$d(x,x^-_{i+2k+2}) \le d(x,x^+_i) + d(x^+_i,\tilde{\tilde{x}}^+_{i+1}) + d(\tilde{\tilde{x}}^+_{i+1},\tilde{x}^+_{i+1}) \tag{9.255}$$
$$+ [d(\tilde{x}^+_{i+1},\tilde{\tilde{x}}^+_{i+2}) + d(\tilde{\tilde{x}}^+_{i+2},\tilde{x}^+_{i+2})] + \dots$$
$$+ \dots + [d(\tilde{x}^+_{i+2k},\tilde{\tilde{x}}^+_{i+2k+1}) + d(\tilde{\tilde{x}}^+_{i+2k+1},\tilde{x}^+_{i+2k+1})]$$
$$+ d(\tilde{x}^+_{i+2k+1},x^-_{i+2k+2}).$$

We decided to denote $x^+_i = \tilde{x}^+_i$ and $x^-_{i+2k+2} = \tilde{x}^-_{i+2k+2}$.
In particular, by (9.250)', we have:

$$d(\tilde{x}^+_{i+2k+1},x^-_{i+2k+2}) = d(\tilde{x}^+_{i+2k+1},\tilde{x}^-_{i+2k+2}) = 0. \tag{9.256}$$

On the other hand, we have:

$$d(\tilde{\tilde{x}}^+_{i+s},\tilde{x}^+_{i+s}) \le K\,\frac{\bar{j}_{i+s}}{\sqrt{\omega}} + \delta t^-_{i+s} + \sqrt{\alpha_1}\,\delta t_{i+s} \quad \text{(by (9.251))} \tag{9.257}$$

Finally, we have:

$$d(\tilde{x}^-_{i+s},\tilde{\tilde{x}}^+_{i+s+1}) = d(\tilde{x}^-_{i+s+1},\tilde{\tilde{x}}^+_{i+s+1}) \quad \text{(by (9.250)')} \tag{9.258}$$
$$\le C_1 a_1 \,\delta t_{i+s+1} \qquad \text{(by (9.252))}.$$

Equations (9.256), (9.257), (9.258) and the fact that $x^+_i = \tilde{x}^+_i$ then imply:

$$d(x,x^-_{i+2k+2}) \le d(x,x^+_i) + C_1 a_1 \sum_{s=1}^{2k+1} \delta t_{i+s} + K\left[\sum_{s=1}^{2k+1} \frac{\bar{j}_{i+s}}{\sqrt{\omega}} + \delta t^-_{i+s} + \sqrt{\alpha_1}\,\delta t_{i+s}\right]$$

$$\le d(x,x^+_i) + (C_1 a_1 + \sqrt{\alpha_1}\,K) \sum_{s=1}^{2k+1} \delta t_{i+s} + K\left[\frac{j_1+j_2}{\sqrt{\omega}} + (\Sigma\,\delta t^-_{i+s})\right]. \tag{9.259}$$

We know that $\frac{j_1+j_2}{\sqrt{\omega}}$ goes to zero, $\Sigma\,\delta t^-_{i+s}$ goes to zero also. We know also that $d(x,x^+_i)$ goes to zero.

We then deduce that if we take δ_1 small enough and if $i+2k+2 \le i+j_0-1$,

x^-_{i+2k+2} is an arbitrary neighbourhood of x, uniformly on every k such that $i+2k+2 \leq i+j_0-1$. (9.260)

By the genericity hypothesis, we have:

$$d\lambda_{2k_0}(\xi(x)) \neq 0 \tag{9.261}$$

as $\xi(x)$ is not tangent to Σ.

Thus with δ_1 small enough, and using (9.260), we have:

$|d\lambda_{2k_0}(\xi(x^-_{i+2k+2}))| \geq \theta > 0$ and $d\lambda_{2k_0}(\xi(x^-_{i+2k+2}))$ is of the sign of $d\lambda_{2k_0}(\xi(x))$ for every k such that $i+2k+2 \leq i+j_0-1$. (9.262)

We then consider a term of (9.218):

$$\lambda_{-2k_0}(x^+_{i+2k+1})\,\lambda_{k_0}(x^+_{i+2k+2}). \tag{9.263}$$

We have, by (9.241):

$$\lambda_{-2k_0}(x^+_{i+2k+1})\,\lambda_{2k_0}(x^+_{i+2k+2}) = \frac{\lambda_{2k_0}(x^+_{i+2k+2})}{\lambda_{2k_0}(x^-_{i+2k+2})}. \tag{9.264}$$

$$\lambda_{-2k_0}(x^+_{i+2k+1})\,\lambda_{2k_0}(x^+_{i+2k+2}) = \tag{9.265}$$

$$\frac{\lambda_{2k_0}(x^-_{i+2k+2}) + d\lambda_{2k_0}(\xi(x^-_{i+2k+2}))\,\delta t_{i+2k+2}}{\lambda_{2k_0}(x^-_{i+2k+2})} + \delta_k,$$

$$\delta_k = \frac{1}{\lambda_{2k_0}(x^-_{i+2k+2})} \tag{9.265)'$$

$$\times \left[\lambda_{2k_0}(x^+_{i+2k+2}) - \lambda_{2k_0}(x^-_{i+2k+2}) - d\lambda_{2k_0}(\xi(x^-_{i+2k+2}))\,a\,\delta t_{i+2k+2}\right].$$

We estimate:

$$|\delta_k| \le \frac{1}{\lambda_{2k_0}(x^-_{i+2k+2})} \Big[|\lambda_{2k_0}(x^+_{i+2k+2}) - \lambda_{2k_0}(\bar{x}^+_{i+2k+2})| + \tag{9.266}$$

$$+ |\lambda_{2k_0}(\bar{x}^+_{i+2k+2}) - \lambda_{2k_0}(x^-_{i+2k+2}) - d\lambda_{2k_0}(\xi(x^-_{i+2k+2}) a \delta t_{i+2k+2}| \Big]$$

by (9.240) and (9.244) $\Big\} \le \dfrac{Ca^2(\delta t_{i+2k+2})^2}{\lambda_{2k_0}(x^-_{i+2k+2})} + \dfrac{K}{\lambda_{2k_0}(x^-_{i+2k+2})} \left[\dfrac{\bar{j}_{i+2k+2}}{\sqrt{\omega}} + \delta t^-_{i+2k+2} + \sqrt{\alpha_1}\, \delta t_{i+2k+2} \right]$

by (A_5) $\le \dfrac{1}{\alpha_3} \left[Ca^2(\delta t_{i+2k+2})^2 + K \left[\dfrac{\bar{j}_{i+2k+2}}{\sqrt{\omega}} + \delta t^-_{i+2k+2} + \sqrt{\alpha_1}\, \delta t_{i+2k+2} \right] \right].$

It follows, by (9.265)':

$$\lambda_{-2k_0}(x^+_{i+2k+1}) \lambda_{2k_0}(x^+_{i+2k+2}) = 1 + \frac{d\lambda_{2k_0}(\xi(x^-_{i+2k+2})) a \delta t_{i+2k+2}}{\lambda_{2k_0}(x^-_{i+2k+2})} + \delta_k, \tag{9.267}$$

with

$$|\delta_k| \le \frac{1}{\alpha_3} \left[Ca^2(\delta t_{i+2k+2})^2 + K \left[\frac{\bar{j}_{i+2k+2}}{\sqrt{\omega}} + \delta t^-_{i+2k+2} + \sqrt{\alpha_1}\, \delta t_{i+2k+2} \right] \right]. \tag{9.268}$$

We denote:

$$\gamma_k = \lambda_{-2k_0}(x^+_{i+2k+1}) \lambda_{2k_0}(x^+_{i+2k+2}); \quad \theta_k = K \left[\frac{\bar{j}_{i+2k+2}}{\sqrt{\omega}} + \delta t^-_{i+2k+2} \right]. \tag{9.269}$$

We thus have:

$$\gamma_k \le 1 + \frac{d\lambda_{2k_0}(\xi(x^-_{i+2k+2})) a \delta t_{i+2k+2}}{\lambda_{2k_0}(x^-_{i+2k+2})} + \tag{9.270}$$

$$+ \frac{1}{\alpha_3} \left[Ca^2(\delta t_{i+2k+2})^2 + K \left[\frac{\bar{j}_{i+2k+2}}{\sqrt{\omega}} + \delta t^-_{i+2k+2} + \sqrt{\alpha_1}\, \delta t_{i+2k+2} \right] \right];$$

thus:

$$\frac{\gamma_k}{1+K\left[\frac{\bar{j}_{i+2k+2}}{\sqrt{\omega}} + \delta t^-_{i+2k+2}\right]} \leq \tag{9.271}$$

$$1 + \frac{a\delta t_{i+2k+2}}{1+\theta_k}\left[\frac{d\lambda_{k_0}(\xi)}{\lambda_{2k_0}(x^-_{i+2k+2})} + \frac{Ca\,\delta t_{i+2k+2}}{\alpha_3} + \frac{1}{a_0}\sqrt{\alpha_1}\right],$$

i.e.

$$\frac{\gamma_k}{1+\theta_k} \leq 1 + \frac{a\delta t_{i+2k+2}}{1+\theta_k}\left[\frac{d\lambda_{2k_0}(\xi)}{\lambda_{2k_0}(x^-_{i+2k+2})} + \frac{Ca\,\delta t_{i+2k+2}}{\alpha_3} + \frac{1}{a_0}\sqrt{\alpha_1}\right].$$

In the same way:

$$\frac{\gamma_k}{1-\theta_k} \geq 1 + \frac{a\delta t_{i+2k+2}}{1-\theta_k}\left[\frac{d\lambda_{2k_0}(\xi)}{\lambda_{2k_0}(x^-_{i+2k+2})} - \frac{Ca\,\delta t_{i+2k+2}}{\alpha_3} - \frac{1}{a_0}\sqrt{\alpha_1}\right]. \tag{9.272}$$

Now we use (9.262).

By (9.262) and (A_3), we know that:

$$\frac{|d\lambda_{2k_0}(\xi(x^-_{i+2k+2}))|}{|\lambda_{2k_0}(x^-_{i+2k+2})|} \geq \frac{\theta}{\alpha_4}, \tag{9.273}$$

for $i+2k+2 \leq j_0-1+i$ and δ_1 small enough.

If δ_1 and ε_1 are small enough, we have:

$$\left[\frac{Ca\,\delta t_{i+2k+2}}{\alpha_3} + \frac{1}{a_0}\sqrt{\alpha_1}\right] \leq \frac{\theta}{2\alpha_4}. \tag{9.274}$$

Indeed, α_1 goes to zero with ε_1 and $\delta t_{i+2k+2} < \delta_1$ as $i+2k+2 \leq i+j_0-1$. (In fact $\sum_{r=1}^{j_0-1} \delta t_{i+r} < \delta_1$.)

To end, we may ensure:

$$\frac{1}{1-\theta_k} \geq \tfrac{1}{2}; \quad \frac{1}{1+\theta_k} \geq \tfrac{1}{2} \tag{9.275}$$

uniformly on every k as $\Sigma\,\theta_k \xrightarrow[\substack{\varepsilon \to 0 \\ \omega \to +\infty}]{} 0 \quad \left(\leq K\left[\frac{j_1+j_2}{\sqrt{\omega}} + \Sigma\,\delta t_j^-\right]\right)$.

We have then, with (9.273), (9.274), (9.275):

$$\begin{cases} \dfrac{\gamma_k}{1+\theta_k} \leq 1 + \alpha_k (a\,\delta t_{i+2k+2}) \\[2ex] \dfrac{\lambda_k}{1-\theta_k} \geq 1 + \beta_k (a\,\delta t_{i+2k+2}) \end{cases} \tag{9.276}$$

with

α_k and β_k keeping a constant sign for k such that (9.277)

$i+2k+2 \leq i+j_0-1$ and $|\alpha_k| \geq \dfrac{\theta}{4\alpha_4}$; $|\beta_k| \geq \dfrac{\theta}{4\alpha_4}$: α_k and

β_k have the same sign.

Finally, let us remark that:

$$\Pi\,(1+\theta_k) \xrightarrow[\substack{\varepsilon_1 \to 0 \\ \omega \to +\infty}]{} 1; \quad \Pi\,(1-\theta_k) \xrightarrow[\substack{\varepsilon_1 \to 0 \\ \omega \to +\infty}]{} 1 \tag{9.278}$$

as $\Sigma\,|\theta_k| = \Sigma\,\theta_k \xrightarrow[\substack{\varepsilon_1 \to 0 \\ \omega \to +\infty}]{} 0$ and as

$\dfrac{\gamma_k}{1-\theta_k}$ and $\dfrac{\gamma_k}{1+\theta_k}$ are positive provided δ_1 is chosen small (9.279)

enough. (Note that $|\alpha_k|$ and $|\beta_k|$ can be easily uniformly upperbounded.)

This implies:

$$\left.\begin{array}{l} \Pi \dfrac{\gamma_k}{1+\theta_k} \le \Pi\,(1+\alpha_k(a\delta t_{i+2k+2})) \\ \Pi \dfrac{\gamma_k}{1-\theta_k} \le \Pi\,(1+\beta_k(a\delta t_{i+2k+2})) \end{array}\right\} \begin{array}{l}\text{the products being taken}\\ \text{on all } k \text{ such that}\\ i+2k+2 \le i+j_0-1.\end{array} \qquad (9.280)$$

By (9.235) and (9.238), the product of the γ_k's goes to 1.
Indeed, taking into account the way we arranged the terms in (9.238), we have, in the product $\Pi\gamma_k$, all the terms which appear in $\sum_{r=1}^{j_0-1} \lambda_{k'_{i+r+1}}(x^+_{i+r})$, but possibly the last one, if j_0-1 is not even.
But we know that $\lambda_{j_0}(x^+_{j_0-1}) \to 1$.
Thus:

$$\Pi\,\gamma_k \to 1. \qquad (9.281)$$

On the other hand, as we have $\theta_k = K\left[\dfrac{\bar{j}_{i+2k+2}}{\sqrt{\omega}} + \delta t^-_{i+2k+2}\right]$,

$$\Sigma\,\theta_k \le K\left[\frac{j_1+j_2}{\sqrt{\omega}} + \Sigma\,\delta t^-_s\right] \to 0. \qquad (9.282)$$

Thus:

$$\Pi(1+\theta_k) \to 1. \qquad (9.283)$$

Hence, by (9.277), $\Pi(1+\alpha_k(a\delta t_{i+2k+2}))$ is lowerbounded by a quantity converging to 1, while $\Pi(1+\beta_k\, a\delta t_{i+2k+2})$ is upperbounded by a quantity also converging to 1. If the α_k's are positive, so are the β_k's. $\Pi(1+\beta_k a\delta t_{i+2k+2})$ is upperbounded by a quantity converging to 1; hence we have:

$$\Sigma\beta_k\, a\delta t_{i+2k+2} \to 0. \qquad (9.284)$$

If, on the contrary, the α_k's are negative, then $\Pi(1+\alpha_k(a\delta t_{i+2k+2}))$ is lowerbounded by a quantity going to 1. Thus:

$$\Sigma\,\alpha_k\, a\delta t_{i+2k+2} \to 0. \qquad (9.285)$$

In any case, given (9.277), we deduce:

$$\Sigma\ \delta t_{i+2k+2} \to 0, \tag{9.286}$$

the sum being taken on all k's such that $i+2k+2 \leq i+j_0-1$.

We now use (9.236) and (9.238) to deduce

$$\Sigma\ \delta t_{i+2k+2} \to 0, \tag{9.287}$$

(the sum being taken on all k's such that $i+2k+1 \leq i+j_0-1$); the proof of which relies on similar arguments, up to minor modifications and also on the fact that $D\psi^{-2k_0}(\xi(\psi^{2k_0}(x)))$ is transverse to Σ at x.

Finally, we thus have:

$$\sum_{r=1}^{j_0-1} \delta t_{i+r} \to 0; \tag{9.288}$$

which proves ②.

To prove ③, we see that:

$$\sum_{r=1}^{j_0} \delta t_{i+r} \geq \delta_1 \tag{9.289}$$

by definition of j_0.

Thus δt_{i+j_0} goes to a quantity greater than δ_1.

Before time δt_{i+j_0}, the curve oscillates as follows, according to the evenness of j_0.

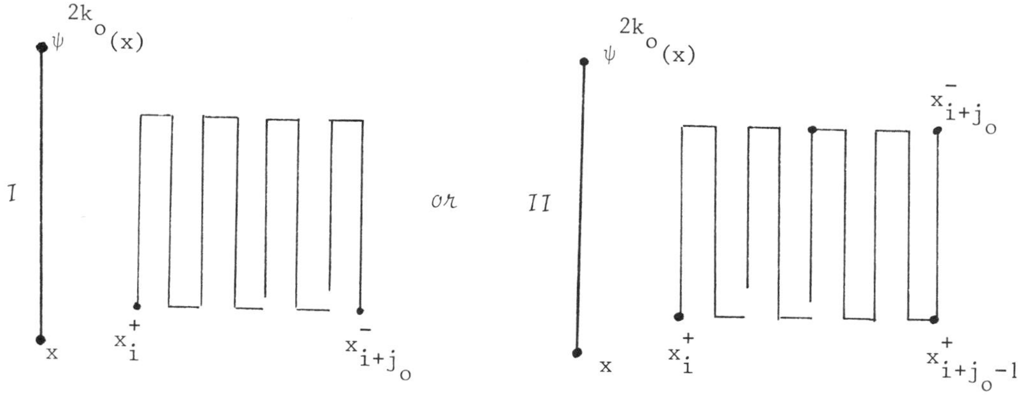

In the first case, using (9.285), we easily see that $d(x_i^+, x_{i+j_0}^-)$ goes to zero, and $d(x, x_{i+j_0}^-)$ also.

In the second case, by (9.285) too, we see that $d(x_i^+, x_{i+j_0}^+)$ goes to zero, and hence $d(x, x_{i+j_0-1}^+)$ goes also to zero. Finally, starting from $x_{i+j_0}^-$, in both cases, the curve remains tangent to ξ during a time $\delta t_{i+j_0} \geq \delta_1/2$; which proves ③ and completes the proof. ■

10 Qualitative explanation of the phenomenon

10a.

We complete now the analysis of the curves x_ε and the "convergence process" by the following theorem which locates the limit curves in some region of the manifold. The complete proof of this theorem will appear later on.

We introduce for $\theta > 0$ the set:

$$A_\theta = \{x \in M \mid \exists\, k \in \mathbb{Z} \text{ satisfying } \lambda_{2k}(x) \leq 1-\theta\}. \tag{10.1}$$

We then have the following results under the sole hypothesis (A_2).

Theorem 8: Let μ^ε_θ the Lebesgue measure of the set of t's such that $x_\varepsilon(t)$ is in A_θ.
If μ^ε_θ, for fixed θ, does not go to zero with ε, the sequence (x_ε) defining the critical point at infinity does not induce a change of topology in the level sets of J.

Corollary 1: If J induces a difference of topology in C_β, one can choose x_ε such that μ^ε_θ goes to zero with ε.
If, then, assuming $(\mathcal{A}_1)$ and $(\mathcal{A}_2)$ (x) hold, $\hat{\mu}^\varepsilon_\theta$ is the total time spent by the straightened curve $\hat{x}_\varepsilon$ in A_θ along the concentration intervals, $\hat{\mu}^\varepsilon_\theta$ goes to zero with ε.

The significance of Theorem 8 and its Corollary 1 is clear: the limit curve is in the subset of M described as follows, during the concentration intervals:

$$B = \{x \in M \mid \forall\, k \in \mathbb{Z},\ \lambda_{2k}(x) \geq 1\}. \tag{10.2}$$

Corollary 2: Under the assumptions of Theorem 7, case I of this theorem holds if $d\lambda_{2k_0}(\xi(x))$ is negative: while case II holds if $d\lambda_{2k_0}(D\psi^{-2k_0}(\xi(\psi^{2k_0}(x))))$ is negative (for x_ε such that it corresponds to a difference of topology in the level sets). There is indetermination if both quantities are positive.

We now give an idea of the proof of Theorem 8.

This proof shows why there are critical points at infinity in this variational problem.

Idea of the proof of Theorem 8: Consider two points of M, x and y, joined by a ξ-orbit.

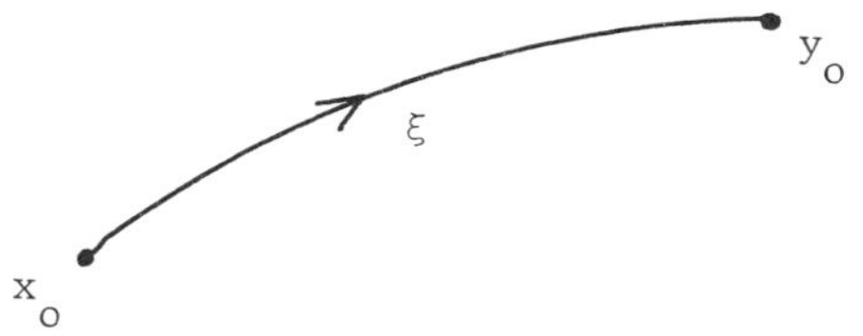

Let $(C_\beta)_{x_0}^{y_0}$ be the space of curves x going from x_0 to y, tangent to β and such that $\alpha(\dot{x}) = \text{Cte} > 0$.

On $(C_\beta)_{x_0}^{y_0}$, we introduce the functional:

$$J(x) = \int_0^1 \alpha(\dot{x})\, dt = \alpha(\dot{x}). \tag{10.3}$$

Denote by x_1 the curve defined by the ξ-orbit from x_0 to y_0, described during time 1.

Let $a_1 = J(x_1) = \int_0^1 \alpha(\dot{x}_1)\, dt.$

x_1 is a critical point to J. Indeed, the first variation of J along a curve x is:

$$\partial J(x) \,.\, z = -\int_0^1 b\eta \; dt \quad (\dot{x} = a\xi + bv) \tag{10.4}$$

where $z = \lambda\xi + \mu v + \eta w$; $\dot{\eta} = \mu a - \lambda b$; $\dot{\overbrace{\lambda + \bar{\mu}\eta}} = b\eta + C$

$$\begin{cases} \lambda(0) = \mu(0) = \eta(0) = 0 \\ \lambda(1) = \mu(1) = \eta(1) = 0 . \end{cases} \tag{10.5}$$

In particular at x_1, we have $\dot{x}_1 = a_1 \xi$ hence $b_1 = 0$ thus:

$$\partial J(x_1) . z = -\int_0^1 b_1 \eta \, dt = 0 \quad \forall z. \tag{10.6}$$

Thus, from the point of view of the calculus of variations, x_1 is critical.
Usually, in the calculus of variations, at a critical point, one computes the Morse index of the functional.
This computation, when carried out here, gives rise to a finite Morse index. One expects then the critical point x_1 to induce a difference of topology in the level sets of J.
This does not hold here, as an essential hypothesis in Morse theory is lacking: the gradient is not Fredholm.
The situation is even more complicated: there exist variations z which are not differentiable and which allow a whole neighbourhood of x_1 in $(C_\beta)_{x_0}^{y_0}$ to decrease; that is a deformation which is only C^0 defined on a neighbourhood of x_1 and along which J decreases.
Here is the deformation:

To define it, we have to leave the space $(C_\beta)_{x_0}^{y_0}$.

Let $y_1 = x_1(t_0)$

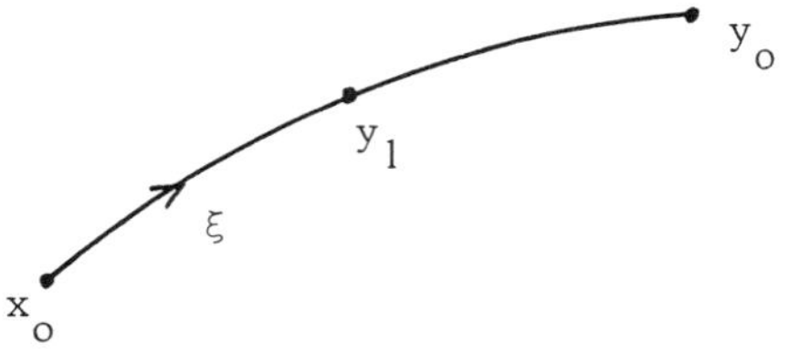

Consider the following curves:
Describe x_1 from x_0 to y_1; then at y_1 describe a piece of v-orbit to a point $\hat{y}_1$. Starting from $\hat{y}_1$, go backwards to y_1. Then continue on x_1 from y_1 to y_0.

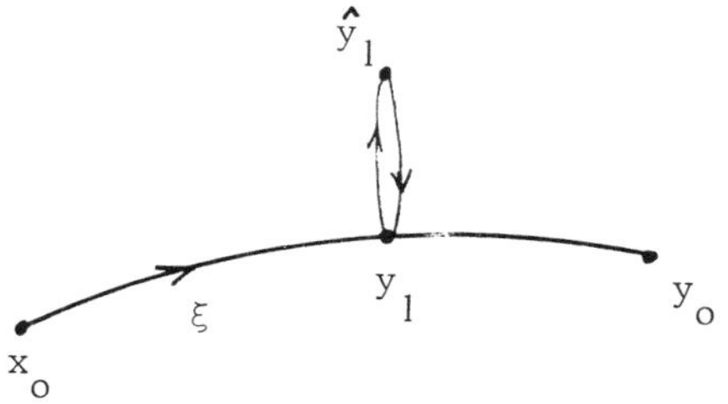

Call $\tilde{x}_s$ the curve obtained this way.
As $\alpha(v)$ is zero, $\int_0^1 \alpha(\dot{\tilde{x}}_1)\,dt = \int_0^1 \alpha(\dot{x}_1)\,dt = a_1$.

Evidently, the curve $\tilde{x}_1$ is not in $(C_\beta)_{x_0}^{y_0}$, as $\alpha(\dot{\tilde{x}}_1)$ cannot be made constant.

However, for $\varepsilon > 0$ given, one can "push" the curve $\tilde{x}_1$ in $(C_\beta)_{x_0}^{y_0}$ on a curve x_1^ε, if one accepts a slight increase in the energy level from $a_1 = \alpha(\dot{x}_1)$ to $a_1^\varepsilon < a_1 + \varepsilon$ (i.e. $J(x_1^\varepsilon) = a_1^\varepsilon < a_1 + \varepsilon$). This will be made clear later on.

Now, notice that the curve $\tilde{x}_1$ obtained this way offers other infinitesimal variations than the curve x_1.
For this, one "opens" along ξ the v-oscillation introduced; that is we consider the following variation:

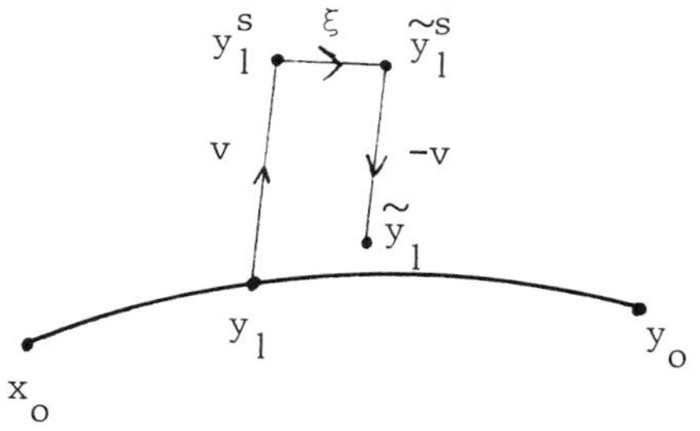

Evidently, we must extend this variation to $[y_1, y_0]$ in a continuous way. For this, let us fix the notations:

y_1^s is obtained from y_1 with pushing through v during the time s:

$$y_1^s = \phi(s, y_1). \tag{10.7}$$

$\tilde{y}_1^s$ is obtained from y_1^s by pushing through ξ during the time ε.

Lastly, $\tilde{y}_1$ is obtained from $\tilde{y}_1^s$ by pushing through v during the time -s:

$$\tilde{y}_1 = \phi(-s, y_1). \tag{10.8}$$

Thus, we go from y_1 to $\tilde{y}_1$ by pushing through $D\phi_{-s}(\xi)$ during the time ε.

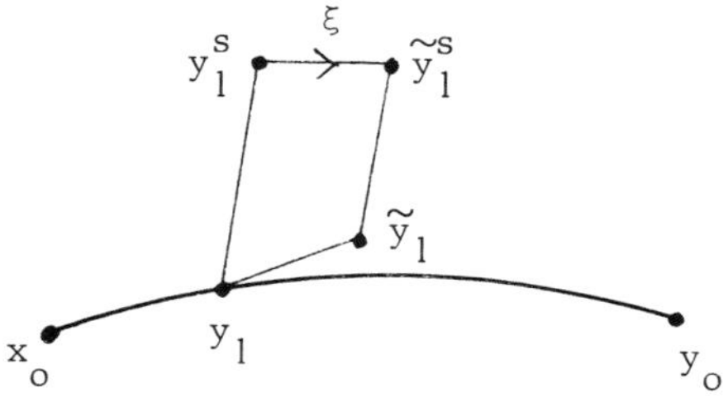

The curve we obtain this way from y_1 to $\tilde{y}_1$ is tangent at first order to $D\phi_{-s}(\xi)(y_1^s) = \lambda_0 \xi(y_1) + \mu_0 v(y_1) + \eta_0 w(y_1)$.
Assume that $y_1 = x_1(t_1)$. We can always find a solution of $[t_1, 1]$ of

$$\dot{\eta} = \mu a_1 - \lambda b_1 \tag{10.9}$$

such that:

$$\begin{cases} \eta(t_1) = \eta_0 & \eta(1) = 0 \\ \mu(t_1) = \mu_0 & \mu(1) = 0 \\ \lambda(t_1) = \lambda_0 & \lambda(1) = 0 \end{cases} \tag{10.10}$$

and which is C^∞ on $[t_1, 1]$.

For this, we consider a C^∞ function μ on $[t_1, 1]$ such that

$$\begin{cases} \mu(1) = 0 & \mu(t_1) = \mu_0 \\ a_1 \int_{t_0}^{1} \mu(t)\,dt = -\eta_0 \end{cases} \tag{10.11}$$

and then we take $\eta = a_1 \int_{t_1}^{t} \mu(\tau)\,d\tau + \eta_0$.

λ can be chosen arbitrary as long as λ is C^{∞} with $\lambda(t_1) = \lambda_0$ and $\lambda(1) = 0$. Let z be the vector thus constructed along the curve interval $[y_1, y_0]$ denoted $y(.)$. We push this piece of curve along the variation z during a small time ε. We are left with a curve (a piece of curve) denoted $l(.)$.

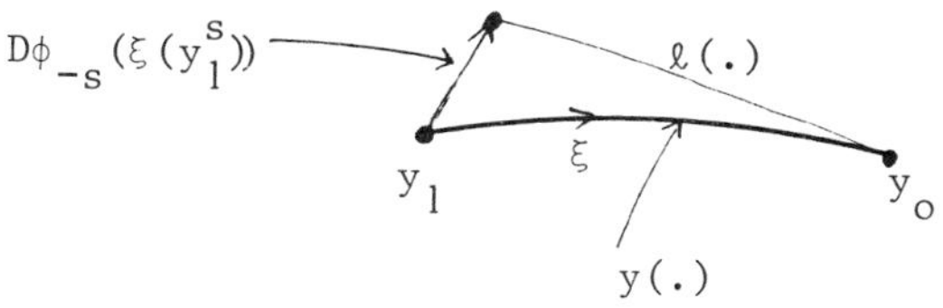

At first order we have:

$$\alpha_{l(t)}(\dot{\widehat{l(t)}}) = \alpha_{y(t)}(\dot{y}(t)) + [\frac{d\alpha}{dt}(z) - d\alpha(\dot{y}, z)]\varepsilon. \qquad (10.12)$$

We indeed know that the first variation of $\alpha_x(\dot{x})$ along a tangent vector z at the curve x is $\frac{d}{dt}\alpha(z) - d\alpha(\dot{x}, z)$ (Proposition 3).

Here $\dot{y} = a_1 \xi$.

Hence:

$$\alpha_{l(t)}(\dot{\widehat{l(t)}}) = a_1 + [\dot{\widehat{\lambda + \bar{\mu}\eta}} - d\alpha(a_1\xi, z)]\varepsilon = a_1 + (\dot{\widehat{\lambda + \bar{\mu}\eta}})\varepsilon \qquad (10.13)$$

as $d\alpha(\xi, .) = 0$ ($z = \lambda\xi + \mu v + \eta w$; $\alpha(\xi) = 1$; $\alpha(v) = 0$; $\alpha(w) = \bar{\mu}$).

a_1 is a positive constant, $\lambda + \bar{\mu}\eta$ is C^{∞}, hence for ε small enough, $\alpha_{l(t)}(\dot{\widehat{l(t)}})$ is strictly positive.

We reparametrize the piece l so that $\alpha(\dot{l}(t))$ is a strictly positive constant. This constant is

$$\frac{1}{1-t_1} \int_{t_1}^{1} \alpha(\dot{l}(t))\, dt. \qquad (10.14)$$

At first order this constant is

$$\frac{1}{1-t_1} \int_{t_1}^{1} [a_1 + (\dot{\widehat{\lambda + \bar{\mu}\eta}})\varepsilon]dt = a_1 + \varepsilon[(\lambda + \bar{\mu}\eta)(1) - (\lambda + \bar{\mu}\eta)(t_1)], \qquad (10.15)$$

Hence:

$$\frac{1}{1-t_1} \int_{t_1}^{1} \alpha(\dot{l}(t))\, dt - a_1 = -\varepsilon\, \alpha_{y_1} (D\phi_{-s}(\xi(y_1^s))) \tag{10.16}$$

at first order.

Indeed $\lambda(1) = \eta(1) = 0$ and $\lambda(1) = \lambda_0$, $\eta(t_1) = \mu_0$.

Lastly, $\lambda(t_1) + \bar{\mu}(y_1)\ \eta(t_1) = \lambda_0 + \mu(y_0)\, \eta_0 = \alpha_{y_1} (D\phi_{-s}(\xi(y_1^s)))$. Using the fact that $\dot{y} = a_1 \xi$, this implies:

$$\frac{1}{1-t_1} \left[\int_{t_1}^{1} \alpha_{l(t)}(\dot{l}(t)) - \alpha_{y(t)}(\dot{y}(t)) \right] = -\varepsilon\, \alpha_{y_1} (D\phi_{-s}(\xi(y_1^s))). \tag{10.17}$$

We extend now the curve l to $[0, t_1]$ as follows. On $[0, t_1[$ we go from x_0 to y_1 along ξ and at time t_1 we plug a variation which consists of describing the piece along $v\ [y_1, y_1^s]$, then a piece along ξ during the time ε (we denote this piece $[y_1, \tilde{y}_1^s]$). Then the backwards piece along v from $\tilde{y}_1^s$ to y_1. As we ask that this variation be made during a time equal to zero, we call it a "Dirac", although this is an abuse of terminology.

At first order, we then have a curve which is "continuous in a geometric sense". Indeed, as we chose $z(t_1) = D\phi_{-s}(\xi(y_1^s))$ and as we push along z during the time ε, $\tilde{y}_1$ is nothing but the point obtained by pushing y_1 along $z(t_1)$ during the time ε, at least at first order.

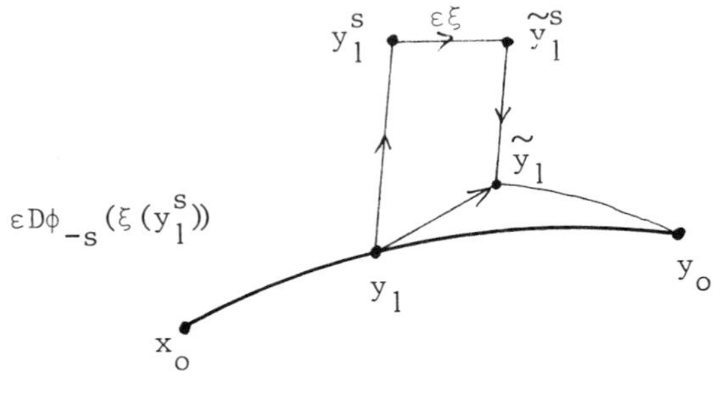

y_1 and $\tilde{y}_1$ are by their very definition joined by a curve tangent to $D_{\phi_s}(\xi(\bullet))$ during the time ε where the $\bullet$ describes the curve tangent to ξ starting at y_1^s and ending at $\tilde{y}_1$ during the time ε. Through $z(t_1)$, y_1 has been pushed in $y_1 + \varepsilon\, d\phi_{-s}(\xi y_1^s))$; hence the two points. ($\tilde{y}_1$ and $y_1 + \varepsilon\, d\phi_{-s}$ are the same at first order.)

We compute the variation along $[\tilde{y}_1, y_0]$.

On what remains of the curve, the variation is 0 on $[x_0, y_1]$ as x_1 and l are the same on this piece.

However, at time t_1, there is a variation of the functional equal at first order to ε. Indeed, there is a contribution of this Dirac we introduced on l; this contribution equals the length described along ξ (indeed the contribution of the pieces tangent to v is 0 as $\alpha(v) \equiv 0$).

Summing up the variations thus are

$$\underbrace{\varepsilon}_{\text{for the Dirac}} \quad \underbrace{-\varepsilon\,\alpha_{y_1}(D\phi_{-s}(\xi(y_1^s)))}_{\text{for the piece } [y, y_0]} = \varepsilon(1-\alpha_{y_1}(D\phi_{-s}(\xi(y_1^s)))). \qquad (10.18)$$

Now, we "push" l in C_β in order to give some consistency to our argument.

We first push l in the set $\{x \in \mathcal{L}_\beta \mid \alpha(\dot{x}) > 0\}$ then we reparametrize l so that on the reparametrized curve x', $\alpha(\dot{x})$ is constant. This is easy: we indeed choose along the pieces tangent to v, which are $[y_1, y_1^s]$ and $[y_1^s, \tilde{y}_1]$, functions λ, μ, η such that

$$\begin{cases} \dot{\overbrace{\lambda + \bar{\mu}\eta}} - b\eta > 0 \\ \dot{\eta} = \mu a - \lambda b = -\lambda b \end{cases} \qquad (10.19)$$

($\dot{l} = a\xi + bv$, l has been reparametrized so that the time spent in describing it previously as a Dirac mass is very small but strictly positive. Notice that on $[y_1, y_1^s]$ and $[\tilde{y}_1^s, \tilde{y}_1]$, $a = 0$.)

We constrain λ, μ, η to be zero at y_1^s and $\tilde{y}_1^s$, and we want to find solutions to these equations. For this we extend λ, μ, η at the whole interval $[0, 1]$ asking that they satisfy on $[x_0, y_1]$ and $[\tilde{y}_1, y_0]$ the equation $\dot{\eta} = \mu a_1 - \lambda b$. Let z_1 be a vector so obtained. If we push by z_1 during the time $\varepsilon_1 = 0(\varepsilon)$, the first variation between the continuous at first order curve l_1 and the curve x_1 is the same as the one between the curve l and x_1, i.e.

$$\varepsilon(1 - \alpha_{y_1}(D\phi_{-s}(\xi(y_1^s)))). \qquad (10.20)$$

Now $\alpha(\dot{l}_1(t))$ is strictly positive. Indeed $\widehat{\lambda + \bar{\mu}\eta - b\eta}^{\bullet}$ is strictly positive along the v pieces. Elsewhere $\alpha(\dot{l}(t))$ is strictly positive. We may thus push during a time ε_1 so small that $\alpha(\dot{l}_1(t))$ remains positive.
We reparametrize l_1 into l_2 so that $\alpha(\dot{l}_2(t))$ equals a constant; l_2 being the curve going from x_0 to y_0 during the time 1.

If we set

$$a_2 = \int_0^1 \alpha(\dot{l}_2(t))\ dt, \tag{10.21}$$

we have:

$$a_2 - a_1 = \varepsilon(1 - \alpha_{y_1}(D\phi_{-s}(\xi(y_1^s)))) \tag{10.22}$$

at first order.

We point out here that these variations are "oriented": indeed, if starting from x_1 we go to l_2 with pushing by z_2, one cannot construct a curve $\tilde{l}_2$ by pushing x_1 by $-z_2$ and which is C_β (this curve is then constrainted to be tangent to $-\xi$ from $\tilde{y}_1$ to y_1^s which forbids this curve from belonging to C_β).

We however may define deformations of x_1 which decrease the energy level if we can find a point y_1 on $x_1(t)$ and a time s such that:

$$1 - \alpha_{y_1}(D\phi_{-s}(\xi(y_1^s))) < 0. \tag{10.23}$$

The same phenomenon takes place on a closed curve. It can be shown that a continuous deformation may be defined on a neighbourhood of a closed curve tangent to ξ on a certain piece, as soon as at a point y_1 of this piece one can find a time s such that $1 - \alpha_{y_1}(D\phi_{-s}(\xi(y_1^s)))$ is strictly negative. This deformation "decreases" the energy levels.
Thus a curve is really critical for J if

$$1 - \alpha_{y_1}(D\phi_{-s}(\xi(y_1^s))) \geq 0 \qquad \forall s; \tag{10.24}$$

and if on another hand, this curve is "essentially" tangent to ξ. Indeed, if on a certain time interval this curve is not tangent to ξ, then the gradient of J allows us to decrease a strictly positively lowerbounded amount on a neighbourhood of this curve.

We consider now the function

$$\omega(s) = 1 - \alpha_{y_1}(D\phi_{-s}(\xi(y_1^s))). \tag{10.25}$$

We look for the extremal points of this function.
For this, we compute $\dot{\omega}(s)$

$$\dot{\omega}(s) = \frac{\partial}{\partial s}(1 - \alpha_{y_1}(D\phi_{-s}(\xi(y_1^s)))) = -\frac{\partial}{\partial s}\alpha_{y_1}(D\phi_{-s}(\xi(y_1^s))). \tag{10.26}$$

Let us consider a chart in a neighbourhood of $y_1^{s_0}$ where v is read as a constant vector. In this chart, taking an s-derivative corresponds to taking a v-derivative. We thus have:

$$\dot{\omega}(s) = -\alpha_{y_1}(D\phi_{-s_0}(\frac{\partial}{\partial s}\xi(y_1^s))(s_0)). \tag{10.27}$$

Now

$$\frac{\partial}{\partial s}\xi(y_1^s)(s_0) = [v, \xi][y_1^{s_0}]. \tag{10.28}$$

Indeed, taking a derivative along $\frac{\partial}{\partial s}$ is taking a v-derivative and the term $\xi . v$ is zero, in that chart (connexion) as v is constant, hence its derivative along any vector, in particular ξ, is zero.
We thus have:

$$\dot{\omega}(s_0) = -\alpha_{y_1}(D\phi_{-s_0}([v, \xi][y_1^{s_0}])) = \alpha_{y_1}(D\phi_{-s}([\xi, v])). \tag{10.29}$$

In particular, $\dot{\omega}(s)$ is zero if and only if $D\phi_{-s}([\xi, v])$ belongs to the kernel of α at y_1. Thus, as $D\phi_{-s_0}(v) = v$, $\dot{\omega}(s_0)$ is zero if and only if

$$D\phi_{-s_0}(\text{plane}(v, [\xi, v])) = D\phi_{-s_0}(\ker \alpha) = \ker \alpha_{y_1}. \tag{10.30}$$

<u>Consequently,</u> ω <u>is extremal when the plane</u> α <u>has come back on itself in the transport by</u> ξ <u>from</u> $y_1^{s_0}$. <u>That is</u> $y_1^{s_0}$ <u>is a coincidence point of</u> y_1.

Thus, if we look for time s such that:

$$1 - \alpha_{y_1}(D\phi_{-s}(\xi(y_1^s))) < 0 \tag{10.31}$$

we have to check if this inequality holds at coincidence points. In fact we can just retain the oriented coincidence points. At the others (oriented in the opposite direction), we have:

$$\alpha_{y_1}(D\phi_{-s}(\xi(y_1^s))) = \lambda\, \alpha_{y_1^s}(\xi(y_1^s)) = \lambda \quad \text{with } \lambda < 0. \tag{10.32}$$

Hence:

$$1 - \alpha_{y_1}(D\phi_{-s}(\xi(y_1^s))) > 0 \quad \text{necessarily.} \tag{10.33}$$

We are then left with the oriented coincidence points governed by ψ^{2k} $(k \in \mathbb{Z})$ at these points:

$$1 - \alpha_{y_1}(D\phi_{-s}(\xi(y_1^s))) = 1 - (\psi^{-2k}) \star \alpha(\xi) = 1 - \frac{\lambda}{\lambda_{2k}}(y_1). \tag{10.34}$$

Indeed we have:

$$((\psi^{2k}) \star \alpha)_{y_1} = \lambda_{2k}(y_1)\, \alpha_{y_1}. \tag{10.35}$$

Hence:

$$\lambda_{2k}(y_1)\, \alpha_{y_1} = \alpha_{\psi^{2k}(y_1)}(D\psi^{2k}(0)). \tag{10.36}$$

Hence:

$$\lambda_{2k}(y_1)\, \alpha_{y_1}(D\psi^{-2k}(\xi(\psi^{2k}(y_1)))) = \alpha_{\psi^{2k}(y_1)}(\xi(\psi^{2k}(y_1))) = 1. \tag{10.37}$$

We thus have if y_1 is on the ξ-piece

$$1 - \frac{1}{\lambda_{2k}(y_1)} \geq 0 \tag{10.38}$$

if no decreasing deformation on a neighbourhood can be defined.

Thus:

$$\lambda_{2k}(y_1) \geq 1 \qquad \forall\ k \in \mathbb{Z}\,. \tag{10.39}$$

This argument holds when y_1 is on a ξ-piece. When we are dealing with those curves x_ε that define a critical point at infinity, we must consider the measure of the time on which

$$\lambda_{2k}(y_1) < 1\,. \tag{10.40}$$

If this measure does not go to zero, it is split on the concentration intervals where the curve is almost tangent to ξ, we may thus define our deformation on this concentration interval decreasing strictly and globally this curve x_ε by a quantity proportional to this measure. Hence if there is a difference in the level set, this measure has to go to zero. Hence our theorem.

Corollary 1 is immediate.

For Corollary 2, we argue by contradiction that if we are in case II of Theorem 7 with $d\lambda_{2k_0}(\xi) < 0$, then the curve is tangent to ξ starting from x if we neglect the initial oscillations and we are left with a piece tangent to ξ where

$$\lambda_{2k_0}(y) < 1. \tag{10.41}$$

Indeed, the limit curve starts at x and is tangent to ξ.

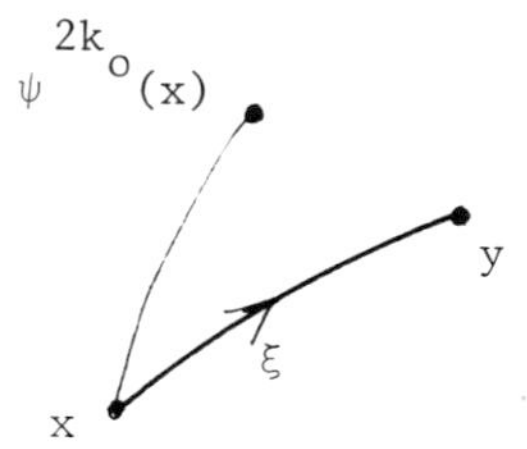

Case II

Hence, in a neighbourhood of x on this curve on the right, we have:

$$\lambda_{2k_0}(y) < \lambda_{2k_0}(x) \quad \text{as} \quad d\lambda_{2k_0}(\xi) < 0. \tag{10.42}$$

Now, $\lambda_{2k_0}(x) = 1$, hence $\lambda_{2k_0}(y) < 1$ on this right piece of curve during a consistent curve (Theorem 7) which is impossible by Theorem 8. ■

10b.

We finish now by noticing that the previous analysis allows us to understand why the curves critical at infinity have the form we have found: where a Dirac of type ψ^{2k_0} is realized on a limit curve at infinity, we must have $\lambda_{2k_0}(x) \geq 1$: otherwise we can decrease the curve in the neighbourhood of the curve. But if we consider now $\psi^{2k_0}(x)$, there is again a point on the curve where a jump along ψ^{-2k_0} is realized. Thus, we have also:

$$\lambda_{-2k_0}(\psi^{2k_0}(x)) \geq 1. \tag{10.43}$$

Now

$$\lambda_{-2k_0}(\psi^{2k_0}(x)) = \frac{1}{\lambda_{2k}(x)} \geq 1. \tag{10.44}$$

Hence:

$$\lambda_{2k_0}(x) = 1 \tag{10.45}$$

which explains Theorem 7.

The fact now that these Diracs are governed by the ψ^{2k} is tied to the fact that $\omega(s)$ is extremal at these points: the deformation which "decreases most" the energy level, or else which optimizes the decreasing speed is related to the coincidence points. Finally, this deformation is no more decreasing (critical points at infinity) if at these extremal points $\lambda^{2k}(x)$ is less than or equal to one, with equality when jumps occur.

11 Expansion of J near infinity. The index of a critical point at infinity

11a. The parametrization normal form

We are then left with these geometric curves made up of ξ-pieces and v-pieces. The v-pieces have been seen to run from a point to a conjugate.

On such a piece, as seen from Chapter 5 onwards, the function $b(t)$ is very particular.

Indeed, as stated in (5. 3), the vector:

$$\begin{cases} A_1^\star = 1 - \dfrac{b^2}{2\omega a} + \dfrac{\int_0^1 b^2 dt}{2\omega a} \\ B_1^\star = - \dfrac{b}{\omega} \\ C_1^\star = - \dfrac{\dot{b}}{\omega a} \end{cases} \tag{11.1}$$

nearly satisfies the transport equation of the forms.

Furthermore, if we are looking at a v-piece between x_{2i} and x_{2i+1}, then nearby x_{2i}, $\begin{bmatrix} A_1^\star \\ B_1^\star \\ C_1^\star \end{bmatrix}$ is nearly $\begin{bmatrix} 1 \\ 0 \\ 0 \end{bmatrix}$

Consequently b has in fact a <u>first normal form</u> on a critical point at infinity:

Namely, we introduce the function φ_i on the v-piece between x_{2i} and x_{2i+1} satisfying:

$$\begin{cases} \dfrac{\partial^2 \varphi_i}{\partial s^2} + \varphi_i + \bar{\mu} \dfrac{\partial \varphi_i}{\partial s} = 0 \\ \varphi_i(0) = \varphi_i(s_i) = 1 \\ \dfrac{\partial \varphi_i}{\partial s}(0) = \dfrac{\partial \varphi_i}{\partial s}(s_i) = 0 \end{cases} \tag{11.2}$$

Here s is the time parameter along the v-orbit from x_{2i} to x_{2i+1}. φ_i exists and is uniquely defined by (11.2) as x_{2i} and x_{2i+1} are conjugate. We thus have:

$$\frac{b(t)}{\sqrt{\omega a}} \sim \pm \sqrt{1 - \varphi_i(s(t))}. \tag{11.3}$$

We state this in:

Proposition 12: Along a near tangent to a v-piece between two nearby conjugate points, $\frac{b(t)}{\sqrt{\omega a}}$ is equivalent to $\pm \sqrt{1 - \varphi_i(s(t))}$ where φ_i satisfies:

$$\begin{cases} \dfrac{\partial^2 \varphi_i}{\partial s^2} + \varphi_i + \bar{\mu} \dfrac{\partial \varphi_i}{\partial s} = 0 \\ \varphi_i(0) = \varphi_i(s_i) = 1 \\ \dfrac{\partial \varphi_i}{\partial s}(0) = \dfrac{\partial \varphi_i}{\partial s}(s_i) = 0 \\ s \in [0, s_i]; \text{ time on the } (\pm) \text{ v-orbit from } x_{2i} \text{ to } x_{2i+1} \\ \text{which are the conjugate points.} \end{cases} \tag{11.4}$$

At first, the question of the sign (11.3) seems to bring in some ambiguity. In fact, in view of the analysis we carried out in Chapters 8, 9 and 10, this ambiguity can be somewhat dropped.

Indeed, the change of sign in (11.3) might occur only when $\varphi_i(s(t)) = 1$. As noticed, a posteriori, in Chapter 10, our curves lie in the region of M where $\lambda^\star(\psi^{2k}(x)) \leq 1$, for any x in this region and $k \in \mathbb{Z}$. ($\lambda^\star$ is the collinearity coefficient of $\phi^\star_{s(\psi^{2k}(x))}\alpha$ on α.)

$\lambda^\star(x_s)$ on the v-orbit through x_{2i} can be identified as $\varphi_i(s)$ where φ_i satisfies:

$$\begin{cases} \dfrac{\partial^2 \varphi_i}{\partial s^2} + \varphi_i + \bar{\mu} \dfrac{\partial \varphi_i}{\partial s} = 0 \\ \varphi_i(0) = 1; \ \dfrac{\partial \varphi_i}{\partial s}(0) = 0 \end{cases} \tag{11.5}$$

and φ_i is extremal only at the coincidence points of x_{2i} (see again Chapter 10). Thus $\varphi_i(s) \leq 1$ and the change of sign can occur only at coincidence points $x_{s_i'}$ such that $\varphi_i(s_i') = 1$; hence at conjugate points.

Therefore, the only possibility for b is to accomplish a piece of v-orbit from x_{2i} to x_{2i+1}, then come back from x_{2i+1} to x_{2i}, etc, following what was stated in Theorem 7 of Chapter 8.

If we consider a deformation line of (4.9) going to a critical point at infinity, these oscillations are in finite number (upperbounded). Otherwise, we leave an L^∞-neighbourhood of this critical point at infinity and one can construct a deformation lemma to move all such curves away from infinity.

As we are dealing with an actual jump, this number is odd.

Thus, we are left, as a model, with only one jump and a definite sign for b on such a piece.

As we wish here to present general ideas rather than justify all the technical details, we will assume for sake of simplicity that whenever a jumps occurs, a single oscillation is associated with it.

So that a critical point at infinity is this geometrical curve, together with a parameter $\sqrt{\omega a} \to +\infty$, the v-pieces being described with $b(t)/\sqrt{\omega a} \sim \pm\sqrt{1-\varphi_i(s(t))}$, where $\pm$ is fixed by the orientation along v of $[x_{2i}, x_{2i+1}]$.

11b. The variations along a critical point at infinity inwards of C_β

There are two kinds of variations along such a geometric curve with this limit parametrization we pointed out in 11a.

The first kind, we will present here, consists of opening up the oscillations in order to see if we are dealing with an actual or a fake critical point at infinity. This will be made clear later on.

These variations are inwards of C_β.

We want to know if a sequence $(\varepsilon \to 0)$ of flow lines of (4.9) arrives or departs from the limit object.

In order to discriminate between these two possibilities, we need a first expansion of J along inwards C_β-variation.

An inward variation has to bring the length along ξ to be a strictly positive constant which is nearly a, a being the length along ξ of the limit object (the curve x).

We are thus led to introduce along a v-piece $[x_{2i}, x_{2i+1}]$ of x, which we will assume for sake of simplicity to be oriented by +v, the differential equation:

$$\begin{cases} \dot{\overbrace{\lambda + \bar{\mu}\eta}} - \eta = \dfrac{a}{\sqrt{\omega a}\sqrt{1-\varphi_i(s)}} & s \in [0, s_i]. \\ \dot{\eta} = -\lambda \end{cases} \tag{11.6}$$

In (11.6), $\cdot$ is a derivation by $\dfrac{\partial}{\partial s} = v$; ω is a large positive parameter and a is the length along ξ of the curve as already stated.

Another way to see (11.6) is to set:

$$\frac{\partial}{\partial t} = b\frac{\partial}{\partial s} = \sqrt{\omega a}\sqrt{1-\varphi_i(s)}\frac{\partial}{\partial s} \tag{11.7}$$

and we then have:

$$\begin{cases} \dfrac{\partial}{\partial t}(\lambda + \bar{\mu}\eta) - b\eta = a \\ \dfrac{\partial \eta}{\partial t} = -\lambda b. \end{cases} \tag{11.8}$$

The homogeneous equation:

$$\begin{cases} \dot{\overbrace{\lambda + \bar{\mu}\eta}} - \eta = 0 & s \in [0, s_i] \\ \dot{\eta} = -\lambda \end{cases} \tag{11.9}$$

has solutions satisfying:

$$(\lambda + \bar{\mu}\eta)(s_i) = (\lambda + \bar{\mu}\eta)(0) \tag{11.9'}$$

as x_{2i} and x_{2i+1} are conjugate points.

Indeed (11.9) expresses the relations which have to be satisfied by a transported vector along a v-piece.

There is thus an indeterminacy in (11.6) which we will discuss later on, when we will introduce the index of a critical point at infinity.

Notice that the parametrization introduced by (11.8), with $b = \sqrt{\omega a}\sqrt{1-\varphi_i(s)}$ corresponds to the first normal form we pointed out in Proposition 12.

In (11.6) there is a problem:

Indeed, $_i$ satisfies on $[x_{2i}, x_{2i+1}]$ parametrized by v:

$$\begin{cases} \dfrac{\partial^2}{\partial s^2}\varphi_i + \varphi_i + \bar{\mu}\,\dfrac{\partial \varphi_i}{\partial s} = 0 \\ \varphi_i(0) = \varphi_i(s_i) = 1 \\ \dfrac{\partial \varphi_i}{\partial s}(0) = \dfrac{\partial \varphi_i}{\partial s}(1) = 0 \,. \end{cases} \tag{11.10}$$

Thus, $1 - \varphi_i$ has a zero of second order at 0 and s_i. Near this point, we have:

$$1 - \varphi_i(s) \sim C_i s^2 \text{ at } 0; \quad 1 - \varphi_i(s) \sim C_i'(s-s_i)^2 \text{ at } s_i. \tag{11.11}$$

Thus:

$$\int_0^{s_i} \frac{1}{\sqrt{1-\varphi_i(s)}}\, ds \quad \text{diverges logarithmically at both ends.} \tag{11.12}$$

This implies, by integration of the first equation in (11.6), that $(\lambda + \bar{\mu}\eta, \eta)$ cannot possibly be L^∞, <u>a fortiori</u> L^∞-small.

We analyse here what is going on in (11.6).

<u>Lemma 21</u>: Consider a solution of (11.6), $(\lambda, \eta)(s)$, and a solution of (11.9), $(\lambda_1, \eta_1)(s)$ taking the same value at a point τ_i in $[0, s_i]$. Then:

$$\begin{cases} |\eta(s) - \eta_1(s)| \le \dfrac{C}{\sqrt{\omega}} \\ \int_0^{s_i} |\lambda(s) - \lambda_1(s)|^2\, ds \le \dfrac{C}{\omega} \\ \int_0^{s_i} |\dot{\eta}(s) - \dot{\eta}_1(s)|^2\, ds \le \dfrac{C}{\omega}\,. \end{cases}$$

Furthermore:

$\lambda + \bar{\mu}\eta(s)$ is equivalent to $-\frac{1}{\sqrt{C_i}}\sqrt{\frac{a}{\omega}}\log s$ near $s = 0$

$\lambda + \bar{\mu}\eta(s)$ is equivalent to $-\frac{1}{\sqrt{C_i'}}\sqrt{\frac{a}{\omega}}\log(s_i - s)$ near $s = s_i$.

Lastly,

$$\Delta_i = \lim_{\substack{\varepsilon\to 0\\ \varepsilon'\to 0}} \left\{ \int_{\varepsilon}^{s_i-\varepsilon'} \frac{ds}{\sqrt{1-\varphi_i(s)}} + \sqrt{\frac{\omega}{a}}\left[(\lambda+\bar{\mu}\eta)(\varepsilon) - (\lambda+\bar{\mu}\eta)(s_i-\varepsilon')\right]\right\} \tag{11.13}$$

exists and is independent of the solution of (11.6) considered as well as on ω and a. This quantity is thus attached to $[x_{2i}, x_{2i+1}]$ and only to it.

<u>Proof of Lemma 21</u>: $(\lambda - \lambda_1, \eta - \eta_1)$ satisfy (11.6) with zero conditions at τ_i:

$$\dot{\overline{(\lambda - \lambda_1) + \bar{\mu}(\eta - \eta_1)}} - (\eta - \eta_1) = \frac{a}{\sqrt{\omega a}\sqrt{1-\varphi_i(s)}} \tag{11.14}$$

$$\dot{\overline{\eta - \eta_1}} = -(\lambda - \lambda_1) \quad s \in [0, s_i], \quad (\lambda-\lambda_i)(\tau_i) = (\eta-\eta_1)(\tau_i) = 0.$$

We thus have:

$$(\eta-\eta_1)(s) = -\int_{\tau_i}^{s}[\lambda-\lambda_1+\bar{\mu}(\eta-\eta_1)\,d\tau + \int_{\tau_i}^{s}\bar{\mu}(\eta-\eta_1)\,d\tau \tag{11.15}$$

$$= -\int_{\tau_i}^{s}\left[(\eta-\eta_1) + \int_{\tau_i}^{\tau}\frac{a}{\sqrt{\omega a}\sqrt{1-\varphi_i(x)}}\,dx\right]d\tau + \int_{\tau_i}^{s}\bar{\mu}(\eta-\eta_1)\,d\tau\,.$$

Thus:

$$(\eta-\eta_1)(s) = \int_{\tau_i}^{s}(\bar{\mu}-1)(\eta-\eta_1)\,d\tau + \sqrt{\frac{a}{\omega}}\int_{\tau_i}^{s}\left(\int_{\tau}^{\tau_i}\frac{1}{\sqrt{1-\varphi_i(x)}}\,dx\right)d\tau. \tag{11.16}$$

As by (11.10), we know that:

$$\sqrt{1-\varphi_i(s)} \sim C_i s \text{ at } 0; \quad \sqrt{1-\varphi_i(s)} \sim C_i'(s_i - s) \text{ at } s_i, \tag{11.17}$$

we have:

$$\int_{\tau_i}^{s} \int_{\tau}^{\tau_i} \frac{1}{\sqrt{1-\varphi_i(x)}} \, dx \, d\tau \leq C_1 \qquad \forall s . \tag{11.18}$$

Thus:

$$\left| (\eta - \eta_1)(s) - \int_{\tau_i}^{s} (\bar{\mu} - 1)(\eta - \eta_1) \, d\tau \right| \leq C_1 \sqrt{\frac{a}{\omega}} . \tag{11.19}$$

This, together with the vanishing of $\eta - \eta_1$ at τ_i, implies the existence of C such that:

$$|\eta(s) - \eta_1(s)| \leq \frac{C}{\sqrt{\omega}} . \tag{11.20}$$

The other inequalities of Lemma 21 follow easily by integration of (11.14), use of (11.20) and use of the vanishing of $\lambda - \lambda_1$ and $\eta - \eta_1$ at τ_i.

We have now to show (11.13):

The existence of a limit to:

$$\int_{\varepsilon}^{s_i - \varepsilon'} \frac{ds}{\sqrt{1-\varphi_i(s)}} + \sqrt{\frac{\omega}{a}} \left[(\lambda + \bar{\mu}\eta)(\varepsilon) - (\lambda + \bar{\mu}\eta)(s_i - \varepsilon') \right] \tag{11.21}$$

follows from the fact that this expression is by integration of the first equation in (11.6):

$$- \sqrt{\frac{\omega}{a}} \int_{\varepsilon}^{s_i - \varepsilon'} \eta(s) \, ds . \tag{11.22}$$

Now, η_1 being a solution of (11.9) is bounded; and $|\eta - \eta_1|$ is also bounded, as we just proved, by $\frac{C}{\sqrt{\omega}}$. Thus η is bounded on $[0, s_i]$ and (11.22), hence (11.21) has a limit, which we call Δ_i. Δ_i does not depend on the particular solution of (11.6) chosen. Indeed, if we consider another solution, (λ', η'), the difference $(\lambda - \lambda', \eta - \eta')$ is a solution of (11.9). Thus, by (11.9)', $[(\lambda - \lambda') + \bar{\mu}(\eta - \eta')](s_i) = [(\lambda - \lambda') + \bar{\mu}(\eta - \eta')](0)$. Hence:

$$\left[(\lambda + \bar{\mu}\eta)(\varepsilon) - (\lambda + \bar{\mu}\eta)(s_i - \varepsilon') \right] - \left[(\lambda' + \bar{\mu}\eta')(\varepsilon) - (\lambda' + \bar{\mu}\eta')(s_i - \varepsilon') \right] \tag{11.23}$$

goes to zero with ε and ε'.

Hence the result.

Δ_i is also independent of ω and a, as, in fact, it can be computed on a solution of:

$$\begin{cases} \overset{\bullet}{\overline{\gamma_i + \bar{\mu}\eta_i}} - \eta_i = \dfrac{1}{\sqrt{1-\varphi_i(s)}} \\ \dot{\eta}_i = -\gamma_i \end{cases} \tag{11.24}$$

as equal to:

$$\Delta_i = \lim_{\substack{\varepsilon\to 0\\ \varepsilon'\to 0}} \left\{ \int_{\varepsilon}^{s_i-\varepsilon'} \frac{ds}{\sqrt{1-\varphi_i(s)}} + [(\gamma_i+\bar{\mu}\eta_i)(\varepsilon)-(\gamma_i+\bar{\mu}\eta_i)(s_i-\varepsilon')] \right\}. \tag{11.25}$$

The proof of Lemma 21 is thereby complete. ■

A variation governed by (11.6) comes now with two problems: the first one, which is not very serious, is due to the indeterminacy in it; one can add to such a variation any variation subject to (11.9). In order to determine once and for all the variation we are looking at, we will impose:

$$\lambda(s_i/2) = \eta(s_i/2) = 0. \tag{11.26}$$

The influence of the solutions of (11.9) is analysed separately in 11c. The solution of (11.6)-(11.26) is denoted:

$$(\tilde{\lambda}, \tilde{\eta}) . \tag{11.27}$$

Notice that, by Lemma 21, and the fact that $(\tilde{\lambda}_1, \tilde{\eta}_1) \equiv (0,0)$, where $(\tilde{\lambda}_1, \tilde{\eta}_1)$ is the solution of (11.9) taking the same value (i.e. (0,0)) then $(\tilde{\lambda}, \tilde{\eta})$ at $\tau_i = s_i/2$, $(\tilde{\lambda}, \tilde{\eta})$ satisfies:

$$\begin{cases} |\tilde{\eta}(s)| \le \dfrac{C}{\sqrt{\omega}} \; ; \; \int_0^{s_i} |\tilde{\lambda}(s)|^2 ds \le \dfrac{C}{\omega} \; ; \; \int_0^{s} |\dot{\tilde{\eta}}|^2 ds \le \dfrac{C}{\omega} \; ; \\ (\tilde{\lambda} + \bar{\mu}\tilde{\eta})(s) \sim -\dfrac{1}{\sqrt{C_i}} \sqrt{\dfrac{a}{\omega}} \log s \qquad \text{at } 0 \\ \qquad\qquad \sim -\dfrac{1}{\sqrt{C_i'}} \sqrt{\dfrac{a}{\omega}} \log(s_i - s) \qquad \text{near } s = s_i . \end{cases} \tag{11.28}$$

The second problem is more serious:

By (11.28) such a variation cannot be made to be L^∞, a fortiori not L^∞ or H^1 small.

Indeed $\tilde{\lambda} + \bar{\mu}\tilde{\eta}(s)$ diverges logarithmically at $s = 0$ and $s = s_i$.
$\tilde{\eta}$ remains meanwhile bounded, and even, by (11.28), going to zero with ω.
Thus at 0 and s_i, the variation is infinite along ξ.
Nevertheless, we can try to extend it along the ξ-pieces, subject to the differential equation:

$$\begin{cases} \dot{\overline{\lambda + \bar{\mu}\eta}} = \varphi(t) \\ \dot{\eta} = \mu a \end{cases} \quad \text{here } \cdot = \xi = \text{derivation along } \xi. \tag{11.29}$$

We will state further on what are the conditions on φ near the points x_{2i+1}.
For the time being, let us try to understand what is going on the v-pieces. The equation (11.6) may be rewritten, in a more intrinsic form:
setting:

$$z(s) = (\tilde{\lambda}\xi + \tilde{\eta}w) \tag{11.30}$$

we have:

$$\left[\frac{\partial}{\partial s} + v; z(s)\right] = \frac{a}{\sqrt{\omega a}} \frac{1}{\sqrt{1-\varphi_i(s)}}\xi + \frac{a}{\sqrt{\omega a}}\theta v; \quad \theta \in C^\infty(\,]0, s_i[\,). \tag{11.31}$$

Indeed, (11.6) and (11.31) are equivalent as can be checked by applying α and β to (11.31).
As we are dealing with a variation along a v-piece and as the functional $J(x) = \int_0^1 \alpha_x(\dot{x})\,dt$ (this is rephrasing of our functional which makes sense not only on C_β or $\mathcal{L}_\beta$, but even on unparametrized, however oriented, curves) is invariant under reparametrization, the θ-term is not relevant. What matters is the variation transverse to $\dot{x}$, i.e. to v along these pieces.
Clearly:

$$z = \sqrt{\frac{a}{\omega}}\, z_0, \tag{11.32}$$

z_0 satisfying:

$$\left[\frac{\partial}{\partial s} + v, z_0\right] = \frac{1}{\sqrt{1-\varphi_i(s)}}\,\xi + \theta v \tag{11.33}$$

$z_0(s_i/2) = 0$; z_0 has no component on v.

If we draw the variation, we have:

(11.34)

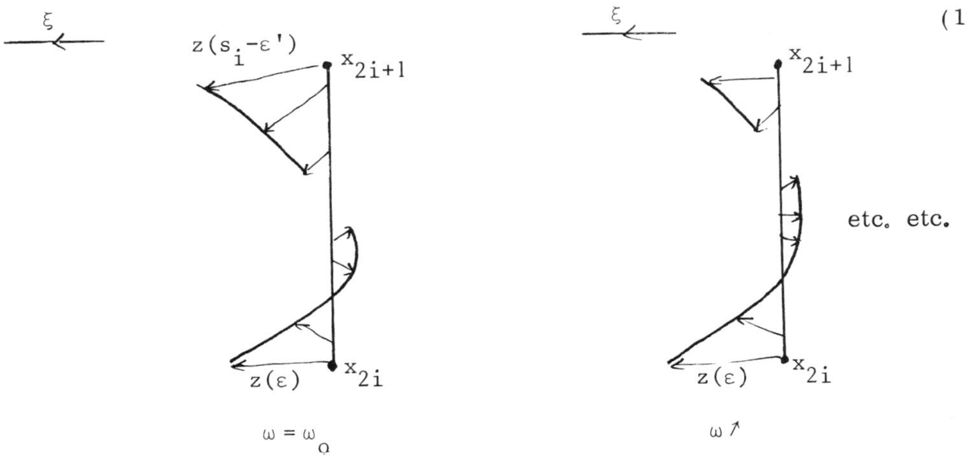

Remark: The linking as shown in (11.34) actually occurs.

There is thus, as $\omega \to +\infty$, more and more control on the variation z which is of the order of $\frac{C}{\sqrt{\omega}}$ on a fixed (when $\omega \to +\infty$) compact set in $]0, s_i[$. Nevertheless, at the ends, 0 and s_i, we always have a logarithmic divergence along ξ.

If we extend now to the ξ-pieces subject to (11.29), we have:

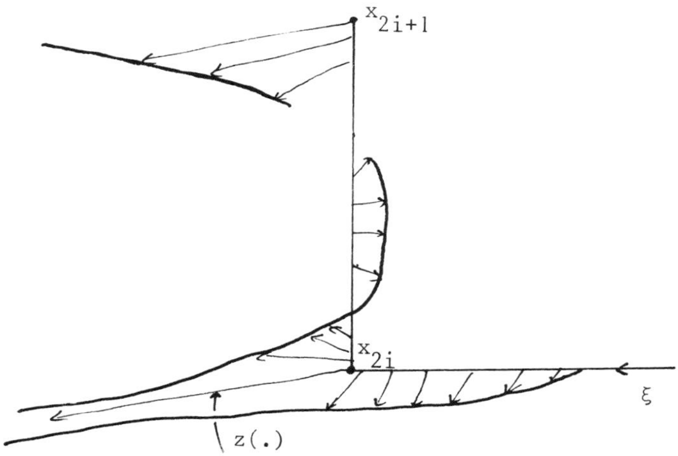

(11.35)

Following a choice of φ in (11.29) behaving as:

$$\begin{cases} \varphi(t) \sim \sqrt{\frac{a}{\omega}} \dfrac{C_i''}{t-t_i} \quad \text{nearby } t = t_i; \\ t_i \text{ corresponding to } x_{2i} \text{ on the } \xi\text{-piece; } C_i'' \text{ bounded constant;} \end{cases} \tag{11.36}$$

one can take care of the logarithmic divergence of $\tilde{\lambda} + \bar{\mu}\tilde{\eta}$ at x_{2i} (also x_{2i+1} on the other ξ-piece) and control in $\sqrt{\frac{a}{\omega}}$ the H_1-norm of the variation on each compact subset of the $]x_{2i-1}, x_{2i}[$.

This will be made precise later on.

The only problem is thus at the <u>corners</u>, i.e. at the points x_{2i}, x_{2i+1}.

There, in fact, following our analysis in Chapters 8 and 9, it is natural that we find some problems, as $\dot{x}_\varepsilon$ changes very rapidly. On the other hand, the curves x_ε stay nevertheless in an L^∞-neighbourhood of x.

We wish thus to cut out the variation on a neighbourhood of the corners (a neighbourhood which we can take to be smaller and smaller as ω goes to $+\infty$) in order to have an L^∞ control on it (going to zero with ω); and then to make a first expansion of J.

Let us first derive what kind of first expansion we may expect, considering the variation as taking place in $\mathcal{L}_\beta$ with the functional $J(y) = \int_0^1 \alpha_y(\dot{y})\,dt$

(which coincides with J on C_β and makes sense on unparametrized oriented curves of $\mathcal{L}_\beta$).

We have:

Lemma 22: Consider x a curve of $\mathcal{L}_\beta$, unparametrized.
Let $J : \mathcal{L}_\beta \to \mathbb{R}$

$$y \to \int_0^1 \alpha_y(\dot{y})\,dt.$$

Consider the variation z defined by (11.30), (11.29), cut out at the corners so that all the (11.28) estimates hold. Call z_1 this new variation. Then,

$$\partial J(x)\,.\,z_1 = \sqrt{\frac{a}{\omega}}\,(\sum_i \Delta_i) + o(\frac{1}{\sqrt{\omega}}).$$

Proof: We have, by the first variation of $\alpha_x(\dot{x})$ given by Proposition 3:
$\partial J(x)\,.\,z_1 = \int [\frac{\partial}{\partial s}\,\alpha(z_1) - d\alpha(\dot{x}, z_1)]ds$ if we parametrize the curve by s
$(\dot{x} = \frac{\partial x}{\partial s})$; which is as well:

$$\partial J(x)\,.\,z_1 = \sum_i \int_{[x_{2i}, x_{2i+1}]} [\overline{\tilde{\lambda}_1 + \bar{\mu}\,\tilde{\eta}_1}^{\bullet} - \tilde{\eta}_1]ds \tag{11.37}$$

$$+ \sum_i \int_{[x_{2i+1}, x_{2i+2}]} \overline{\tilde{\lambda}_1 + \bar{\mu}\,\tilde{\eta}_1}^{\bullet}\, ds$$

where $\tilde{\lambda}_1$ and $\tilde{\eta}_1$ are the components of z_1 and the parametrization is the one of (11.30) and (11.29) on the corresponding v and ξ-pieces.

Thus:

$$\partial J(x)\,.\,z_1 = -\sum_i \int_{[x_{2i}, x_{2i+1}]} \tilde{\eta}_1\, ds \tag{11.38}$$

the other terms indeed cancel in (11.37).

Now:

$$-\sum_i \int_{[x_{2i}, x_{2i+1}]} \tilde{\eta}_1\, ds = -\sum_i \int_{\varepsilon}^{s_i - \varepsilon'} \tilde{\eta}_1\, ds + o(\frac{1}{\sqrt{\omega}}) \tag{11.39}$$

as z_1 satisfies (11.28).

But $z = z_1$ outside a small neighbourhood of the corners. Thus:

$$-\int_{\varepsilon}^{s_i-\varepsilon'} \tilde{\eta}_1 \, ds = -\int_{\varepsilon}^{s_i-\varepsilon'} \tilde{\eta} \, ds = \tag{11.40}$$

$$= \sqrt{\frac{a}{\omega}} \left[\int_{\varepsilon}^{s_i-\varepsilon'} \frac{ds}{\sqrt{1-\varphi_i(s)}} + \sqrt{\frac{\omega}{a}} \left((\tilde{\lambda} + \bar{\mu}\tilde{\eta})(\varepsilon) - (\tilde{\lambda} + \bar{\mu}\tilde{\eta})(s_i - \varepsilon') \right) \right].$$

By Lemma 21, this has a limit, when ε and ε' go to zero, equal to $\sqrt{\frac{a}{\omega}}$.

Thus:

$$\partial J(x) \cdot z_1 = -\sum_i \int_{[x_{2i}, x_{2i+1}]} \tilde{\eta}_1 \, ds = \sqrt{\frac{a}{\omega}} \left(\sum_i \Delta_i\right) + o\left(\frac{1}{\sqrt{\omega}}\right). \tag{11.41}$$

Hence Lemma 22. ■

11c. The variations tangent to the border-line. The index of a critical point at infinity

We are now left with indeterminancies.

Namely, we are considering variations such that on the v-pieces, we have:

$$\begin{cases} \dot{\overline{\lambda + \bar{\mu}\eta}} - \eta = 0 \\ \dot{\eta} = -\lambda \end{cases} \qquad \cdot = \frac{\partial}{\partial s} = \pm v; \ \mu \text{ arbitrary, } s \in [0, s_i] \tag{11.42}$$

On the ξ-pieces we have:

$$\begin{cases} \dot{\overline{\lambda + \bar{\mu}\eta}} = \varphi \\ \dot{\eta} = \mu a \end{cases} \tag{11.43}$$

$\cdot = \frac{\partial}{\partial s} = \xi$; φ L^{∞}-small.

(11.20) defines, up to the μ-indeterminacy, the equation of a transported vector (by v) along the v-piece $[x_{2i}, x_{2i+1}]$. This can be easily checked by applying α and β to the transport equation:

$$\left[\frac{\partial}{\partial s} + v, \ \lambda\xi + \mu v + \eta w\right] = \psi(s) v, \quad \psi \text{ arbitrary.} \tag{11.44}$$

Thus, calling:

$z_i(s)$ the variation subject to (11.20) on $[0, s_i]$, (11.45)

we have:

$$z_i(s_i) = D\phi_{s_i}(z_i(0)) + \delta s_i v; \qquad \delta s_i \in \mathbb{R}. \tag{11.46}$$

Consequently, in order to compute $J(x+z)$, we may always see $x + z$ as being geometrically realized as follows:

(11.47)

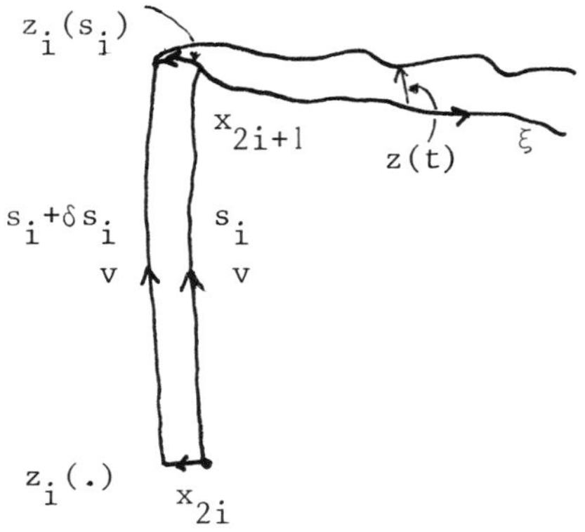

i.e., along the v-pieces, we just push by v during a time $s_i + \delta s_i$ starting at $x_{2i} + z_i(0)$; along the ξ-pieces, we have as usual a tangential variation. In this way, $J(x + z) - J(x)$ comes only from the ξ-pieces and is thus equal to the variation of $J(x)$ along these pieces, which is of second order (the first variation in zero. Indeed, by (11.20), (11.21), this first variation is $\sum_i [(\lambda + \bar{\mu}\eta)(x_{2i+2}) - (\lambda + \bar{\mu}\eta)(x_{2i+1})] = 0$ as x_{2i} is conjugate to x_{2i+1}).

The expansion of $J(x + z) - J(x)$ is:

$$J(x + z) - J(x) = z \cdot (\partial J(x) \cdot z) = \tag{11.48}$$

$$= z \cdot \sum_i \left\{ [(\lambda + \bar{\mu}\eta)(x_{2i+2}) - (\lambda + \bar{\mu}\eta)(x_{2i+1})] - (\int_{x_{2-+1}}^{x_{2i+2}} b\eta) \right\}$$

$$= z \cdot \left(\sum_i [(\lambda + \bar{\mu}\eta)(x_{2i}) - (\lambda + \bar{\mu}\eta)(x_{2i+1})] - \int_{x_{2i+1}}^{x_{2i+2}} b\eta \right) .$$

We first compute:

$$z\ .\ \left(\int_{x_{2i+1}}^{x_{2i+2}} b\eta\right) \tag{11.49}$$

b is, as noted previously, equal to $d\alpha(\dot{x}, w) = \gamma(\dot{x})$.

Thus, the first variation of b along z is:

$$\frac{d}{dt}\gamma(z) - d\gamma(\dot{x}, z) = \dot{\mu} + a\eta\,\tau - b\eta\,\bar{\mu}_{\xi} \tag{11.50}$$

On a ξ-piece, b is zero. Thus this variation is:

$$\dot{\mu} + a\eta\,\tau\ . \tag{11.51}$$

Thus, using again the fact that b is zero on such a piece (thus $bz\ .\ \eta = 0$ and $b\eta(x_{2i+2}) = b\eta(x_{2i+1}) = 0)$, we derive:

$$z\ .\ \left[-\int_{x_{2i+1}}^{x_{2i+2}} b\eta\right] = -\int_{x_{2i+1}}^{x_{2i+2}} (\dot{\mu} + a\eta\,\tau)\,\eta\ . \tag{11.52}$$

We are thus left with:

$$z\ .\ \left[(\lambda + \bar{\mu}\,\eta)(x_{2i}) - (\lambda + \bar{\mu}\,\eta)(x_{2i+1})\right] \tag{11.53}$$

which has the following simple interpretation:

Consider the differential equations:

$$\begin{cases} \dot{\overline{\lambda + \bar{\mu}\,\eta}} = \eta \\ \dot{\eta} = -(\lambda + \bar{\mu}\,\eta) + \bar{\mu}\,\eta \quad \text{initial data given by } z_i(0) \\ \cdot = \dfrac{\partial}{\partial s} = \pm v \quad \text{on the } \pm v\text{-piece from } x_{2i}+z_i(0) \text{ to} \\ x_{2i+1}+z_i(s_i) \quad \text{during the time } [0, s_i+\delta s_i] \text{ and } (11.20) \\ \text{with the same initial data.} \end{cases} \tag{11.54}$$

Then, with evident notations:

$$z\ .\ \left[(\lambda + \bar{\mu}\,\eta)(x_{2i}) - (\lambda + \bar{\mu}\,\eta)(x_{2i+1})\right] = \tag{11.55}$$

$$(\lambda+\bar{\mu}\,\eta)(x_{2i}+z_i(0)) - (\lambda+\bar{\mu}\,\eta))(x_{2i+1}+z_i(s_i)) - (\lambda+\bar{\mu}\,\eta)(x_{2i}) + (\lambda+\bar{\mu}\,\eta)(x_{2i+1})$$

at first order.

To give an intrinsic form to this expression, we come back to Proposition 7. A transported vector $Z = \begin{bmatrix} A \\ B \\ C \end{bmatrix}(s)$ along v satisfies the differential equation:

$$\dot{Z} = \Gamma Z \quad \text{in the basis } (\xi, v, [\xi, v]). \tag{11.56}$$

Let

$$\begin{cases} V(s) = \begin{bmatrix} a_0(s) & b_0(s) & c_0(s) \\ a_1(s) & b_1(s) & c_1(s) \\ a_2(s) & b_2(s) & c_2(s) \end{bmatrix} & \text{be the resolvant matrix of } (11.34) \\ V(0) = \mathrm{Id}. \end{cases} \tag{11.57}$$

Then:

$$\alpha_{x_s}(D\phi_s(Z(0))) = [1, 0, 0]\, V(s) \begin{bmatrix} A(0) \\ B(0) \\ C(0) \end{bmatrix} \tag{11.58}$$

$$= a_0(s) A(0) + b_0(s) B(0) + c_0(s) C(0).$$

and

$$\alpha_{x_s}(D\phi_s(Z(0))) - \alpha_{x_0}(Z(0)) = (a_0(s) - 1) A(0) + b_0(s) B(0) + c_0(s) C(0). \tag{11.59}$$

Setting:

$$z(s) = \lambda \xi + \mu v + \eta w, \tag{11.60}$$

we thus have:

$$(\lambda + \bar{\mu}\eta)(s) - (\lambda + \bar{\mu}\eta)(0) = (a_0(s) - 1)\lambda(0) + b_0(s)\mu(0) + c_0(s)\eta(0) \tag{11.61}$$

In fact V depends also on the starting point of the transport differential equations, which we will denote y:

$$V(s) = V(s,y). \tag{11.62}$$

Thus

$$(\lambda+\bar{\mu}\,\eta)(s)-(\lambda+\bar{\mu}\,\eta)(0) = (a_0(s,y)-1)\lambda(0)+b_0(s,y)\mu(0)+c_0(s,y)\,\eta(0). \tag{11.63}$$

With $y = x_{2i}$ and $s = s_i$, we have:

$$a_0(s_i,x_{2i}) = 1;\; b_0(s_i,x_{2i}) = c_0(s_i,x_{2i}) = 0 \tag{11.64}$$

as x_{2i+1} is conjugate to x_{2i}.

Now, by (11.24), the variation z we are considering, under (11.20) and (11.21) is defined by:

$$\begin{cases} \lambda_i,\ \mu_i,\ \eta_i: & \text{coordinates of } z_i(0) \text{ at } x_{2i} \text{ and} \\ \delta s_i & \text{such that } z_i(s_i) = D\phi_{s_i}(z_i(0)) + \delta s_i v \end{cases} \tag{11.65}$$

and we may view the variation along this v-piece as pushing along the transport vector of $z_i(0)$ along this v-piece during the time $s_i + \delta s_i$. The variation (11.33), in view of (11.41), (11.42) and (11.43) is thus:

$$\begin{aligned} & z\,.\,\left[(\lambda+\bar{\mu}\,\eta)(x_{2i}) - (\lambda+\bar{\mu}\,\eta)(x_{2i+1})\right] \\ &= -\left\{\left[\frac{\partial a_0}{\partial s}(s_i,x_{2i})\,\lambda_i + \frac{\partial b_0}{\partial s}(s_i,x_{2i})\,\mu_i + \frac{\partial c_0}{\partial s}(s_i,x_{2i})\,\eta_i\right]\delta s_i \right.\\ &\quad + \left(\frac{\partial a_0}{\partial y}\bigg|_{(s_i,x_{2i})} \cdot z_i(0)\right)\lambda_i + \left(\frac{\partial b_0}{\partial y}\bigg|_{(s_i,x_{2i})} \cdot z_i(0)\right)\mu_i \\ &\quad \left. + \left(\frac{\partial c_0}{\partial y}\bigg|_{(s_i,x_{2i})} \cdot z_i(0)\right)\eta_i\right\} \end{aligned} \tag{11.66}$$

We are now ready to define the index of a critical point at infinity. Consider:

$$a \pm \text{ v-piece } [x_{2i}, x_{2i+1}]. \tag{11.67}$$

Let

$V(s, y)$ be the resolvent matrix of the transport equation for the vectors starting at y along the v-orbit, in the basis $(\xi, v, [\xi, v])$, (11.68)

$$\text{i.e. } \frac{\partial V}{\partial s} = \Gamma V; \quad \Gamma = \begin{bmatrix} 0 & 0 & -1 \\ 0 & 0 & \bar{\mu}_{\xi}(s,y) \\ 1 & 0 & \bar{\mu}(s,y) \end{bmatrix}.$$

Let

$$V(s,y) = \begin{bmatrix} a_0(s,y) & b_0(s,y) & c_0(s,y) \\ a_1(s,y) & b_1(s,y) & c_1(s,y) \\ a_2(s,y) & b_2(s,y) & c_2(s,y) \end{bmatrix} . \tag{11.69}$$

Let z be a variation of the critical point at infinity x satisfying:

$$\begin{cases} \dot{\overline{\lambda + \bar{\mu}\eta}} - \eta = 0 \\ \dot{\eta} = -\lambda \\ \cdot = \dfrac{\partial}{\partial s} = \pm v; \ \mu \text{ arbitrary, } s \in [0, s_i] \text{ on the piece } [x_{2i}, x_{2i+1}]; \end{cases} \tag{11.20}$$

$$\begin{cases} \dot{\overline{\lambda + \bar{\mu}\eta}} = \varphi(s) \\ \dot{\eta} = \mu a \\ \cdot = \dfrac{\partial}{\partial s} = \xi \text{ on a } \xi\text{-piece; } \varphi \text{ is } L^{\infty}\text{-small.} \end{cases} \tag{11.21}$$

Let

$z_i(0) = \lambda_i \xi + \mu_i v + \eta_i [\xi, v]$ be the value of this variation at x_{2i}; (11.70)

$z_i(s_i)$ = value of z at $x_{2i+1} = D\phi_{s_i}(z_i(0)) + \delta s_i v$; (11.71)

$$\hat{\mu}_i = a_1(s_i, x_{2i})\lambda_i + b_1(s_i, x_{2i})\mu_i + c_1(s_i, x_{2i})\eta_i ; \tag{11.72}$$

$$\hat{\eta}_i = a_2(s_i, x_{2i})\lambda_i + b_2(s_i, x_{2i})\mu_i + c_2(s_i, x_{2i})\eta_i ; \tag{11.73}$$

$$|z|^2_{H^1(x_{2i+1}, x_{2i+2})} = \int_{[x_{2i+1}, x_{2i+2}]} (\dot{\lambda}^2 + \dot{\mu}^2 + \dot{\eta}^2)\, ds. \tag{11.74}$$

We then have:

<u>Proposition 14</u>: $J(x + z) = a + \sum_i \left[(\hat{\mu}_i + \delta s_i)\, \hat{\eta}_i - \mu_i \eta_i - \delta s_i \left[\frac{\partial a_0}{\partial s}(s_i, x_{2i}) \lambda_i + \right.\right.$

$$\left. + \frac{\partial b_0}{\partial s}(s_i, x_{2i}) \mu_i + \frac{\partial c_0}{\partial s}(s_i, x_{2i}) \eta_i \right] - \left(\frac{\partial a_0}{\partial y}\Big|_{(s_i, x_{2i})} \cdot z_i(0) \right) \lambda_i -$$

$$- \left(\frac{\partial b_0}{\partial y}\Big|_{(s_i, x_{2i})} \cdot z_i(0) \right) \mu_i - \left(\frac{\partial c_0}{\partial s}\Big|_{(s_i, x_{2i})} \cdot z_i(0) \right) \eta_i$$

$$\left. + \frac{1}{a} \int_{[x_{2i}, x_{2i+2}]} (\dot{\eta}^2 - a^2 \eta^2 \tau)\, ds \right] + o(|z|^2),$$

where η is an H^1-arbitrary function equal to η_i at x_{2i} and $\hat{\eta}_i$ at x_{2i+1}. This formula gives the index of the critical point at infinity. <u>Notice that it does not depend on φ.</u> Here

$$|z| = \sum_i |\delta s_i| + \sum_i [\,|\mu_i| + |\eta_i| + |\lambda_i|\,] + \sum_i |z|_{H^1(x_{2i+1}, x_{2i+2})}.$$

<u>Proof</u>: It follows from (11.48), (11.66) and (11.52).

The second variation is the sum of the expressions in (11.66) and (11.52). In (11.52), we have $\dot{\eta} = \mu a$. We thus integrate by part:

$$-\int_{[x_{2i+1}, x_{2i+2}]} (\dot{\mu} + a\eta\tau)\eta = \mu(x_{2i+1})\, \eta(x_{2i+1}) - \mu(x_{2i+2})\, \eta(x_{2i+2}) \tag{11.75}$$

$$+ \int_{[x_{2i+1}, x_{2i+2}]} a\mu^2 - a\eta^2 \tau$$

$$= \mu(x_{2i+1})\, \eta(x_{2i+1}) - \mu(x_{2i+2})\, \eta(x_{2i+2})$$

$$+ \frac{1}{a} \int_{[x_{2i+1}, x_{2i+2}]} [\dot{\eta}^2 - a\eta^2 \tau].$$

Now

$$\mu(x_{2i+2}) = \mu_{i+1}; \quad \eta(x_{2i+2}) = \eta_{i+1} \quad \text{by (11.70)} \tag{11.76}$$

and by (11.71), (11.68) and (11.72), (11.73):

$$\mu(x_{2i+1}) = \hat{\mu}_i + \delta s_i; \quad \eta(x_{2i+1}) = \hat{\eta}_i. \tag{11.77}$$

This yields the result.

The fact that the remainder term is $o(|z|^2)$ with the definition stated of $|z|$, retaining the H^1-norm only on the ξ-pieces is just due to the invariance of J under reparametrization, in particular along the v-pieces where the variation is L^∞ in λ and η which are governed by (11.20), hence are L^∞-bounded by $\sum_i |\lambda_i| + |\eta_i|$; while the μ-variation along these pieces can be absorbed through a reparametrization, yielding only a $\sum_i |\delta s_i|$ term. Hence the proof of Proposition 14. ■

12 Foliations transverse to contact forms

Throughout this work, we assumed hypothesis (A_1) to hold.

This hypothesis gives some kind of <u>elliptic</u> structure to the transport equation along v. The critical points at infinity, as seen, are linked to the conjugate points, which might exist because of (A_1).

There is an opposite situation, when (A_1) does not hold:

This is an actual case.

Indeed, we have the following:

<u>Theorem 9</u>: Let α_0 be the standard contact structure on S^3. There exists a codimension-one non singular foliation $\tilde{\gamma}$ transverse to α_0.

If v is the vector field $\tilde{\gamma} \cap \alpha_0$, then any point of S^3 has no coincidence point other than itself (<u>a fortiori</u>, no conjugate point).

The same result holds for all contact forms on three-dimensional manifolds constructed by W. B. Lickorish [A foliation for 3-manifolds, Annals of Mathematics, 82 (1965), pp. 414-420].

<u>Proof</u>: It is sufficient to construct a foliation transverse to the contact form $\alpha_0 = (y-x)\,dz + dx$ on the standard torus $x^2 + y^2 \leq 1$; $z \in [0,1]$.
The global construction follows by gluing up the tori together along the trace of $\tilde{\gamma} \cap \alpha_0$ which is easily seen to be possible.

The foliation is:

$$\tilde{\gamma} = x\,dx + y\,dy + (1 - x^2 - y^2)\,dz. \quad \blacksquare$$

<u>Remark</u>: This explicit example has been provided to me by D. Bennequin.

Let us consider the vector field in $\tilde{\gamma} \cap \alpha_0$.

In the transport along v, any contact form α having α_0 as a contact structure, rotates following Proposition 11.

$\tilde{\gamma}$ being a foliation does not rotate. As α and $\tilde{\gamma}$ are transverse, α cannot possibly make a complete revolution and no point of S^3 has a coincidence point distinct from itself.

The same construction and arguments readily generalize to all the contact forms constructed by Lickorish.

Hence the proof of Theorem 9.

Remark: It is likely that the same result holds for every contact form.

The explicit examples provided in the proof of Theorem 9 are a model of what happens in the general case of a contact form on the torus tangent to the core of this torus and transverse to a foliation. Let:

$$v = \tilde{\gamma} \cap \alpha = \begin{cases} -(y-x)y \\ x(y-x) + x^2 + y^2 - 1 \\ y \end{cases} \tag{12.1}$$

v has two periodic orbits lying on the boundary of the solid torus; one is attractive, the other is repulsive.

These periodic orbits are:

$$0_1 : \begin{cases} x^2 + y^2 = 1 \\ y = x = \frac{1}{\sqrt{2}} \end{cases} \qquad 0_2 : \begin{cases} x^2 + y^2 = 1 \\ y = x = -\frac{1}{\sqrt{2}} \end{cases} . \tag{12.2}$$

We will analyse further on more this explicit example.

It is quite easy in the situation of a contact form α transverse to a foliation $\tilde{\gamma}$ to compute the transport equations; and they are nearly explicit. Indeed we know that in the transport along v the form $\tilde{\gamma}$ remains collinear to itself. We thus have only a v-orbit x_s:

$$\phi^\star_s \tilde{\gamma} = a^\star(s)\,\tilde{\gamma} . \tag{12.3}$$

Let

$$\beta = d\alpha(v, .) . \tag{12.4}$$

We split β on the basis of the forms tangent to v provided by $\tilde{\gamma}$ and α :

$$\beta = a_1 \alpha + b_1 \tilde{\gamma}, \tag{12.5}$$

$a\star$, a_1 and b_1 allow us to compute the transport equations.

Indeed, we wish to compute, given x_0:

$$\beta_{x_0}(D\phi_{-s}(.)); \quad \alpha_{x_0}(D\phi_{-s}(.)); \quad \gamma_{x_0}(D\phi_{-s}(.)) . \tag{12.6}$$

We first compute:

$$\beta_{x_s}(D\phi_s(.)); \quad \alpha_{x_s}(D\phi_s(.)); \quad \gamma_{x_s}(D\phi_s(.)). \tag{12.7}$$

We have, for one differential form δ:

$$\frac{\partial}{\partial s}\delta_{x_s}(D\phi_s(.)) = d\delta_{x_s}(v, D\phi_s(.)) . \tag{12.8}$$

Thus:

$$\frac{\partial}{\partial s}\alpha_{x_s}(D\phi_s(.)) = \beta_{x_s}(D\phi_s(.)) \tag{12.9}$$

$$\frac{\partial}{\partial s}\beta_{x_s}(D\phi_s(.)) = d\beta_{x_s}(v, D\phi_s(.)) = d(a_1\alpha+b_1\tilde{\gamma})(v, D\phi_s(.)) \tag{12.10}$$

$$= \frac{\partial a_1}{\partial v}\alpha(D\phi_s(.)) + a_1\beta_{x_s}(D\phi_s(.)) + \frac{\partial b_1}{\partial v}\tilde{\gamma}(D\phi_s(.))$$

$$+ b_1 d\tilde{\gamma}(v, D\phi_s(.)) .$$

As $\tilde{\gamma}$ is a foliation, we have:

$$d\tilde{\gamma}(v, .) = \lambda\star\tilde{\gamma} . \tag{12.11}$$

$\lambda\star$ is easy to compute.

Indeed, by (12.3), we have:

$$\frac{\partial}{\partial s}\gamma_{x_s}(D\phi_s(.)) = \dot{a}\star(s)\tilde{\gamma}_{x_0} = d\tilde{\gamma}_{x_s}(v, D\phi_s(.)) = \lambda\star(x_s)\tilde{\gamma}_{x_s}(D\phi_s(.)). \tag{12.12}$$

Thus:

$$\gamma_{x_s}(D\phi_s(\cdot)) = e^{\int_0^s \lambda^\star(x_\tau)\,d\tau}\,\tilde{\gamma}_{x_0} \tag{12.13}$$

and

$$a^\star(s) = e^{\int_0^s \lambda^\star(x_\tau)\,d\tau}\,. \tag{12.14}$$

Rather than keeping $a^\star$ as data, we will keep $\lambda^\star$ which takes a more intrinsic form through (12.11).

Thus:

$$\frac{\partial}{\partial s}\beta_{x_s}(D\phi_s(\cdot)) = a_1\beta_{x_s}(D\phi_s(\cdot)) + \frac{\partial a_1}{\partial v}\alpha_{x_s}(D\phi_s(\cdot)) + \frac{\partial b_1}{\partial v}e^{\int_0^s \lambda^\star(x_\tau)\,d\tau}\,\tilde{\gamma}_{x_0}(\cdot) + b_1\lambda^\star e^{\int_0^s \lambda^\star d\tau}\,\tilde{\gamma}_{x_0}\,. \tag{12.15}$$

$$\frac{\partial}{\partial s}\alpha_{x_s}(D\phi_s(\cdot)) = \beta_{x_s}(D\phi_s(\cdot)) = a_1\,\alpha_{x_s}(D\phi_s(\cdot)) + b_1 e^{\int_0^s \lambda^\star(x_\tau)\,d\tau}\,\tilde{\gamma}_{x_0}\,. \tag{12.16}$$

(12.16) yields:

$$\alpha_{x_s}(D\phi_s(\cdot)) = e^{\int_0^s a_1\,d\tau}\alpha_{x_0} + e^{\int_0^s a_1\,d\tau}\left[\int_0^s b_1 e^{\int_0^\tau (\lambda^\star - a_1)\,dx}\,d\tau\right]\tilde{\gamma}_{x_0} \tag{12.17}$$

or else

$$\begin{aligned}\alpha_{x_0}(D\phi_{-s}(\cdot)) &= e^{-\int_0^s a_1\,d\tau}\alpha_{x_s} - \int_0^s b_1 e^{\int_0^\tau (\lambda^\star - a_1)\,dx}\,d\tau\;\tilde{\gamma}_{x_0}(D\phi_{-s}(\cdot))\\ &= e^{-\int_0^s a_1\,d\tau}\alpha_{x_s} - e^{-\int_0^s \lambda^\star(x_\tau)\,d\tau}\left[\int_0^s b_1 e^{\int_0^\tau (\lambda^\star - a_1)\,dx}\,d\tau\right]\tilde{\gamma}_{x_s}\end{aligned} \tag{12.18}$$

$$= \left[e^{-\int_0^s a_1 d\tau} + \left(e^{-\int_0^s \lambda^\star(x_\tau) d\tau} \right) \frac{a_1}{b_1}(x_s) \int_0^s b_1 e^{\int_0^\tau (\lambda^\star - a_1) dx} d\tau \right] \alpha_{x_s} -$$

$$- \frac{e^{-\int_0^s \lambda^\star(x_\tau) d\tau}}{b_1(x_s)} \int_0^s b_1 e^{\int_0^\tau (\lambda^\star - a_1) dx} d\tau \, \beta_{x_s} .$$

We then derive using (12.5)

$$\beta_{x_0}(D\phi_{-s}(\cdot)) = a_1(x_s) e^{-\int_0^s a_1 d\tau} \alpha_{x_s} + e^{-\int_0^s \lambda^\star d\tau} \times \tag{12.19}$$

$$\times \left(b_1(x_0) - \int_0^s b_1 e^{\int_0^\tau (\lambda^\star - a_1) dx} d\tau \right) \tilde{\gamma}_{x_s}$$

$$= \frac{a_1(x_s)}{b_1(x_s)} \left(\int_0^s b_1 e^{\int_0^\tau (\lambda^\star - a_1) dx} d\tau \right) e^{-\int_0^s \lambda^\star d\tau} \alpha_{x_s}$$

$$+ e^{\int_0^s \lambda^\star d\tau} \left(1 - \frac{1}{b_1(x_s)} \int_0^s b_1 e^{\int_0^\tau (\lambda^\star - a_1) dx} d\tau \right) \beta_{x_s} .$$

We thus have computed the transport equation along v, in the (α, β) frame:

Proposition 15: Let $\tilde{\gamma}$ be a foliation transverse to α. Let $v = \tilde{\gamma} \cap \alpha$,

$\beta = d\alpha(v,.)$; $\beta = a_1 \alpha + b_1 \tilde{\gamma}$; $\phi_s^\star \tilde{\gamma} = \tilde{\gamma}_{x_s}(D\phi_s(\cdot)) = e^{\int_0^s \lambda^\star(x_\tau) d\tau} \tilde{\gamma}_{x_0}$. Then:

$$\alpha_{x_0}(D\phi_{-s}(\cdot)) = \left[e^{-\int_0^s a_1 d\tau} + e^{-\int_0^s \lambda^\star(x_\tau) d\tau} \frac{a_1}{b_1}(x_s) \int_0^s b_1 e^{\int_0^\tau (\lambda^\star - a_1) dx} \right] d\tau \alpha_{x_s}$$

$$- \frac{e^{-\int_0^s \lambda^\star(x_\tau) d\tau}}{b_1(x_s)} \int_0^s b_1 e^{\int_0^\tau (\lambda^\star - a_1) dx} d\tau \, \beta_{x_s}$$

$$\beta_{x_0}(D\phi_{-s}(\cdot)) = \frac{a_1}{b_1}(x_s) \, e^{-\int_0^s \lambda^\star d\tau} \left(\int_0^s b_1 e^{\int_0^\tau (\lambda^\star - a_1) dx} d\tau \right) \alpha_{x_s} +$$

$$+ e^{-\int_0^s \lambda^\star d\tau} \left(1 - \frac{1}{b_1(x_s)} \int_0^s b_1 e^{\int_0^\tau (\lambda^\star - a_1) dx} d\tau \right) \beta_{x_s} .$$

Having studied the transport equations, the next step is to examine the variational problem we have been studying, when considered with such a v. This variational problem bears some drastically different features.

This will appear in a later publication.

References to the main text

[1] A. Weinstein, On the hypotheses of Rabinowitz' periodic orbit theorems, J. Diff. Eq. 33, 1979, p.353-358.

[2] S. Smale, Regular curves on Riemannian manifolds, Trans. Amer. Math. Soc. 87, 1958, p.492-512.

[3] W. Boothby, On the integral curves of a linear differential form of maximum rank, Math. Ann., 177, 1968, p.1-104.